Urban Drainage

Urban Drainage
2nd Edition

David Butler[†] and John W. Davies[††]

[†]*Professor of Water Engineering*
Department of Civil and Environmental Engineering
Imperial College London
[††]*Head of Civil Engineering*
School of Science and the Environment
Coventry University

Spon Press
Taylor & Francis Group

LONDON AND NEW YORK

First published 2000 by E & FN Spon
2 Park Square, Milton Park, Abingdon, Oxon OX14 4RN

Simultaneously published in the USA and Canada
by E & FN Spon
270 Madison Avenue, New York, NY 10016

Second Edition published 2004 by Spon Press
2 Park Square, Milton Park, Abingdon, Oxon OX14 4RN

Simultaneously published in the USA and Canada
by Spon Press
270 Madison Avenue, New York, NY 10016

Reprinted 2006

Spon Press is an imprint of the Taylor & Francis Group

© 2000, 2004 David Butler and John W. Davies

Typeset in Sabon by Wearset Ltd, Boldon, Tyne and Wear
Printed and bound in Great Britain by MPG Books Ltd, Bodmin

British Library Cataloguing in Publication Data
A catalogue record for this book is available from the British Library

Library of Congress Cataloging in Publication Data
Butler, David.
  Urban drainage / David Butler and John W. Davies. – 2nd ed.
    p.   cm.
  1. Urban runoff.   I. Davies, John W.   II. Title
TD657. B88   2004
628'.21—dc22

                                              2003025636

ISBN 0–415–30607–8 (pbk)
ISBN 0–415–30606–X (hbk)

Contents

Readership

In this book, we cover engineering and environmental aspects of the drainage of rainwater and wastewater from areas of human development. We present basic principles and engineering best practice. The principles are essentially universal but, in this book, are mainly illustrated by UK practice. We have also included introductions to current developments and recent research.

The book is primarily intended as a text for students on undergraduate and postgraduate courses in Civil or Environmental Engineering and researchers in related fields. We hope engineering aspects are treated with sufficient rigour and thoroughness to be of value to practising engineers as well as students, though the book does not take the place of an engineering manual.

The basic principles of drainage include wider environmental issues, and these are of significance not only to engineers, but to all with a serious interest in the urban environment, such as students, researchers and practitioners in environmental science, technology, policy and planning, geography and health studies. These wider issues are covered in particular parts of the book, deliberately written for a wide readership (indicated in the table opposite). The material makes up a significant portion of the book, and if these sections are read together, they should provide a coherent and substantial insight into a fascinating and important environmental topic.

The book is divided into twenty-four chapters, with numerical examples throughout, and problems at the end of each chapter. Comprehensive reference lists that point the way to further, more detailed information, support the text. Our aim has been to produce a book that is both comprehensive and accessible, and to share our conviction with all our readers that urban drainage is a subject of extraordinary variety and interest.

Chapter	Coverage of wider issues
1	All
2	All
3	3.5, 3.6
12	12.1, 12.2, 12.3
16	16.1, 16.2
17	17.1, 17.2
18	18.1
19	19.1, 19.2, 19.3
20	20.1, 20.2
21	21.1, 21.2, 21.3, 21.6, 21.7
22	22.1, 22.2
23	All
24	All

Acknowledgements

Many colleagues and friends have helped in the writing of this book. We are particularly grateful to Dr Dick Fenner of University of Cambridge for his encouragement and many useful comments. We would also like to acknowledge the helpful comments of John Ackers, Black & Veatch; Professor Bob Andoh, Hydro International; Emeritus Professor Bryan Ellis, Middlesex University; Andrew Hagger, Thames Water; Brian Hughes; Dr Pete Kolsky, Water and Sanitation Programme, World Bank; Professor Duncan Mara, University of Leeds; Nick Orman, WRc; Martin Osborne, BGP Reid Crowther; Sandra Rolfe and Professor David Balmforth, MWH Europe. We thank colleagues at Imperial College: Professor Nigel Graham, Professor Ĉedo Maksimović, Professor Howard Wheater, and current and former researchers Dr Maria do Céu Almeida, David Brown, Dr Eran Friedler, Dr Kim Littlewood, Dr Fayyaz Memon, Dr Jonathan Parkinson and Dr Manfred Schütze. At Coventry University, we thank Professor Chris Pratt.

Clearly, many people have helped with the preparation of this book, but the opinions expressed, statements made and any inadvertent errors are our sole responsibility.

Thanks most of all to:
Tricia, Claire, Simon, Amy
Ruth, Molly, Jack

Notation list

a	constant
a_{50}	effective surface area for infiltration
A	catchment area
	cross-sectional area
	plan area
A_b	area of base
A_D	impermeable area from which runoff received
A_{gr}	sediment mobility parameter
A_i	impervious area
A_o	area of orifice
A_p	gully pot cross-sectional area
$API5$	FSR 5-day antecedent precipitation index
ARF	FSR rainfall areal reduction factor
b	width of weir
	sediment removal constant
	constant
b_p	width of Preissman slot
b_r	sediment removal constant (runoff)
b_s	sediment removal constant (sweeping)
B	flow width
B_c	outside diameter of pipe
B_d	downstream chamber width (high side weir)
	width of trench at top of pipe
B_u	upstream chamber width (high side weir)
c	concentration
	channel criterion
	design number of appliances
	wave speed
c_0	dissolved oxygen concentration
c_{0s}	saturation dissolved oxygen concentration
c_v	volumetric sediment concentration
C	runoff coefficient
C_d	coefficient of discharge
C_v	volumetric runoff coefficient

C_R	dimensionless routing coefficient
d	depth of flow
d'	sediment particle size
d_c	critical depth
d_m	hydraulic mean depth
d_1	depth upstream of hydraulic jump
d_2	depth downstream of hydraulic jump
d_{50}	sediment particle size larger than 50% of all particles
D	internal pipe diameter
	rainfall duration
	wave diffusion coefficient
	longitudinal dispersion coefficient
D_o	orifice diameter
D_{gr}	sediment dimensionless grain size
D_p	gully pot diameter
DWF	dry weather flow
e	voids ratio
	sediment accumulation rate in gully
E	specific energy
	gully hydraulic capture efficiency
	industrial effluent flow-rate
$EBOD$	Effective BOD_5
f	soil infiltration rate
	potency factor
f_c	soil infiltration capacity
f_o	soil initial infiltration rate
f_s	number of sweeps per week
f_t	soil infiltration rate at time t
F_m	bedding factor
F_r	Froude number
F_{se}	factor of safety
g	acceleration due to gravity
G	water consumption per person
G'	wastewater generated per person
h	head
h_a	acceleration head
h_f	head loss due to friction
h_L	total head loss
h_{local}	local head loss
h_{max}	depth of water
	gully pot trap depth
H	total head
	difference in water level
	height of water surface above weir crest
	depth of cover to crown of pipe

H_{min}	minimum difference in water level for non-drowned orifice
i	rainfall intensity
i_e	effective rainfall intensity
i_n	net rainfall intensity
I	inflow rate
	pipe infiltration rate
	rainfall depth
j	time
J	housing density
	criterion of satisfactory service
	empirical coefficient
k	constant
k_b	effective roughness value of sediment dunes
k_{DU}	dimensionless frequency factor
k_L	local head loss constant
k_s	pipe roughness
k_T	constant at $T°C$
k_1	depression storage constant
k_2	Horton's decay constant
k_3	unit hydrograph exponential decay constant
k_4	pollutant washoff constant
k_5	amended pollutant washoff constant
k_{20}	constant at 20°C
K	routing constant
	constant in CSO design (Table 12.6)
	Rankine's coefficient
	empirical coefficient
K_{LA}	volumetric reaeration coefficient
L	length
	load-rate
	gully spacing
L_E	equivalent pipe length for local losses
L_1	initial gully spacing
m	Weibull's event rank number
	reservoir outflow exponent
M	mass
	empirical coefficient
M_s	mass of pollutant on surface
$MT\text{-}D$	FSR rainfall depth of duration D with a return period T
n	number
	Manning's roughness coefficient
	porosity
n_{DU}	number of discharge units
N	total number
	Bilham's number of rainfall events in 10 years

O	outflow rate
p	pressure
	probability of appliance discharge
	BOD test sample dilution
	projection ratio
P	wetted perimeter
	perimeter of infiltration device
	population
	power
	probability
	height of weir crest above channel bed
P_d	downstream weir height (high side weir)
P_F	peak factor
P_s	surcharge pressure
P_u	upstream weir height (high side weir)
$PIMP$	FSR percentage imperviousness
PR	WP percentage runoff
q	flow per unit width
	appliance flow-rate
Q	flow-rate
Q_{av}	average flow-rate
Q_b	gully bypass flow-rate
Q_c	gully capacity
Q_d	continuation flow-rate (high side weir)
Q_f	pipe-full flow-rate
Q_{min}	minimum flow
Q_o	wastewater baseflow
Q_p	peak flow-rate
Q_r	runoff flow-rate
Q_u	inflow (high side weir)
$\overline{Q}$	gully approach flow
$\overline{Q}_L$	limiting gully approach flow
r	risk
	number of appliances discharging simultaneously
	FSR ratio of 60 min to 2 day 5 year return period rainfall
	discount rate
r_b	oxygen consumption rate in the biofilm
r_s	oxygen consumption rate in the sediment
r_{sd}	settlement deflection ratio
r_w	oxygen consumption rate in the bulk water
R	hydraulic radius
	ratio of drained area to infiltration area
	runoff depth
R_e	Reynolds number
$RMED$	FEH median of annual rainfall maxima

s	ground slope
S	storage volume
	soil storage depth
S_c	critical slope
S_d	sediment dry density
S_f	hydraulic gradient or friction slope
S_G	specific gravity
S_o	pipe, or channel bed, slope
$SAAR$	FSR standard average annual rainfall
SMD	FSR soil moisture deficit
$SOIL$	FSR soil index
t	time
	pipe wall thickness
t'	duration of appliance discharge
t_c	time of concentration
t_e	time of entry
t_f	time of flow
t_p	time to peak
T	rainfall event return period
	wastewater temperature
	pump cycle (time between starts)
T'	mean interval between appliance use
T_a	approach time
T_c	time between gully pot cleans
u	unit hydrograph ordinate
U_*	shear velocity
$UCWI$	FSR urban catchment wetness index
v	mean velocity
v_c	critical velocity
v_f	pipe-full flow velocity
v_{GS}	gross solid velocity
v_L	limiting velocity without deposition
v_{max}	maximum flow velocity
v_{min}	minimum flow velocity
v_t	threshold velocity required to initiate movement
V	volume
V_f	volume of first flush
V_I	inflow volume
V_O	outflow volume
	baseflow volume in approach time
V_t	basic treatment volume
w	channel bottom width
	pollutant-specific exponent
W	width of drainage area
	pollutant washoff rate

W_b	sediment bed width
W_c	soil load per unit length of pipe
W_{csu}	concentrated surcharge load per unit length of pipe
W_e	effective sediment bed width
	external load per unit length of pipe
W_s	settling velocity
W_t	crushing strength per unit length of pipe
W_w	liquid load per unit length of pipe
x	longitudinal distance
	return factor
X	chemical compound
y	depth
Y	chemical element
Y_d	downstream water depth (high side weir)
Y_u	upstream water depth (high side weir)
z	potential head
	side slope
Z_1	pollutant-specific constant
	FSR growth factor
Z_2	pollutant-specific constant
	FSR growth factor
α	channel side slope angle to horizontal
	number of reservoirs
	turbulence correction factor
	empirical coefficient
β	empirical coefficient
γ	empirical coefficient
ε	empirical coefficient
	gully pot sediment retention efficiency
ε'	gully pot cleaning efficiency
ζ	sediment washoff rate
η	sediment transport parameter
	pump efficiency
θ	transition coefficient for particle Reynolds number
	angle subtended by water surface at centre of pipe
	Arrhenius temperature correction factor
κ	sediment supply rate
λ	friction factor
λ_b	friction factor corresponding to the sediment bed
λ_c	friction factor corresponding to the pipe and sediment bed
λ_g	friction factor corresponding to the grain shear factor
μ	coefficient of friction
μ'	coefficient of sliding friction
ν	kinematic viscosity
ρ	density

τ_b	critical bed shear stress
τ_o	boundary shear stress
γ	unit weight
	temperature correction factor
χ	surface sediment load
χ_u	ultimate (equilibrium) surface sediment load
ω	counter
ψ	shape correction factor for part-full pipe

Units are not specifically included in this notation list, but have been included in the text.

Abbreviations

AMP	Asset management planning
AOD	Above ordnance datum
ARF	Areal reduction factor
ASCE	American Society of Civil Engineers
ATU	Allylthiourea
BHRA	British Hydrodynamics Research Association
BOD	Biochemical oxygen demand
BRE	Building Research Establishment
BS	British Standard
CAD	Computer aided drawing/design
CARP	Comparative acceptable river pollution procedure
CBOD	Carbonaceous biochemical oxygen demand
CCTV	Closed-circuit television
CEC	Council of European Communities
CEN	European Committee for Standardisation
CFD	Computational fluid dynamics
CIWEM	Chartered Institution of Water and Environmental Management
CIRIA	Construction Industry Research and Information Association
COD	Chemical oxygen demand
CSO	Combined sewer overflow
DG5	OFWAT performance indicator
DO	Dissolved oxygen
DoE	Department of the Environment
DOT	Department of Transport
DN	Nominal diameter
DU	Discharge unit
EA	Environment Agency
EC	Escherichia coli
EGL	Energy grade line
EMC	Event mean concentration
EN	European Standard

EPA	Environmental Protection Agency (US)
EQO	Environmental quality objectives
EQS	Environmental quality standards
EWPCA	European Water Pollution Control Association
FC	Faecal coliform
FEH	Flood Estimation Handbook
FOG	Fats, oils and grease
FORGEX	FEH focused rainfall growth curve extension method
FS	Faecal streptococci
FSR	Flood Studies Report
FWR	Foundation for Water Research
GL	Ground level
GMT	Greenwich Mean Time
GRP	Glass reinforced plastic
HDPE	High density polyethylene
HGL	Hydraulic grade line
HMSO	Her Majesty's Stationery Office
HR	Hydraulics Research
HRS	Hydraulics Research Station
IAWPRC	International Association on Water Pollution Research and Control
IAWQ	International Association on Water Quality
ICE	Institution of Civil Engineers
ICP	Inductively coupled plasma
IDF	Intensity – duration – frequency
IE	Intestinal enterococci
IL	Invert level
IoH	Institute of Hydrology
IWEM	Institution of Water and Environmental Management
LC50	Lethal concentration to 50% of sample organisms
LOD	Limit of deposition
MAFF	Ministry of Agriculture, Fisheries and Food
MDPE	Medium density polyethylene
MH	Manhole
MPN	Most probable number
NERC	Natural Environment Research Council
NOD	Nitrogenous oxygen demand
NRA	National Rivers Authority
NWC	National Water Council
OFWAT	Office of Water Services
OD	Outside diameter
OS	Ordnance Survey
PAH	Polyaromatic hydrocarbons
PCB	Polychlorinated biphenyl
PID	proportional–integral–derivative

PVC-U	Unplasticised polyvinylchloride
QUALSOC	Quality impacts of storm overflows: consent procedure
RRL	Road Research Laboratory
RTC	Real-time control
SAAR	Standard annual average rainfall
SDD	Scottish Development Department
SEPA	Scottish Environmental Protection Agency
SOD	Sediment oxygen demand
SRM	Sewerage Rehabilitation Manual
SG	Specific gravity
SS	Suspended solids
STC	Standing Technical Committee
SUDS	Sustainable (urban) drainage systems
SWO	Stormwater outfall
TBC	Toxicity-based consents
TKN	Total Kjeldahl nitrogen
TOC	Total organic carbon
TRRL	Transport & Road Research Laboratory
TWL	Top water level
UKWIR	United Kingdom Water Industry Research
UPM	Urban pollution management
WAA	Water Authorities Association
WaPUG	Wastewater Planning User Group
WC	Water closet (toilet)
WEF	Water Environment Federation (US)
WFD	Water Framework Directive
WMO	World Meteorological Organisation
WP	Wallingford Procedure
WPCF	Water Pollution Control Federation (US)
WSA	Water Services Association
WTP	Wastewater treatment plant
WO	Welsh Office
WRc	Water Research Centre

1 Introduction

1.1 What is urban drainage?

Drainage systems are needed in developed urban areas because of the interaction between human activity and the natural water cycle. This interaction has two main forms: the abstraction of water from the natural cycle to provide a water supply for human life, and the covering of land with impermeable surfaces that divert rainwater away from the local natural system of drainage. These two types of interaction give rise to two types of water that require drainage.

The first type, *wastewater*, is water that has been supplied to support life, maintain a standard of living and satisfy the needs of industry. After use, if not drained properly, it could cause pollution and create health risks. Wastewater contains dissolved material, fine solids and larger solids, originating from WCs, from washing of various sorts, from industry and from other water uses.

The second type of water requiring drainage, *stormwater*, is rainwater (or water resulting from any form of precipitation) that has fallen on a built-up area. If stormwater were not drained properly, it would cause inconvenience, damage, flooding and further health risks. It contains some pollutants, originating from rain, the air or the catchment surface.

Urban drainage systems handle these two types of water with the aim of minimising the problems caused to human life and the environment. Thus urban drainage has two major interfaces: with the public and with the environment (Fig. 1.1). The public is usually on the transmitting rather than receiving end of services from urban drainage ('flush and forget'), and this may partly explain the lack of public awareness and appreciation of a vital urban service.

Fig. 1.1 Interfaces with the public and the environment

In many urban areas, drainage is based on a completely artificial system of sewers: pipes and structures that collect and dispose of this water. In contrast, isolated or low-income communities normally have no main drainage. Wastewater is treated locally (or not at all) and stormwater is drained naturally into the ground. These sorts of arrangements have generally existed when the extent of urbanisation has been limited. However, as will be discussed later in the book, recent thinking – towards more sustainable drainage practices – is encouraging the use of more natural drainage arrangements wherever possible.

So there is far more to urban drainage than the process of getting the flow from one place to another via a system of sewers (which a non-specialist could be forgiven for finding untempting as a topic for general reading). For example, there is a complex and fascinating relationship between wastewater and stormwater as they pass through the system, partly as a result of the historical development of urban drainage. When wastewater and stormwater become mixed, in what are called 'combined sewers', the disposal of neither is 'efficient' in terms of environmental impact or sustainability. Also, while the flow is being conveyed in sewers, it undergoes transformation in a number of ways (to be considered in detail in later chapters). Another critical aspect is the fact that sewer systems may cure certain problems, for example health risks or flooding, only to create others in the form of environmental disruption to natural watercourses elsewhere.

Overall, urban drainage presents a classic set of modern environmental challenges: the need for cost-effective and socially acceptable technical improvements in existing systems, the need for assessment of the impact of those systems, and the need to search for sustainable solutions. As in all other areas of environmental concern, these challenges cannot be considered to be the responsibility of one profession alone. Policy-makers, engineers, environment specialists, together with all citizens, have a role. And these roles must be played in partnership. Engineers must understand the wider issues, while those who seek to influence policy must have some understanding of the technical problems. This is the reasoning behind the format of this book, as explained in the Preface. It is intended as a source of information for all those with a serious interest in the urban environment.

1.2 Effects of urbanisation on drainage

Let us consider further the effects of human development on the passage of rainwater. Urban drainage replaces one part of the natural water cycle and, as with any artificial system that takes the place of a natural one, it is important that the full effects are understood.

In nature, when rainwater falls on a natural surface, some water returns to the atmosphere through evaporation, or transpiration by plants; some infiltrates the surface and becomes groundwater; and some runs off the surface (Fig. 1.2(a)). The relative proportions depend on the nature of the surface, and vary with time during the storm. (Surface runoff tends to increase as the ground becomes saturated.) Both groundwater and surface runoff are likely to find their way to a river, but surface runoff arrives much faster. The groundwater will become a contribution to the river's general baseflow rather than being part of the increase in flow due to any particular rainfall.

Development of an urban area, involving covering the ground with artificial surfaces, has a significant effect on these processes. The artificial surfaces increase the amount of surface runoff in relation to infiltration, and therefore increase the total volume of water reaching the river during or soon after the rain (Fig. 1.2(b)). Surface runoff travels quicker over hard surfaces and through sewers than it does over natural surfaces and along natural streams. This means that the flow will both arrive and die away faster, and therefore the peak flow will be greater (see Fig. 1.3). (In addition, reduced infiltration means poorer recharge of groundwater reserves.)

This obviously increases the danger of sudden flooding of the river. It also has strong implications for water quality. The rapid runoff of stormwater is likely to cause pollutants and sediments to be washed off the surface or scoured by the river. In an artificial environment, there are likely to be more pollutants on the catchment surface and in the air than there would be in a natural environment. Also, drainage systems in which there is mixing of wastewater and stormwater may allow pollutants from the wastewater to enter the river.

The existence of wastewater in significant quantities is itself a consequence of urbanisation. Much of this water has not been made particularly 'dirty' by

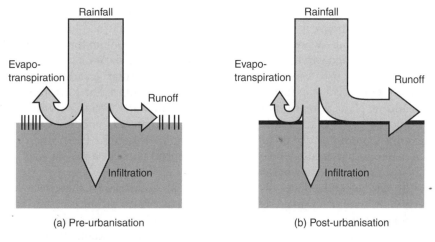

Fig. 1.2 Effect of urbanisation on fate of rainfall

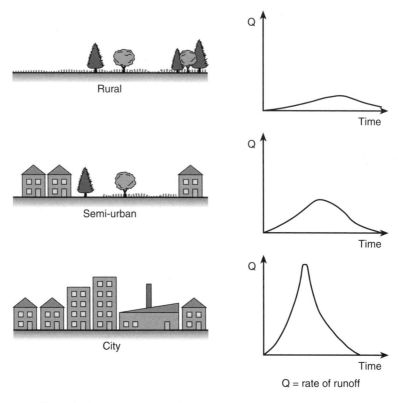

Fig. 1.3 Effect of urbanisation on peak rate of runoff

its use. Just as it is a standard convenience in a developed country to turn on a tap to fill a basin, it is a standard convenience to pull the plug to let the water 'disappear'. Water is also used as the principal medium for disposal of bodily waste, and varying amounts of bathroom litter, via WCs.

In a developed system, much of the material that is added to the water while it is being turned into wastewater is removed at a wastewater treatment plant prior to its return to the urban water cycle. Nature itself would be capable of treating some types of material, bodily waste for example, but not in the quantities created by urbanisation. The proportion of material that needs to be removed will depend in part on the capacity of the river to assimilate what remains.

So the general effects of urbanisation on drainage, or the effects of replacing natural drainage by urban drainage, are to produce higher and more sudden peaks in river flow, to introduce pollutants, and to create the need for artificial wastewater treatment. While to some extent impersonating nature, urban drainage also imposes heavily upon it.

1.3 Urban drainage and public health

In human terms, the most valuable benefit of an effective urban drainage system is the maintenance of public health. This particular objective is often overlooked in modern practice and yet is of extreme importance, particularly in protection against the spread of diseases.

Despite the fact that some vague association between disease and water had been known for centuries, it was only comparatively recently (1855) that a precise link was demonstrated. This came about as a result of the classic studies of Dr John Snow in London concerning the cholera epidemic sweeping the city at the time. That diseases such as cholera are almost unknown in the industrialised world today is in major part due to the provision of centralised urban drainage (along with the provision of a microbiologically safe, potable supply of water).

Urban drainage has a number of major roles in maintaining public health and safety. Human excreta (particularly faeces) are the principal vector for the transmission of many communicable diseases. Urban drainage has a direct role in effectively removing excreta from the immediate vicinity of habitation. However, there are further potential problems in large river basins in which the downstream discharges of one settlement may become the upstream abstraction of another. In the UK, some 30% of water supplies are so affected. This clearly indicates the vital importance of disinfection of water supplies as a public health measure.

Also, of particular importance in tropical countries, standing water after rainfall can be largely avoided by effective drainage. This reduces the mosquito habitat and hence the spread of malaria and other diseases.

Whilst many of these problems have apparently been solved, it is essential that in industrialised countries, as we look for ever more innovative sanitation techniques, we do not lose ground in controlling serious diseases. Sadly, whilst we may know much about waterborne and water-related diseases, some rank among the largest killers in societies where poverty and malnutrition are widespread. Millions of people around the world still lack any hygienic and acceptable method of excreta disposal. The issues associated with urban drainage in low-income communities are returned to in more detail in Chapter 23.

1.4 History of urban drainage engineering

Early history

Several thousand years BC may seem a long way to go back to trace the history of urban drainage, but it is a useful starting point. In many parts of the world, we can imagine animals living wild in their natural habitat and humans living in small groups making very little impact on their environment. Natural hydrological processes would have prevailed; there might have been floods in extreme conditions, but these would not have been made worse by human alteration of the surface of the ground. Bodily wastes would have been 'treated' by natural processes.

Artificial drainage systems were developed as soon as humans attempted to control their environment. Archaeological evidence reveals that drainage was provided to the buildings of many ancient civilisations such as the Mesopotamians, the Minoans (Crete) and the Greeks (Athens). The Romans are well known for their public health engineering feats, particularly the impressive aqueducts bringing water *into* the city; less spectacular, but equally vital, were the artificial *drains* they built, of which the most well known is the *cloaca maxima*, built to drain the Roman Forum (and still in use today).

The English word *sewer* is derived from an Old French word, *essever*, meaning 'to drain off', related to the Latin *ex-* (out) and *aqua* (water). The Oxford English Dictionary gives the earliest meaning as 'an artificial water-course for draining marshy land and carrying off surface water into a river or the sea'. Before 1600, the word was not associated with wastewater.

London

The development of drainage in London provides a good example of how the association between wastewater and stormwater arose. Sewers originally had the meaning given above and their alignment was loosely based on the natural network of streams and ditches that preceded them. In a quite unconnected arrangement, bodily waste was generally disposed of into cesspits (under the residence floor), which were periodically emptied. Flush toilets (discharging to cesspits) became common around 1770–1780, but it remained illegal until 1815 to connect the overflow from cesspits to the sewers. This was a time of rapid population growth and, by 1817, when the population of London exceeded one million, the only solution to the problem of under-capacity was to allow cesspit overflow to be connected to the sewers. Even then, the cesspits continued to be a serious health problem in poor areas, and, in 1847, 200 000 of them were eliminated completely by requiring houses to be connected directly to the sewers.

This moved the problem elsewhere – namely, the River Thames. By the 1850s, the river was filthy and stinking (Box 1.1) and directly implicated in the spread of deadly cholera.

There were cholera epidemics in 1848–1849, 1854 and 1867, killing tens of thousands of Londoners. The Victorian sanitary reformer Edwin Chadwick passionately argued for a dual system of drainage, one for human waste and one for rainwater: 'the rain to the river and the sewage to the soil'. He also argued for small-bore, inexpensive, self-cleansing sewer pipes in preference to the large brick-lined tunnels of the day. However, the complexity and cost of engineering two separate systems prevented his ideas from being put into practice. The solution was eventually found in a plan by Joseph Bazalgette to construct a number of 'combined' interceptor sewers on the north and the south of the river to carry

---

Box 1.1 Michael Faraday's abridged letter to The Times *of 7th July 1855*

I traversed this day by steamboat the space between London and Hungerford Bridges [*on the River Thames*], between half-past one and two o'clock. The appearance and smell of water forced themselves on my attention. The whole of the river was an opaque pale brown fluid. The smell was very bad, and common to the whole of the water. The whole river was for the time a real sewer.

If there be sufficient authority to remove a putrescent pond from the neighbourhood of a few simple dwellings, surely the river which flows for so many miles through London ought not be allowed to become a fermenting sewer. If we neglect this subject, we cannot expect to do so with impunity; nor ought we to be surprised if, ere many years are over, a season give us sad proof of the folly of our carelessness.

---

the contents of the sewers to the east of London. The scheme, an engineering marvel (Fig. 1.4), was mostly constructed by 1875, and much of it is still in use today.

Again, though, the problem had simply been moved elsewhere. This time, it was the Thames estuary, which received huge discharges of wastewater. Storage was provided to allow release on the ebb tide only, but there was no treatment. Downstream of the outfalls, the estuary and its banks were disgustingly polluted. By 1890, some separation of solids was carried out at works on the north and south banks, with the sludge dumped at sea. Biological treatment was introduced in the 1920s, and further improvements followed. However, it was not until the 1970s that the quality of the Thames was such that salmon were commonplace and porpoises could be seen under Blackfriars Bridge.

UK generally

After the Second World War, many parts of the UK had effective wastewater treatment facilities, but there could still be significant wastewater pollution during wet weather. Most areas were drained by combined sewers, carrying wastewater and stormwater in the same pipe. (The first origins of this system can be found in the connection of wastewater to stormwater sewers, as described above.) Such a system must include combined sewer overflows (CSOs) to provide relief during rain storms, allowing excess flows to escape to a nearby river or stream. As we will discover, CSOs remain a problem today.

During the 1950s and 1960s, there was significant research effort on improving CSO design. This led to a number of innovative new arrangements, and to general recommendations for reducing pollution. Most

Fig. 1.4 Construction of Bazalgette's sewers in London (from *The Illustrated London News*, 27 August 1859, reproduced with permission of The Illustrated London News Picture Library)

sewer systems in the UK today are still combined, even though from 1945 it had become the norm for newly-constructed developments to be drained by a separate system of sewers (one pipe for wastewater, one for stormwater). These issues will be explored further in Chapters 2 and 12.

However, in some parts of the UK, particularly around industrial estuaries like the Mersey and the Tyne, there were far more serious problems of wastewater pollution than those caused by CSOs. In those areas all wastewater, in wet and dry weather, was discharged directly to the estuary without any treatment at all. Box 1.2 considers the Tyne, and the work that was done to improve matters.

The water industry

In 1974, the water industry in England and Wales was reorganised, and water authorities were formed. These were public authorities that controlled most aspects of the water cycle, including water supply (except in areas where private water companies existed). However, most new water authorities allowed local authorities to remain in charge of sewerage,

Box 1.2 Tyneside interceptor sewer scheme

Tyneside had undergone rapid development during the industrial revolution, and those providing housing for the rapidly expanding workforce had not felt it necessary to look further than the conveniently placed Tyne for disposal of stormwater and untreated wastewater. The area was drained by a multitude of main sewers running roughly perpendicular to the river, discharging untreated wastewater along the length of the north and south banks even in dry weather. This unpleasant situation had existed for many years. The sewer systems were the responsibility of a number of different local authorities and, since pollution was considered to have low political priority, the effort to find a comprehensive solution was not made until the 1960s with the formation of an overall sewerage authority. This authority drew up plans for interceptor sewers running along both sides of the Tyne picking up the flows from each main sewer and taking them to a treatment works. A tunnel under the Tyne was needed to bring flows from the south (Fig. 1.5).

The Tyneside scheme also included provision for intercepting wastewater from a coastal strip to the north of the Tyne. Here,

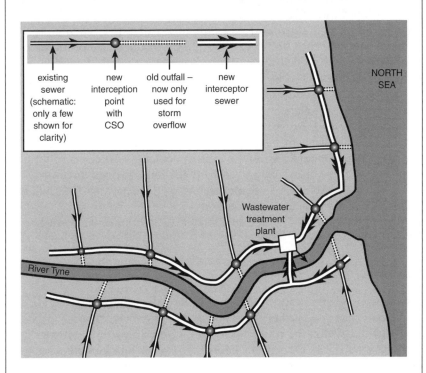

Fig. 1.5 Tyneside interceptor sewer scheme (schematic plan)

again, wastewater had received no treatment and was discharged via sea outfalls that barely reached the low tide mark. The area was drained by combined sewers, and some overflows had consisted simply of outlet relief pipes discharging from holes in the seawall at the top of the beach, so that in wet weather the overflow from the combined sewer flowed across the popular beach to the sea.

acting as agents. The overall control of the water authorities generally allowed more regional planning and application of overall principles. This was helped by the expanded Water Research Centre, whose pragmatic, common-sense approaches encouraged improvement in the operation of sewer systems. However, drainage engineering remained a fairly low-tech business, with drainage engineers generally rather conservative, relying on experience rather than specialised technology to solve problems.

Modelling and rehabilitation

A change came in the early 1980s, with the introduction of computer modelling of sewer systems. Such models had been available in the US for a while, but the first modelling package written for UK conditions, *WASSP* (Wallingford Storm Sewer Package), which was based on a set of calculations covering rainfall, runoff and pipe flow called the Wallingford Procedure, was launched in 1981. The first version was not particularly user-friendly and needed a mainframe computer to run on, but later the software was developed in response to the development of computers and the demand for a good user interface. The tool had a profound effect on the attitudes and practices of drainage engineers. To model a system, its physical data had to be known; creating computer models therefore demanded improvement in sewer records. The use of models encouraged far more understanding of how a system actually worked. A philosophy that high-tech problem analysis could make huge savings in construction costs became established, and was set out in the *Sewerage Rehabilitation Manual* of the Water Research Centre.

Rehabilitation is considered in Chapter 18, and modelling in Chapters 19 and 20.

The 1990s

As drainage engineers in the UK moved into the 1990s, they experienced two major changes. The first was that the industry was reorganised again. In England and Wales, the water authorities were privatised. Regulatory functions that had been carried out internally, like pollution-monitoring, were moved to a new organisation: the National Rivers Authority, which, in turn, became part of the Environment Agency in 1996. Later, in Scotland, three large water authorities took over water functions from local authorities (and were merged into one large authority in 2002).

The other big change was the gradual application of much more stringent pollution regulations set by the European Union. The Bathing Water Directive (CEC, 1976) required 'bathing waters' to be designated, and for their quality to comply with bacterial standards. Huge investment in coastal wastewater disposal schemes was carried out in response. For example, in the south-west of England, the 'clean sweep' programme was developed to improve the sea water quality at *eighty-one* beaches and their surroundings. This was based on thirty-two engineering schemes valued at £900 million (Brokenshire, 1995).

In Brighton and Hastings on England's south coast, huge combined sewer storage tunnels were constructed to avoid CSO spills onto local beaches during storm events. And in the north-east of England, similar major investment was made along the route of the coastal interceptor sewer constructed in the 1970s, already described in Box 1.2. So, on that length of coast, there was a great deal of change in twenty years: from the contents of combined sewers overflowing all over the beach, to massive storage tunnels satisfying strict limits on storm discharges to the sea (Firth and Staples, 1995).

The Urban Waste Water Treatment Directive (CEC, 1991) also had far-reaching effects. This specified a minimum level of wastewater treatment, based on the urban population size and the receiving water type, to be achieved by 2005. Sea disposal of sludge was completely banned by the end of 1998. Pollution standards are considered in Chapter 3.

Current challenges

The twenty-first century brings fresh challenges to the field of urban drainage. In the arena of legislation, the EU Water Framework Directive (CEC, 2000) seeks to maintain and improve the quality of Europe's surface and ground waters. Whilst this may not have a direct impact on drainage design or operation, it will exert pressure to further upgrade the performance of system discharge points such as combined sewer overflows and will influence the types of substances that may be discharged to sewer systems. Further details can be found in Chapter 3.

An emerging, if controversial, threat is that of climate change. The anthropogenic impact on our global climate now seems to have been demonstrated conclusively, but the implications are not fully understood. Our best predictions indicate that there will be significant changes to the rainfall regime, and these are discussed in Chapter 5. These changes must, in turn, be taken into account in new drainage design. The implications for existing systems are a matter for research (Evans *et al.*, 2003).

One of the most serious implications is the increased potential for sewer (pluvial) flooding. External or, even worse, internal flooding with sewage is considered to be wholly unacceptable in the twenty-first century according to some sources (WaterVoice Yorkshire, 2002). Given the stochastic nature of rainfall and the potential for more extreme events in the future, this is an area that is likely to require careful attention by urban drainage researchers and practitioners (as considered further in Section 11.2.2).

Changing aims

It has already been stated that the basic function of urban drainage is to collect and convey wastewater and stormwater. In the UK and other developed countries, this has generally been taken to cover all wastewater, and all it contains (subject to legislation about hazardous chemicals and industrial effluents). For stormwater, the aim has been to remove rainwater (for storms up to a particular severity) with the minimum of inconvenience to activities on the surface.

Most people would see the efficient removal of stormwater as part of 'progress'. In a developing country, they might imagine a heavy rainstorm slowing down the movement of people and goods in a sea of mud, whereas in a city in a developed country they would probably consider that it should take more than mere rainfall to stop transport systems and businesses from running smoothly. Nowadays, however, as with other aspects of the environment, the nature of progress in relation to urban drainage, its consequences, desirability and limits, are being closely reassessed.

The traditional aim in providing storm drainage has been to remove water from surfaces, especially roads, as quickly as possible. It is then disposed of, usually via a pipe system, to the nearest watercourse. This, as we have discovered in Section 1.2, can cause damage to the environment and increase the risk of flooding elsewhere. So, while a prime purpose of drainage is still to protect people and property from stormwater, attention is now being paid not only to the surface being drained but also to the impact of the drained flow on the receiving water. Consequently, interest in more natural methods of disposing of stormwater is increasing. These include infiltration and storage (to be discussed in full in Chapter 21), and the general intention is to attempt to reverse the trend illustrated in Fig. 1.3: to decrease the peak flow of runoff and increase the time it takes to reach the watercourse.

Another way in which attempts are being made to reverse the effects of urbanisation on drainage described in Section 1.2 is to reduce the non-biodegradable content in wastewater. Public campaigns with slogans like 'bag it and bin it, don't flush it' or 'think before you flush' have been mounted to persuade people not to treat the WC as a rubbish bin.

These tendencies towards reducing the dependence on 'hard' engineering solutions to solve the problems created by urbanisation, and the philosophy that goes with them, are associated with the word 'sustainability' and are further considered in Chapter 24.

1.5 Geography of urban drainage

The main factors that determine the extent and nature of urban drainage provision in a particular region are:

- wealth
- climate and other natural characteristics
- intensity of urbanisation
- history and politics.

The greatest differences are the result of differences in wealth. Most of this book concentrates on urban drainage practices in countries that can afford fully engineered systems. The differences in countries that cannot will be apparent from Chapter 23 where we consider low-income communities.

Countries in which rainfall tends to be occasional and heavy have naturally adopted different practices from those in which it is frequent and generally light. For example, it is common in Australia to provide 'minor' (underground, piped) systems to cope with low quantities of stormwater, together with 'major' (overground) systems for larger quantities. Other natural characteristics have a significant effect. Sewers in the Netherlands, for example, must often be laid in flat, low-lying areas and, therefore, must be designed to run frequently in a pressurised condition.

Intensity of urbanisation has a strong influence on the percentage of the population connected to a main sewer system. Table 1.1 gives percentages in a number of European countries.

Historical and political factors determine the age of the system (which is likely to have been constructed during a period of significant development and industrialisation), characteristics of operation such as whether or not the water/wastewater industry is publicly or privately financed, and strictness of statutory requirements for pollution control and the manner in which they are enforced. Countries in the European Union are subject to common requirements, as described in Section 1.4.

Boxes 1.3 to 1.5 present a selection of examples to give an idea of the wide range of different urban drainage problems throughout the world.

Table 1.1 Percentage of population connected to main sewers in selected European countries (1997 figures)

Country	% population connected to sewer
Germany	92
Greece	58
Italy	82
Netherlands	97
Portugal	57
UK	96

Box 1.3 Orangi, Karachi, Pakistan

The squatter settlement of Orangi in Karachi (*New Scientist*, 1 June 1996) has a population of about 1 million. It has some piped water supplies but, until the 1980s, had no sewers. People had to empty bucket latrines into the narrow alleys. In a special self-help programme, quite different from government-sponsored improvement schemes, the community has built its own sewers, with no outside contractors. A small septic tank is placed between the toilet and the sewer to reduce the entry of solids into the pipe. The system itself has a simplified design. The wastewater is carried to local rivers and is discharged untreated. The system is being built up alley-by-alley, as the people make the commitment to the improvements. This is a great success for community action, and has created major improvements in the immediate environment. But problems seem certain to occur elsewhere in the form of pollution in the receiving river, until treatment, which would have to be provided by the central authorities, is sufficient.

Box 1.4 Villages in Hong Kong

A scheme in Hong Kong (Lei *et al.*, 1996) has provided sewers for previously unsewered villages. Here residents had 'discharged their toilet waste into septic tanks which very often overflowed due to improper maintenance, while their domestic sullage is discharged into the surface drains'. This had caused pollution of streams and rivers, and contributed to pollution of coastal waters (causing 'red tides'). A new scheme provides sewers to remove the need for the septic tanks and carry the wastewater to existing treatment facilities. One problem during construction was 'Fung Shui', the traditional Chinese belief that the orientation of features in the urban landscape may affect the health and good luck of the people living there. When carrying out sewer construction within traditional Chinese villages, engineers had to take great care over these issues, by consultation with residents.

Box 1.5 Jakarta, Indonesia

Indonesia has a territory of over 1.9 million km^2 for its 200 million inhabitants (with the population currently growing at 3 million per annum). Approximately 110 million live on the island of Java which has an area of only 127000 km^2, making it one of the most densely populated parts of the world. The largest city is Jakarta, with an official population of 10 million but probably much larger. Jakarta has many transient settlements. Over 20% of the housing could be classed as temporary and 40% is semi-permanent. About 60% of the population live in settlements called *kampungs* that now have a semi-legitimate status. Housing programmes divide kampungs into two categories: 'never-to-be-improved' and those 'to-be-improved'. Residents of the first category are encouraged to return to their villages, move away from Java or select a permanent housing area in Jakarta. The 'to-be-improved' category *kampungs* are upgraded by introducing some basic services. By 1984, the housing improvement programmes had reached 3.8 million inhabitants, yet it has been estimated that 50% of the population within these settlements has yet to be served.

Incredibly, for a city of its size, Jakarta has no urban drainage system. So, for example, most of the 700000 m^3 of wastewater produced daily goes directly to dikes, canals and rivers. Just a small proportion is pre-treated by septic tanks. The area is prone to seasonal flooding of streets, commercial properties and homes. As a response, existing drains have been re-aligned in some locations to route the stormwater more directly and more quickly to the sea. Sewerage pilot-schemes have been constructed, but finance is in short supply (Varis and Somlyody, 1997).

Problems

1.1 Do you think urban drainage is taken for granted by most people in developed countries? Why? Is this a good or bad thing?

1.2 How does urbanisation affect the natural water cycle?

1.3 Some claim that urban drainage engineers, throughout history, have saved more lives than doctors and nurses. Can that be justified, nationally and internationally?

1.4 Pollution from urban discharges to the water environment should be controlled in some way. What are the reasons for this? How should the limits be determined? Could there be such a thing as a requirement that is too strict? If so, why?

1.5 What have been the main influences on urban drainage engineers since the start of their profession?

References

Brokenshire, C.A. (1995) South West Water's 'clean sweep' programme: some engineering and environmental aspects. *Journal of the Institution of Water and Environmental Management*, **9**(6), December, 602–613.

CEC (2000) *Directive Establishing a Framework for Community Action in the Field of Water Policy*, 2000/60/EC.

Council of European Communities (1976) Directive concerning the quality of bathing water (76/160/EEC).

Council of European Communities (1991) Directive concerning urban waste water treatment (91/271/EEC).

Evans, E.P., Thorne, C.R., Saul, A., Ashley, R., Sayers, P.N., Watkinson, A., Penning-Rowsell, E.C. and Hall, J.W. (2003) *An Analysis of Future Risks of Flooding and Coastal Erosion for the UK Between 2030–2100. Overview of Phase 2*. Foresight Flood and Coastal Defence Project, Office of Science and Technology.

Firth, S.J. and Staples, K.D. (1995) North Tyneside bathing waters scheme. *Journal of the Institution of Water and Environmental Management*, **9**(1), February, 55–63.

Lei, P.C.K., Wong, H.Y., Liu, P.H. and Tang, D.S.W. (1996) Tackling sewage pollution in the unsewered villages of Hong Kong. *International Conference on Environmental Pollution, ICEP.3*, **1**, Budapest, April, European Centre for Pollution Research, 334–341.

Varis, O. and Somlyody, L. (1997) Global urbanisation and urban water: can sustainability be afforded? *Water Science and Technology*, 35(9), 21–32.

WaterVoice Yorkshire (2002) *WaterVoice calls for action to put an end to sewer flooding*. Press Release, June. www.watervoice.org.uk.

2   Approaches to urban drainage

2.1 Types of system: piped or natural

Development of an urban area can have a huge impact on drainage, as discussed in Section 1.2 and represented on Figs 1.2 and 1.3. Rain that has run off impermeable surfaces and travelled via a piped drainage system reaches a river far more rapidly than it did when the land and its drainage was in a natural state, and the result can be flooding and increased pollution. Rather than rely on 'end of pipe solutions' to these problems, the recent trend has been to try to move to a more natural means of drainage, using the infiltration and storage properties of semi-natural features.

Of course, artificial drainage systems are not universal anyway. Some isolated communities in developed countries, and many other areas throughout the world, have never had main drainage.

So, the first distinction between types of urban drainage system should be between those that are based fundamentally on pipe networks and those that are not.

Much of this chapter, and of this book, is devoted to piped systems, so let us now consider the alternatives to piped systems.

The movement towards making better use of natural drainage mechanisms has been given different names in different countries. In the US and other countries, the techniques tend to be called 'best management practices', or BMPs. In Australia the general expression 'water sensitive urban design' communicates a philosophy for water engineering in which water use, re-use and drainage, and their impacts on the natural and urban environments, are considered holistically. In the UK, since the mid-1990s, the label has been SUDS (Sustainable Urban Drainage Systems, or SUstainable Drainage Systems).

These techniques – including soakaways, infiltration trenches, swales, water butts, green roofs and ponds – concentrate on stormwater. They are considered in more detail in Chapter 21. Some schemes for reducing dependence on main drainage also involve more localised collection and treatment of *wastewater*. However, movements in this direction, while of great significance, are only in their early stages (as described in Chapter 24).

2.2 Types of piped system: combined or separate

Urban drainage systems handle two types of flow: wastewater and stormwater. An important stage in the history of urban drainage was the connection of wastewater to ditches and natural streams whose original function had been to carry stormwater. The relationship between the conveyance of wastewater and stormwater has remained a complex one; indeed, there are very few systems in which it is simple or ideal.

Piped systems consist of *drains* carrying flow from individual properties, and *sewers* carrying flow from groups of properties or larger areas. The word *sewerage* refers to the whole infrastructure system: pipes, manholes, structures, pumping stations and so on.

There are basically two types of conventional sewerage system: a *combined* system in which wastewater and stormwater flow together in the same pipe, and a *separate* system in which wastewater and stormwater are kept in separate pipes.

Some towns include hybrid systems, for example a 'partially-separate' system, in which wastewater is mixed with some stormwater, while the majority of stormwater is conveyed by a separate pipe. Many other towns have hybrid systems for more accidental reasons: for example, because a new town drained by a separate system includes a small old part drained by a combined system, or because wrong connections resulting from ignorance or malpractice have caused unintended mixing of the two types of flow.

We will now consider the characteristics of the two main types of sewerage system. Other types of drainage will be considered in Chapters 21, 23 and 24.

2.3 Combined system

In the UK, most of the older sewerage systems are combined and this accounts for about 70% by total length. Many other countries have a significant proportion of combined sewers: in France and Germany, for example, the figure is also around 70%, and in Denmark it is 45%.

A sewer network is a complex branching system, and Fig. 2.1 presents an extreme simplification of a typical arrangement, showing a very small proportion of the branches. The figure is a plan of a town located beside a natural water system of some sort: a river or estuary, for example. The combined sewers carry both wastewater and stormwater together in the same pipe, and the ultimate destination is the wastewater treatment plant (WTP), located, in this case, a short distance out of the town.

In dry weather, the system carries wastewater flow. During rainfall, the flow in the sewers increases as a result of the addition of stormwater. Even in quite light rainfall, the stormwater flows will predominate, and in heavy falls the stormwater could be fifty or even one hundred times the average wastewater flow.

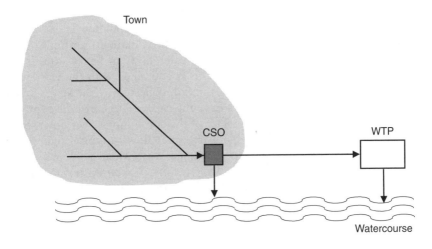

Fig. 2.1 Combined system (schematic plan)

It is simply not economically feasible to provide capacity for this flow along the full length of the sewers – which would, by implication, carry only a tiny proportion of the capacity most of the time. At the treatment plant, it would also be unfeasible to provide this capacity in the treatment processes. The solution is to provide structures in the sewer system which, during medium or heavy rainfall, divert flows above a certain level out of the sewer system and into a natural watercourse. These structures are called combined sewer overflows, or CSOs. A typically-located CSO is included in Fig. 2.1.

The basic function of a CSO is illustrated in Fig. 2.2. It receives inflow, which, during rainfall, consists of stormwater mixed with wastewater. Some flow is retained in the sewer system and continues to the treatment works – the continuation flow. The amount of this flow is an important

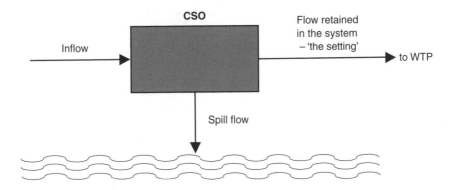

Fig. 2.2 CSO inflow and outflow

characteristic of the CSO, and is referred to as the 'setting'. The remainder is overflowed to the watercourse – the overflow or 'spill flow'.

It is useful at this point to consider the approximate proportions of flow involved. Let us assume that the stormwater flow, in heavy rain, is fifty times the average wastewater flow. This is combined with the wastewater flow that would exist regardless of rainfall, collected by the sewer system upstream of the CSO (which *does* have the capacity to carry the combined flow). Let us assume that the capacity of the continuing sewer downstream of the CSO is 8 times the average wastewater flow (a typical figure). The inflow is therefore fifty-one times average wastewater flow ($51 \times av$), made up of $50 \times av$ stormwater, plus, typically, $1 \times av$ wastewater. In this case the flow diverted to the river will therefore be $51 - 8 = 43 \times av$.

This diverted flow would seem to be a highly dilute mixture of rain-water and wastewater (ostensibly in the proportions 50 to 1). Also, CSOs are designed with the intention of retaining as many solids as possible in the sewer system, rather than allowing them to enter the watercourse. Therefore, the impact on the environment of this untreated discharge might appear to be slight. However, storm flows can be highly polluted, especially early in the storm when the increased flows have a 'flushing' effect in the sewers. There are also limits on the effectiveness of CSOs in retaining solids. And the figures speak for themselves! Most of the flow in this case is going straight into the watercourse, not onto the treatment works. To put it simply: CSOs cause pollution, and this is a significant drawback of the combined system of sewerage. The design of CSOs is con-sidered further in Chapter 12.

2.4 Separate system

Most sewerage systems constructed in the UK since 1945 are separate (about 30%, by total length). Fig. 2.3 is a sketch plan of the same town as shown on Fig. 2.1, but this time sewered using the separate system. Wastewater and stormwater are carried in separate pipes, usually laid side-by-side. Waste-water flows vary during the day, but the pipes are designed to carry the maximum flow all the way to the wastewater treatment plant. The storm-water is not mixed with wastewater and can be discharged to the water-course at a convenient point. The first obvious advantage of the separate system is that CSOs, and the pollution associated with them, are avoided.

An obvious disadvantage might be cost. It is true that the pipework in separate systems is more expensive to construct, but constructing two pipes instead of one does not cost twice as much. The pipes are usually constructed together in the same excavation. The stormwater pipe (the larger of the two) may be about the same size as the equivalent com-bined sewer, and the wastewater pipe will be smaller. So the additional costs are due to a slightly wider excavation and an additional, relatively small pipe.

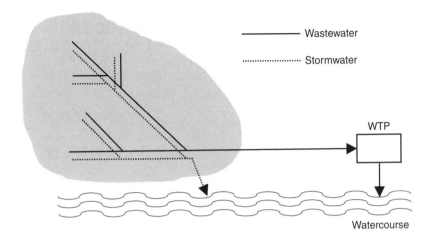

Fig. 2.3 Separate system (schematic plan)

Separate systems do have drawbacks of their own, and we must consider them now. The drawbacks relate to the fact that perfect separation is effectively impossible to achieve. First, it is difficult to ensure that polluted flow is carried only in the wastewater pipe. Stormwater can be polluted for many reasons, including the washing-off of pollutants from the catchment surface. This will be considered in more detail in Chapter 6. Second, it is very hard to ensure that no rainwater finds its way into the wastewater pipe. Rainwater enters the wastewater pipe by two main mechanisms: infiltration and direct inflow.

Infiltration

Infiltration to a pipe takes place when groundwater seeps in via imperfections: for example, cracks or damage from tree-roots or poor joints. It can take place in all types of sewer but is likely to cause the most problems in the wastewater pipe of a separate system because the extra water will have the most impact on the remaining pipe capacity. (Exfiltration, the leaking of liquid out of a sewer, can also be a problem, particularly in areas of sensitive groundwater. This will be considered in Chapter 4.)

Inflow

Direct inflow usually results from wrong connections. These may arise out of ignorance or deliberate malpractice. A typical example, which might belong to either category, is the connection of a home-made garden drain into the wastewater manhole at the back of the house. A survey of one separate system (Inman, 1975) found that 40% of all houses had some arrangement whereby stormwater could enter the wastewater sewer. It may at first sight seem absurd that a perfectly good infrastructure system can be put at risk by such mismanagement and human weakness, but it is a very real problem. Since a drainage system does not run under pressure, and is not 'secure', it is hard to stop people damaging the way it operates. In the USA, 'I and I' (infiltration and inflow) surveys can involve injecting smoke into a manhole of the wastewater system and looking out for smoke rising from the surface or roof drainage of guilty residents!

2.5 Which sewer system is better?

This obvious question does not have a simple answer. In the UK, new developments are normally given separate sewer systems, even when the new system discharges to an existing combined system. As has been described in Chapter 1, during the 1950s, engineers started to pay particular attention to the pollution caused by CSOs, and this highlighted the potential advantages of eliminating them by using separate systems. It was quite common for consulting engineers, when asked to investigate problems with a combined sewer system, to recommend in their report a solution like the rebuilding of a CSO, but to conclude with a sentence like, 'Of course the long-term aim should be the replacement of the entire combined system by a separate one; however this is not considered economically feasible at present'. No wonder it wasn't considered feasible! The expense and inconvenience of a large-scale excavation in every single street in the town, together with all the problems of coping with the flows during construction and reconnecting every property, would have been a major discouragement, to say the least.

As the philosophy of sewer rehabilitation took hold in the 1980s, this vague ideal for the future was replaced by the more pragmatic approach of 'make best use of what's there already'. Many engineers reassessed the automatic assumption that the separate system was the better choice. This was partly a result of increasing experience of separate systems and the problems that go with them. One of the main problems – the difficulty of keeping the system separate – tends to get worse with time, as more and more incorrect connections are made. Theoretical studies have shown that only about one in a hundred wrong connections would nullify any pollution advantage of separate sewers over combined ones (Nicholl, 1988). There was also increasing awareness that stormwater is not 'clean'. The application of new techniques for improving CSOs, combined with the use of sewer system computer models to fine-tune proposals for rehabilitation works, led to significant reductions in the pollution caused by many existing combined sewer systems. So, by the early 1990s, while few were proposing that all new systems should be combined, the fact that there were a large number of existing combined systems was not, in itself, a major source of concern.

Recently, the goal of more sustainable urban drainage has drawn new attention to particular shortcomings of combined systems: the unnatural mixing of waterborne waste with stormwater, leading to the expensive and energy-demanding need for re-separation, and the risk of environmental pollution. So current thinking suggests that while existing systems – combined or separate – may continue to be improved and developed, it is most unlikely that they would be converted wholesale from one type to the other. If drainage practices for new developments change, it is likely to be in the direction of increased use of source control (non-piped) methods of handling stormwater, to be described in Chapter 21, and certainly not a return to combined sewers.

All this suggests that there is no need to answer the question 'Which system is better?', but it is still worthwhile reflecting in some detail on the advantages and disadvantages of separate and combined systems, in order to highlight the operational differences between existing systems of the two types.

First we should consider some typical characteristics. Maximum flow of wastewater in a separate system, as a multiple of the average wastewater flow, depends on the size and layout of the catchment. Typically the maximum is 3 times the average. In a combined system, the traditional capacity at the inlet to a wastewater treatment plant (in the UK) is 6 times average wastewater flow; of this, 3 times the average is diverted to storm tanks and 3 times is given full treatment. Therefore during rainfall, a combined sewer (downstream of a CSO) is likely to be carrying at least 6 times average wastewater flow, whereas the wastewater pipe in a separate system is likely to carry no more than 3 times the average.

This, together with the construction methods outlined in Section 2.3, and the obvious fact that, during rain, combined sewers carry a mixture of two types of flow, give rise to a number of differences between combined and separate systems. One interesting advantage of the combined system is that, if the wastewater flow is low, and, in light rain, the combined flow does not exceed 3 times average wastewater flow, all the stormwater (which may be polluted) is treated. In a separate system, none of that stormwater would receive treatment.

A list of advantages and disadvantages is given in Table 2.1.

2.6 Urban water system

As described, the most common types of sewerage system are combined, separate and hybrid. In this section we will look at how these pipe networks fit within the whole urban water system. Figs 2.4 and 2.5 are diagrammatic representations of the system. They do not show individual pipes, structures or processes, but a general representation of the flow paths and the interrelationship of the main elements. Solid arrows represent intentional flows and dotted arrows unintentional ones.

Table 2.1 Separate and combined system, advantages and disadvantages

Separate system	Combined system
Advantages	**Disadvantages**
No CSOs – potentially less pollution of watercourses.	CSOs necessary to keep main sewers and treatment works to feasible size. May cause serious pollution of watercourses.
Smaller wastewater treatment works.	Larger treatment works inlets necessary, probably with provision for stormwater diversion and storage.
Stormwater pumped only if necessary.	Higher pumping costs if pumping of flow to treatment is necessary.
Wastewater and storm sewers may follow own optimum line and depth (for example, stormwater to nearby outfall).	Line is a compromise, and may necessitate long branch connections. Optimum depth for stormwater collection may not suit wastewater.
Wastewater sewer small, and greater velocities maintained at low flows.	Slow, shallow flow in large sewers in dry weather flow may cause deposition and decomposition of solids.
Less variation in flow and strength of wastewater.	Wide variation in flow to pumps, and in flow and strength of wastewater to treatment works.
No road grit in wastewater sewers.	Grit removal necessary.
Any flooding will be by stormwater only.	If flooding and surcharge of manholes occurs, foul conditions will be caused.
Disadvantages	**Advantages**
Extra cost of two pipes.	Lower pipe construction costs.
Additional space occupied in narrow streets in built-up areas.	Economical in space.
More house drains, with risk of wrong connections.	House drainage simpler and cheaper.
No flushing of deposited wastewater solids by stormwater.	Deposited wastewater solids flushed out in times of storm.
No treatment of stormwater.	Some treatment of stormwater.

Heavy-bordered boxes indicate 'sources' and dashed, heavy-bordered boxes show 'sinks'.

Combined

Figure 2.4 shows this system for a combined sewer network. There are two main inflows. The first is rainfall that falls on to catchment surfaces such as 'impervious' roofs and paved areas and 'pervious' vegetation and soil. It is at this point that the quality of the flow is degraded as pollutants on the

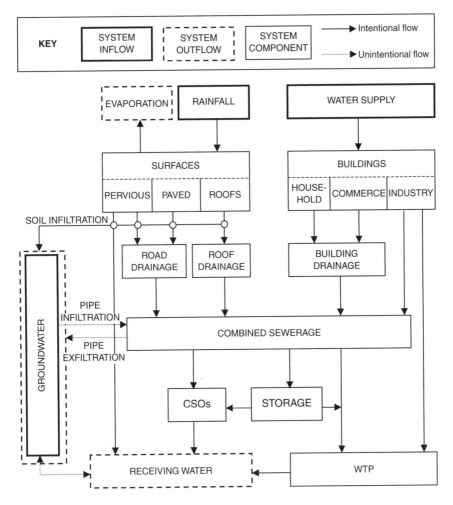

Fig. 2.4 Urban water system: combined sewerage

catchment surfaces are washed off. This is a highly variable input that can only be properly described in statistical terms (as will be considered in Chapter 5). The resulting runoff retains similar statistical properties to rainfall (Chapter 6). There is also the associated outflow of evaporation, whereby water is removed from the system. This is a relatively minor effect in built-up, urban areas. Rainfall that does not run off will find its way into the ground and eventually the receiving water. The component that runs off is conveyed by the roof and highway drainage as stormwater directly into the combined sewer.

The second inflow is water supply. Water consumption is more regular than rainfall, although even here there is some variability (Chapter 4). The resulting wastewater is closely related in timing and magnitude to the water supply. The wastewater is conveyed by the building drainage directly to the combined sewer. An exception is where industry treats its own waste separately and then discharges treated effluent directly to the receiving water. The quality of the water (originally potable) deteriorates during usage.

The combined sewers collect stormwater and wastewater and convey them to the wastewater treatment plant. Unintentional flow may leave the pipes via exfiltration to the ground. At other locations, groundwater may act as a source and add water into the system via pipe infiltration. This is of relatively good quality and dilutes the normal flow. In dry weather, the flow moves directly to the treatment plant with patterns related to the water consumption. During significant rainfall, much of the flow will discharge directly to the receiving water at CSOs (Chapter 12). Discharges are intermittent and are statistically related to the rainfall inputs. If storage is provided, some of the flow may be temporarily detained prior to subsequent discharge either via the CSO or to the treatment plant. The treatment plant will, in turn, discharge to the receiving water.

Separate

The diagram shown in Fig. 2.5 is similar to Fig. 2.4, except that it depicts a separate system with two pipes: one for stormwater and one for wastewater. The separate storm sewers normally discharge directly to a receiving water. The separate wastewater sewers convey the wastewater directly to the treatment plant. As with combined sewers, both types of pipe are subject to infiltration and exfiltration. In addition, as has been discussed, wrong connections and cross-connections at various points can cause unintentional mixing of the stormwater and wastewater in either pipe.

Hybrid

Many older cities in the UK have a hybrid urban drainage system that consists of a combined system at its core (often in the oldest areas) with separate systems at the suburban periphery. The separate wastewater sewers discharge their effluent to the core combined system, but the storm sewers discharge locally to receiving waters. This arrangement has prolonged the life of the urban wastewater system as the older core section is only subjected to the relatively small extra wastewater flows whilst the larger storm flows are handled locally.

Problems

2.1 'Mixing of wastewater and stormwater (in combined sewer systems) is fundamentally irrational. It is the consequence of historical accident,

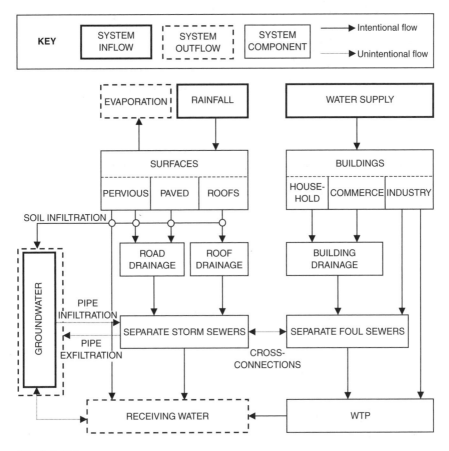

Fig. 2.5 Urban water system: separate sewerage

and remains a cause of significant damage to the water environment.'
Explain and discuss this statement.

2.2 Explain the characteristics of the combined and separate systems of
sewerage. Discuss the advantages and disadvantages of both.

2.3 There are two main types of sewerage system: combined and separate.
Is one system better than the other? Should we change what already
exists?

2.4 Why is it hard to keep separate systems separate? What causes the
problems and what are the consequences?

2.5 Describe how combined and separate sewer systems interact with the
overall urban *water* system. (Use diagrams.)

Key sources

Marsalek, J., Barnwell, T.O., Geiger, W., Grottker, M., Huber, W.C., Saul, A.J., Schilling, W. and Torno, H.C. (1993) Urban drainage systems: design and operation. *Water Science and Technology*, 27(12), 31–70.

Van de Ven, F.H.M., Nelen, A.J.M. and Geldof, G.D. (1992) Urban drainage, in *Drainage Design* (eds P. Smart & J.G. Herbertson). Blackie & Sons.

References

Inman, J. (1975) Civil engineering aspects of sewage treatment works design. *Proceedings of the Institution of Civil Engineers, Part 1*, 58, May, 195–204, discussion, 669–672.

Nicholl, E.H. (1988) *Small Water Pollution Control Works: Design and Practice*, Ellis Horwood, Chichester.

3   Water quality

3.1 Introduction

In the past, there has been a tendency amongst civil engineers not to concern themselves in any detail with the quality aspects of wastewater and stormwater which is conveyed in the systems they design and operate. This is a mistake for several reasons.

- Significant quality changes can occur in the drainage system.
- Decisions made in the sewer system have significant effects on the WTP performance.
- Direct discharges from drainage systems (e.g. combined sewer overflows, stormwater outfalls) can have a serious pollutional impact on receiving waters.

Therefore, this chapter looks at the basic approaches to characterising wastewater and stormwater including outlines of the main water quality tests used in practice. Typical test data is given in Chapters 6 and 7. It considers water quality impacts of discharges from urban drainage systems, and relevant legislation and water quality standards.

3.2 Basics

3.2.1 Strength

Water has been called the 'universal solvent' because of its ability to dissolve numerous substances. The term 'water quality' relates to all the constituents of water, including both dissolved substances and any other substances carried by the water.

The *strength* of polluted liquid containing a constituent of mass M in water of volume V is its *concentration* given by $c = M/V$, usually expressed in mg/l. This is numerically equivalent to parts per million (ppm) assuming the density of the mixture is equal to the density of water (1000 kg/m^3). The plot of concentration c as a function of time t is known as a *pollutograph*

Example 3.1

A laboratory test has determined the mass of constituent in a 2 litre waste-water sample to be 0.75 g. What is its concentration (c) in mg/l and ppm? If the wastewater discharges at a rate of 600 l/s, what is the pollutant load-rate (L)?

Solution

$$c = \frac{M}{V} = \frac{750}{2} = 375 \text{ mg/l} = 375 \text{ ppm}$$

$$L = cQ = 0.375 \times 600 = 225 \text{ g/s}$$

(see Fig. 12.9 for an example). Pollutant mass-flow or flux is given by its *load-rate* $L = M/t = cQ$ where Q is the liquid flow-rate.

In order to calculate the average concentration, either of wastewater during the day or of stormwater during a rain event, the *event mean concentration* (EMC) can be calculated as a flow weighted concentration c_{av}:

$$c_{av} = \frac{\Sigma Q_i c_i}{Q_{av}} \qquad (3.1)$$

c_i concentration of each sample i (mg/l)
Q_i flow rate at the time the sample was taken (l/s)
Q_{av} average flow-rate (l/s).

3.2.2 Equivalent concentrations

It is common practice when dealing with a pollutant (X) that is a compound to express its concentration in relation to the parent element (Y). This can be done as follows:

*Concentration of compound X **as element Y** =*

$$\textit{concentration of compound X} \times \frac{\textit{atomic weight of element Y}}{\textit{molecular weight of compound X}} \qquad (3.2)$$

The conversion of concentrations is based on the gram molecular weight of the compound and the gram atomic weight of the element. Atomic weights for common elements are given in standard texts (e.g. Droste, 1997).

Expressing substances in this way allows easier comparison between different compounds of the same element, and more straightforward calculation of totals. Of course, it also means care needs to be taken in noting in which form compounds are reported (see Example 3.2).

Example 3.2

A laboratory test has determined the mass of orthophosphate (PO_4^{3-}) in a 1 l stormwater sample to be 56 mg. Express this in terms of phosphorus (P).

Solution

Gram atomic weight of P is 31.0 g
Gram atomic weight of O is 16.0 g
Gram molecular weight of orthophosphate is $31 + (4 \times 16) = 95$ g
Hence from equation 3.2:

$$56 \text{ mg } PO_4^{3-}/l = 56 \times \frac{31 \text{ g}P}{95 \text{ g } PO_4^{3-}} = 18.3 \ PO_4^{3-} - P/l$$

3.3 Parameters

There is a wide range of quality parameters used to characterise wastewater and these are described in the following section. Further details on these and many other water quality parameters and their methods of measurement can be found elsewhere (e.g. DoE various; AWWA, 1992). Specific information on the range of concentrations and loads encountered in practice is given in Chapters 6 (wastewater) and 7 (stormwater).

3.3.1 Sampling and analysis

There are three main methods of sampling: grab, composite and continuous. *Grab* samples are simply discrete samples of fixed volume taken to represent local conditions in the flow. They may be taken manually or extracted by an automatic sampler. A *composite* sample consists of a mixture of a number of grab samples taken over a period of time or at specific locations, taken to more fully represent the composition of the flow. *Continuous* sampling consists of diverting a small fraction of the flow over a period of time. This is useful for instruments that give almost instantaneous measurements, e.g. pH, temperature.

In sewers, where flow may be stratified, samples need to be taken throughout the depth of flow if a true representation is required. Mean concentrations can then be calculated by weighting with respect to the local velocity and area of flow.

In all of the tests available to characterise wastewater and stormwater, it is necessary to distinguish between precision and accuracy. In the context of laboratory measurements, *precision* is the term used to describe how well the analytical procedure produces the same result on the same sample when the test is repeated. *Accuracy* refers to how well the test reproduces the actual value. It is possible, for example, for a test to be very precise, but very inaccurate with all values closely grouped, but around the wrong value! Techniques that are both precise and accurate are required.

3.3.2 Solids

Solid types of concern in wastewater and stormwater can broadly be categorised into four classes: gross, grit, suspended and dissolved (see Table 3.1). Gross and suspended solids may be further sub-divided according to their origin as wastewater and stormwater.

Gross solids

There is no standard test for the gross solids found in wastewater and stormwater, but they are usually defined as solids (specific gravity (SG) = 0.9–1.2) captured by a 6 mm mesh screen (i.e. solids >6 mm in two dimensions). Gross sanitary solids (also variously known as aesthetic, refractory or intractable solids) include faecal stools, toilet paper and 'sanitary refuse' such as women's sanitary protection, condoms, bathroom litter, etc. Faecal solids and toilet paper break up readily and may not travel far in the system as gross solids. Gross stormwater solids consist of debris such as bricks, wood, cans, paper, etc.

The particular concern about these solids is their 'aesthetic impact' when they are discharged to the aqueous environment and find their way onto riverbanks and beaches. They can also cause maintenance problems by deposition and blockage, and can cause blinding of screens at WTPs, particularly during storm flows.

Grit

Again, there is no standard test for determination of grit, but it may be defined as the inert, granular material (SG ≈ 2.6) retained on a 150 μm sieve. Grit forms the bulk of what is termed *sewer sediment* and the nature and problems associated with this material will be returned to in Chapter 16.

Table 3.1 Basic classification of solids

Solid type	Size (μm)	SG (–)
Gross	>6000	0.9–1.2
Grit	>150	2.6
Suspended	≥0.45	1.4–2.0
Dissolved	<0.45	–

Suspended solids

The suspended solids (SS) content is the solid matter (both organic and inorganic) maintained in suspension, and retained when a sample is filtered (0.45 μm pore size). In the SS test, the residue is washed, dried and weighed under standard conditions and expressed as a concentration. The accuracy of the SS test is approximately ±15%.

The finer fractions of suspended solids (<63 μm) are extremely efficient carriers of pollutants, carrying greater than their proportionate share (see Sections 6.4.2 and 16.5.2). High concentrations may have a number of adverse effects on the receiving water, including increased turbidity, reduced light penetration, blanketing of the bed, and interference with many types of fish and aquatic invertebrates. Even after deposition, the pollutants attached to these enriched sediments still present a risk, since they can cause a 'delayed' sediment oxygen demand (see Section 3.4.3), or may be resuspended at high flows.

By definition, solids not in suspension (i.e. with a diameter <0.45 μm) are dissolved (see Example 3.3).

Volatile solids

The solids retained during the SS test can be ignited at 550 °C in a muffle furnace. The residue is known as non-volatile or fixed material. The volatilised fraction (the volatile solids) gives an indication of the organic content of the SS.

3.3.3 Oxygen

A key to understanding the reactions occurring anywhere within the urban drainage system is measurement and prediction of the oxygen levels in the aqueous phase. Dissolved oxygen (DO) levels depend on physical, chemical and biochemical activities in the system.

Dissolved oxygen

Oxygen in water is only sparingly soluble. In equilibrium with air, the solubility of DO in water is referred to as its *saturation* value, and it decreases with the increase of both temperature and purity (salinity, solids content) and with the decrease in atmospheric pressure (Table 3.2). Hence, warm water (even with no impurities) is, in effect, a pollutant source.

Dissolved oxygen (DO) can be measured analytically using the Winkler titration method. Titration is a laboratory technique where measured volumes of a reagent (the titrant) are incrementally added to a sample up to the equivalent amount of the constituent being analysed. Membrane electrodes are now more commonly and conveniently used both in the laboratory and for *in situ* measurements.

Example 3.3

In a standard laboratory test, a crucible and filter pad are dried and their combined mass measured at 64.592 g. A 250 ml wastewater sample is drawn through the filter under vacuum. The filter and residue are then placed on the crucible in an oven at 104 °C for drying. The new combined mass is 64.673 g. The crucible and its contents are next placed in a muffle furnace at 550 °C. After cooling, the combined mass is measured as 64.631 g. Determine (a) the suspended solids concentration of the sample, (b) the volatile fraction of the suspended solids.

Solution

Mass of suspended solids removed:
Crucible + filter + solids	= 64.673 g
Crucible + filter	= 64.592 g
Mass of suspended solids	= 0.081 g

Concentration of SS:
81 (mg)/0.250 (l)	= 324 mg/l

Mass of volatile suspended solids removed:
Initial crucible + filter + solids	= 64.673 g
Final crucible + filter + solids	= 64.631 g
Mass of volatile solids	= 0.042 g

Volatile fraction of suspended solids:
42 (mg)/81 (mg)	= 0.52

The concentration of DO is an excellent indicator of the 'health' of a receiving water. All the higher forms of river life require oxygen. Coarse fish, for example, require in excess of 3 mg/l (see Table 3.3). In the absence of toxic impurities, there is a close correlation between DO and biodiversity.

Table 3.2 Dissolved oxygen concentration (under standard conditions) in water as function of temperature

Temperature (°C)	DO (mg/l)
0	14.62
5	12.80
10	11.33
15	10.15
20	9.17
25	8.38
30	7.63

Table 3.3 Oxygen requirements of fish species (adapted from Gray, 1999)

Characteristic species	Minimum DO concentration (mg/l)	Minimum saturation (%)	Comment
Trout, bullhead	7–8	100	Fish require much oxygen
Perch, minnow	6–7	<100	Need more oxygen for active life
Roach, pike, chub	3	60–80	Can live for long periods at this level
Carp, tench, bream	<1	30–40	Can live for short periods at this level

3.3.4 Organic compounds

Wastewater and stormwater contain significant quantities of organic matter in both particulate and soluble form. Organic compounds in water are unstable and are readily oxidised either biologically or chemically to stable, relatively inert, end products such as carbon dioxide, nitrates, sulphates and water. There are three main categories of biodegradable organics present:

• carbohydrates such as sugars, starch and cellulose
• proteins which are complex molecules built up of amino acids and urea
• lipids and fats.

Decomposition of organic matter by micro-organisms consumes DO. In urban drainage systems the main implication is oxygen depletion in:

• sewers, resulting in an anaerobic environment (considered in Chapter 17)
• receiving waters (considered later in this chapter).

An indirect indication of the amount of organic material in a wastewater can be derived from one of two tests: the biochemical oxygen demand value (BOD) or the chemical oxygen demand (COD). A third option is the total organic carbon (TOC) test which gives a more direct measure of the carbon content of the sample under test.

Biochemical oxygen demand (BOD_5)

This test is a laboratory simulation of the microbial processes occurring in water contaminated with organic compounds. It measures the DO consumed in a sample diluted in a 300 ml bottle during a specified incubation period (usually 5 days at a temperature of 20 °C in darkness). The DO is used by micro-organisms as they break down organic material and certain inorganic compounds. Thus:

$$BOD_5 = (c_{DOI} - c_{DOF})/p \tag{3.3}$$

p sample dilution = volume of sample/volume of bottle
c_{DOI} initial dissolved oxygen concentration (mg/l)
c_{DOF} final dissolved oxygen concentration (mg/l)

Measured amounts of the wastewater sample are diluted with prepared water containing nutrients and DO. Seed micro-organisms are added if insufficient are available in the sample itself. Equation 3.3 assumes the dilution water has negligible BOD_5 (see Example 3.4). The test may also measure the oxygen used to oxidise reduced forms of nitrogen (nitrogenous demand – NBOD) unless an inhibitor (e.g. allylthiourea (ATU)) is used. The evolution of BOD with time is shown in Fig. 3.1.

The test does not give a measure of the *total* oxidisable organic matter because of the presence of considerable quantities of carbonaceous matter resistant to biological oxidation. Over 5 days, only the readily biodegradable fraction of organic material present in the water will be broken down.

The test can be extended up to 10–20 days to reach the ultimate carbonaceous CBOD ≈ 1.5 BOD_5. BOD tests are subject to inhibition if the wastewater contains any toxic components (e.g. trace metals in runoff) and they should be seen as an indication rather than as an accurate determination.

BOD_5 is a very common parameter used in the control of treated wastewater effluent quality through the setting and monitoring of discharge consent standards (e.g. 25 mg/l (CEC, 1991a)). Rivers are considered to be

---

Example 3.4

A laboratory test for BOD_5 (ATU) is carried out by mixing a 5 ml sample with distilled water into a 300 ml bottle. Prior to the test the DO concentration of the mixture was 7.45 mg/l and after 5 days it had reduced to 1.40 mg/l. What is the BOD_5 concentration of the sample?

Solution

     Dilution, $p = 5/300 = 0.0167$

Equation 3.3:

     $BOD_5 = (7.45 - 1.40)/0.0167 = 363$ mg/l

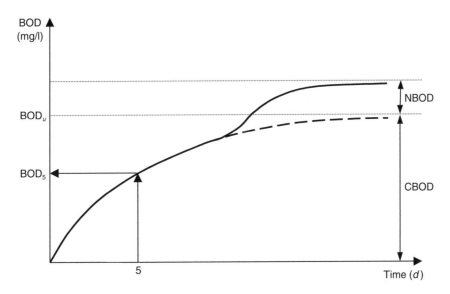

Fig. 3.1 Development of biochemical oxygen demand (BOD) with time

polluted if their BOD_5 exceeds 5 mg/l, and 3 mg/l is required for salmonid rivers (CEC, 1978).

Chemical oxygen demand (COD)

The COD test measures the oxygen equivalent of the organic matter that can be oxidised by a strong chemical oxidising agent (potassium dichromate) in an acidic medium. The standard test lasts for 3 hours using either a titrimetric or colorimetric method. Colorimetric test methods rely on measuring the intensity of light from colour changes in the reaction. Almost all organic compounds are oxidised. Some inorganics are also oxidised but ammonium and ammonia are not. Thus, this measurement is a good estimation of the total content of organic matter.

One of the main limitations of the COD test is its inability to differentiate between biodegradable and biologically inert organic matter. However, if sufficient data is available, it is often possible to empirically relate the COD with BOD values, such as:

$$c_{BOD} \approx a \times c_{COD} \qquad (3.4)$$

c_{BOD} biochemical oxygen demand concentration (mg/l)
c_{COD} chemical oxygen concentration (mg/l)
a 0.4–0.8 (–)

It should be stressed, however, that '*a*' will vary between wastewaters, and no universal relationship has been found between the two parameters. Nevertheless it is a good indicator of the wastewater treatability.

The COD of a sample can be further differentiated into several forms. The first major category is inert (suspended or soluble) material that is non-biodegradable within the timescale associated with an urban drainage system. The second is biodegradable matter, which in turn can be divided into readily and slowly degradable material. The former refers to material that can be immediately oxidised by micro-organisms and the latter to matter which is degraded more slowly. The approximate relationship between the COD fractions and BOD is summarised in Fig. 3.2. The methods for characterising the various COD fractions are still under development and are not yet standardised (Henze, 1992).

Total organic carbon (TOC)

Unlike the BOD and COD tests, the TOC test directly measures the total organic carbon content of a sample. The test is based on the fact that carbon dioxide (CO_2) is a product of combustion. It is carried out in an instrument containing a small furnace at 950 °C in the presence of a catalyst (having first removed the *inorganic* carbon). After this, the CO_2 released is measured for the known volume of sample. It can be performed very rapidly using a single instrument and is especially applicable in the analysis of small concentrations of organic matter.

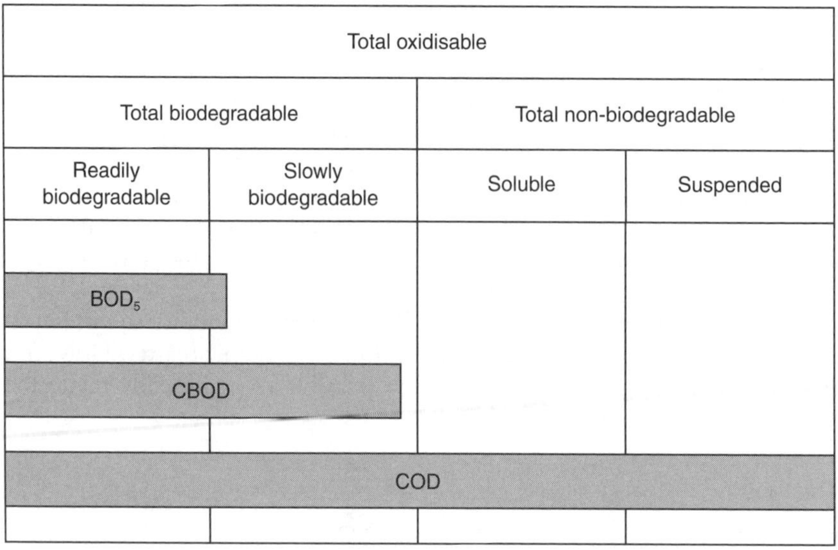

Total oxidisable			
Total biodegradable		Total non-biodegradable	
Readily biodegradable	Slowly biodegradable	Soluble	Suspended
BOD$_5$			
CBOD			
COD			

Fig. 3.2 Relationship between BOD and COD for organic matter

COD is approximately related to TOC as follows (see Example 3.5):

$$COD \approx 2.5 \times TOC \qquad (3.5)$$

Example 3.5

If organic material can be represented by the chemical formula $C_6H_{12}O_6$ (glucose), calculate the *theoretical* relationship between COD and TOC.

Solution

In the COD and TOC tests, organic material is oxidised to carbon dioxide and water:

$$C_6H_{12}O_6 + 6O_2 \rightarrow 6CO_2 + 6H_2O$$

From the above equation, each mole of organic material (molecular weight = 180) requires 6 moles of oxygen (molecular weight 32) for oxidation. As the weight in grams of a substance is the number of moles × molecular weight:

$$COD = \frac{6 \text{ moles } O_2}{1 \text{ mole } C_6H_{12}O_6} = \frac{6 \times 32}{1 \times 180} = 1.067 \text{ gO}_2/\text{gC}_6H_{12}O_6$$

Also, each mole of organic material contains 6 atoms of carbon, so:

$$TOC = \frac{6 \text{ atoms } C}{1 \text{ mole } C_6H_{12}O_6} = \frac{6 \times 12}{1 \times 180} = 0.4 \text{ gC/gC}_6H_{12}O_6$$

$$\therefore \quad COD = \frac{1.067}{0.4} \times TOC = 2.67 TOC$$

3.3.5 Nitrogen

Nitrogen exists in four main forms: organic (in the protein that makes up much matter), ammonia (or ammonia salts), nitrite and nitrate. Total nitrogen is the sum of all forms although, in wastewater and stormwater, organic and ammonia nitrogen make up most of the total. The concentration of nitrogen in domestic wastewater is usually related to the BOD_5.

Excessive levels of nitrogen discharged to receiving waters can promote the growth of undesirable aquatic plants such as algae and floating macrophytes. In severe cases, the receiving water can experience *eutrophic* symptoms such as water discoloration, odours and depressed oxygen levels (considered in more detail later in this chapter).

Organic nitrogen (org.N)

Organic nitrogen includes such natural materials as proteins and peptides, nucleic acids and urea, and numerous synthetic organic materials although not *all* organic nitrogen compounds.

Analytically, organic nitrogen and ammonia are determined together using the Kjeldahl method. In this test, the aqueous sample is boiled to remove any pre-existing ammonia and then digested, during which the organic nitrogen is converted to ammonia. The amount of ammonia produced is then determined as detailed in the next section.

Ammonia nitrogen (NH₃–N)

Ammonia nitrogen exists in solution in two forms, as the ammonium ion (NH_4^+) and as ammonia gas (NH_3) depending on the pH and temperature of the wastewater.

$$NH_3 + H^+ \rightleftharpoons NH_4^+ \tag{3.6}$$

At values of pH ≤ 7, virtually all the ammonia is present as ammonium. At pH 9, for example, 35% is present as NH_3.

Ammonia nitrogen is determined analytically by raising the pH, and using a distillation process. The final measurement is then made by titration or by colorimetry. Sometimes organic nitrogen and ammonia nitrogen are determined together in the total Kjeldahl nitrogen (TKN) test, which is similar to the basic Kjeldahl method except any pre-existing ammonia is *not* removed.

Any unionised ammonia (NH_3) present in a wastewater discharged to the environment is particularly toxic to fish, depending on the dissolved oxygen of the receiving water. Ammonia also exerts an oxygen demand during its conversion to nitrite and subsequently to nitrate.

Nitrite and nitrate nitrogen (NO₂⁻–N, NO₃⁻–N)

Nitrite is the intermediate oxidation state of nitrogen. It is relatively unstable, easily oxidised and its presence shows that oxidation of nitrogenous matter is taking place. Nitrate forms the most highly oxidised state of nitrogen found in wastewater. Determination is usually by colorimetric methods.

3.3.6 Phosphorus

Phosphorus can be expressed as total, organic or inorganic (ortho- and poly-) phosphorus. Organic phosphate is a minor constituent of wastewater and stormwater, with most phosphorus being in the inorganic form. Polyphosphates consist of combinations of phosphorus, oxygen and hydrogen atoms. Orthophosphates (e.g. PO_4^{3-}, HPO_4^{2-}, $H_2PO_4^-$, H_3PO_4) are simpler compounds and may be in solution or attached to particles. Orthophosphates can be determined directly, but poly- and organic phosphate must first be converted to orthophosphates before determination.

Phosphorus-containing compounds are also implicated in receiving water eutrophication. Generally, phosphorus is the controlling nutrient in urban freshwater systems. Salmonid rivers have upper limits of 0.065 mg P/l (CEC, 1978).

3.3.7 Sulphur

Sulphurous compounds are found mainly in wastewater in the form of organic compounds and sulphates (SO_4^{2-}). Under anaerobic conditions, these are reduced to sulphides (S^-), mercaptans and certain other compounds. The principal product, hydrogen sulphide (H_2S), is formed mainly by biofilms on the walls of sewers and in sediment deposits.

Hydrogen sulphide is a flammable and very poisonous gas and, when escaping into the atmosphere, can cause serious odour nuisance. It is acutely toxic to aquatic organisms and could be a factor in fish kills near CSOs. Hydrogen sulphide in damp conditions can be oxidised biologically to sulphuric acid (H_2SO_4) which may cause serious damage to sewer materials, especially concrete (see Chapter 17).

3.3.8 Hydrocarbons and FOG

Hydrocarbons are organic compounds containing only carbon and hydrogen. They are classified into four groups based on molecular structure: aliphatic or straight-chained, branch-chained, aromatic (based on the benzene ring) and alicyclic. In this book, we are mainly concerned with the petroleum-derived group commonly found in stormwater, which includes petrol, lubricating and road oils. They are among the more stable organic compounds and are not easily biodegraded. Most have a strong affinity for suspended particulate matter. They are determined by extraction with carbon tetrachloride.

Hydrocarbons are lighter than water, and virtually insoluble, causing films and emulsions on the water surface and reducing atmospheric re-aeration. Those in accumulated sediments can persist for long periods and exert a chronic impact on bottom-dwelling organisms, as well as being remobilised by subsequent storm events.

FOG is a general term used to include the fats, oils, greases and waxes of plant or food-based origin present in wastewater. They are determined gravimetrically by extraction with trichlorotrifluroethane (Freon).

Fats in sewer systems can cause blockages. When discharged to the environment they cause films and sheens on the water surface.

3.3.9 Heavy metals and synthetic compounds

A considerable number of heavy metals and synthetic organic and inorganic chemicals can be found in wastewater and stormwater. Among the many constituents of concern are metal species such as arsenic, cyanide, lead, cadmium, iron, copper, zinc and mercury. Metals can exist in particulate, colloidal and dissolved (labile) phases depending mainly on the prevailing redox and pH conditions. The concentration of individual metals is often determined by atomic absorption spectrophotometry, although newer, multi-element equipment (e.g. ICP) is becoming more widespread.

Metals in stormwater are predominantly in the particulate phase. This is important because the environmental mobility and bioavailability (and hence toxicity) of metals is highly related to their concentration in solution. Many (particularly the more soluble forms of zinc and copper) are known to have toxic effects on aquatic life and can inhibit biological processes at the WTP.

Herbicides and pesticides can be toxic to a variety of aquatic life at very low concentrations. Some of the more toxic varieties (e.g. chlorinated organics, DDT and PCBs) are no longer used but their residues can still be found in the environment.

3.3.10 Micro-organisms

Direct determination of the presence of pathogenic micro-organisms (e.g. salmonella, enteroviruses) in wastewater and stormwater is not normally carried out. Indeed, even microbiological indicator tests are not routinely undertaken. However, bacterial indicator organisms such as total coliforms and faecal coliforms (*E.coli.*), and faecal streptococci (FS) are known to occur in large numbers in both wastewater and stormwater.

One of the major objectives of urban drainage systems (as discussed in Chapter 1) is the protection of public health, particularly reducing the risk associated with human contact with excreta. Wastewater, even when treated at WTPs, is not routinely disinfected before discharge to a receiving water unless it is classified as a bathing water (see below). Discharges from CSOs and SWOs are not disinfected either, and bacterial standards for recreational activity can be violated during even modest storm events.

3.4 Processes

3.4.1 Hydrolysis

Hydrolysis is an important precursor to both aerobic and anaerobic transformations. It consists of the natural reaction of large organic molecules with water in the presence of enzymes to produce smaller molecules that are potentially available for utilisation by bacteria. It is temperature dependent.

Suspended organic particles are hydrolysed, bringing them into solution. For example, insoluble cellulose is slowly hydrolysed in two stages to form dextrose (Inman, 1979):

$$(C_6H_{10}O_5)n + nH_2O \rightarrow nC_6H_{12}O_6$$

Urea represents 80% of the nitrogen content of fresh wastewater. It is relatively rapidly hydrolysed to ammonia-nitrogen (by the enzyme *urease*) at a rate of 3 mg.N/l per hour at 12 °C in stored samples (Painter, 1958). Polyphosphates will slowly hydrolyse to the orthophosphate form.

3.4.2 Aerobic degradation

Aerobic processes are those carried out by aerobic bacteria in the presence of free oxygen, whereby larger, but soluble, organic molecules are degraded to simple and stable end-products. Such micro-organisms may be freely suspended individually or in flocs in the flow, or be attached to pipe walls or sediment beds as biofilms.

The particular class of organism that carries out this reaction is aerobic *heterotrophic* bacteria. These take organic material as a 'food' source to provide energy or to synthesise new bacterial cells. End-products of this reaction are CO_2, H_2O and other oxidised forms such as nitrate, phosphate and sulphate. This can be simply represented as:

$$C,H,O,N,P,S + O_2 \rightarrow H_2O + CO_2 + NO_3^- + PO_4^{3-} + SO_4^{2-} + \text{new cells} + \text{energy}$$

Unless it is replaced, the oxygen in water can quickly be used up producing an environment hostile to aerobic micro-organisms. Conditions in the bulk flow in gravity sewers are likely to be aerobic.

Nitrification

In an aerobic environment with low levels of organic material, heterotrophic micro-organisms will no longer be able to thrive. However, another group of organisms known as *autotrophs* (nitrosomas and nitrobacter) can thrive, provided there is a sufficient source of oxygen. These utilise inorganic nutrients as an energy source (carbon dioxide and oxidised forms of nitrogen, phosphorus and sulphur). Nitrification is the biological process by which ammonia (an inorganic nutrient) is converted first to nitrite and then to nitrate. This can be summarised as:

$$NH_4^+ + 2O_2 \rightarrow NO_3^- + 2H^+ + H_2O + \text{new cells} + \text{energy}$$

The nitrification reaction consumes large quantities of oxygen and alkalinity.

It is unusual for nitrification to occur naturally in urban drainage networks, but it is possible towards the end of long, well-aerated outfall sewers, particularly in warmer climates.

3.4.3 Denitrification

In the denitrification process, nitrate is reduced to nitrogen gas by the same heterotrophic bacteria responsible for carbonaceous oxidation. Reduction occurs when dissolved oxygen levels are at or near zero. A carbon source must be available to the bacteria.

Denitrification can be stimulated in anaerobic sewer environments by adding a source of nitrate (see Chapter 17).

3.4.4 Anaerobic degradation

Anaerobic processes are those carried out by anaerobic bacteria in the absence of oxygen whereby large organic molecules are degraded to simple organic gases as end products, resulting in a partial breakdown of the substrate.

The class of organisms known as anaerobic heterotrophic bacteria carries out this reaction. These take organic material as a food source to provide energy or to synthesise new bacterial cells. They must take their oxygen from dissolved inorganic salts, therefore they produce reduced forms of end-products together with methane and carbon dioxide. Although this is a three or more stage process it can be simply represented as:

$$C,H,O,N,P,S + H_2O \rightarrow CO_2 + CH_4 + NH_3 + H_2S + \text{new cells} + \text{energy}$$

The products of anaerobic processes are more objectionable and sometimes more dangerous than those of aerobic processes. For example, hydrogen sulphide is malodorous, and methane is combustible. Anaerobic conditions are likely to occur when sediment beds form in sewers, and are common in pressurised rising mains.

3.5 Receiving water impacts

All receiving waters can assimilate wastes to some extent, depending on their natural self-purification capacity. Problems arise when pollutant loads exceed this capacity, thus harming the aquatic ecology and restricting the potential use of the water (e.g. water supply, recreation, fisheries). Discharges in urban areas can be continuous or intermittent, depending on their source. The aim of the urban drainage designer (in pollution terms) is, therefore, to balance the effects of these discharges against the assimilation capacity of the receiving water, so as to optimise water quality and minimise treatment costs.

3.5.1 Emissions

Urban drainage emissions can be categorised as direct and indirect, listed as below.

Direct – from the sewer system

- Intermittent discharges from CSOs consisting of a mixture of stormwater, domestic, commercial and industrial wastewater with groundwater and sewer deposits.
- Intermittent discharges from separate storm sewer outfalls or direct stormwater discharges consisting mainly of runoff from urban surfaces.

Indirect – via the treatment plant

- Continuous low level inputs from normally functioning WTPs.
- Intermittent shock loads (of suspended solids and/or ammonia) from WTPs disturbed by wet weather transient loads.

Intermittent discharges are particularly difficult to quantify and regulate because of their nature. Their acute (immediate) impact can only be measured during a spill event, and their chronic (long-term) effects are often difficult to isolate from background pollution. Therefore, design standards and performance criteria specifically tailored to intermittent discharges are needed. These are discussed in more detail later in the following section.

3.5.2 Processes

The processes occurring in receiving waters subject to discharges from urban drainage systems include (House *et al.*, 1993):

- Physical: transport, mixing, dilution, flocculation, erosion, sedimentation, thermal effects and re-aeration
- Biochemical: aerobic and anaerobic oxidation, nitrification, adsorption and desorption of metals and other toxic compounds
- Microbiological: growth and die-off, toxicant accumulation.

The extent and importance of individual processes will depend on the temporal and spatial scales as shown in Fig. 3.3. For example, a shock load into flowing water in a river will transport downstream relatively quickly, interacting with the water column as it progresses. Thus, a significant length of water will be exposed to contamination for a short period. On the other hand, a load discharged to stagnant water in a lake will disperse more slowly and will generally persist for a longer period. We can identify three relevant timescales:

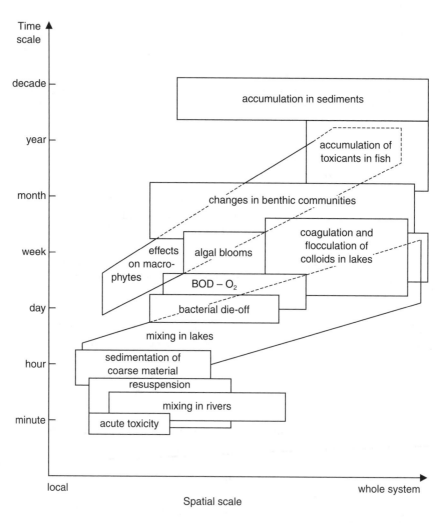

Fig. 3.3 Time and spatial scales for receiving water impacts (based on Aalderink and Lijklema [1985] with permission of the authors)

- short-term (acute)
- medium-term (delayed)
- long-term (chronic, cumulative).

3.5.3 Impacts

Impacts can be divided into direct water quality effects (DO depletion, eutrophication, toxics), public health issues and aesthetic influences. These are summarised for various determinands and receiving water types in Table 3.4.

DO depletion

The most important phenomena caused by intermittent discharges (particularly CSOs) are the following:

- Mixing of low DO spills with the receiving water
- Degradation of discharged (dissolved and particulate) organic matter that exerts an *immediate* oxygen demand on the receiving water. In the case of a river, these occur in a plug moving downstream
- *Delayed* sediment oxygen demand (SOD) caused by the deposited sediment and the scouring effect of discharges after the polluted plug has passed. Typical undisturbed SOD levels are 0.15–2.75 $g/m^2.d$, elevated to 240–1500 $g/m^2.d$ during storm flow conditions (House *et al.*, 1993). As

Table 3.4 Qualitative assessment of receiving water impacts of urban discharges (after House *et al.*, 1993)

Receiving water	Water quality				Public health	Aesthetics	
	Dissolved oxygen	Nutrients	Sediments	Toxics	Microbials	Clarity	Sanitary debris
Streams							
• Steep	–	–	–	x	xx	–	xx
• Slack	x	–	x	x	xx	–	xx
Rivers							
• Small	xx	–	x	x	xx	–	xx
• Large	x	–	x	x	xx	x	xx
Estuaries							
• Small	x	x	x	x	xx	x	xx
• Large	–	–	x	–	xx	x	xx
Lakes							
• Shallow	x	xx	x	x	xx	x	xx
• Deep	x	x	x	x	xx	x	xx

xx Probable   x Possible   – Unlikely

the rate of decomposition is low, the area affected can be extensive. Sensitive benthic organisms are rapidly eliminated but are soon replaced by high population densities of a few species tolerant of the low oxygen silty conditions.

The relative magnitude of these effects will depend on the specific circumstances of the discharge and the recipient. Hvitved-Jacobsen (1986) indicates that, in larger rivers, immediate consumption dominates, whereas small rivers with flows <0.5 m³/s often show depletion due to delayed consumption.

The most apparent consequences of reduced DO levels are fish kills. Additionally, odour problems may be experienced due to putrefaction.

Eutrophication

If large quantities of nutrients such as nitrogen or phosphorus are discharged into receiving waters, excessive growth of aquatic weeds and algae may occur. This can lead to:

- oxygen depletion
- anaerobic conditions in bottom muds
- fish kills
- aesthetic problems.

This is a long-term problem usually associated with shallow, stagnant waters such as lakes, estuaries and the coastal zone, but rivers may also be affected. Intermittent discharges are usually a relatively small constituent of the total nutrient load.

Toxics

Intermittent discharges are a significant source of elevated levels of ammonia (toxic to fish), chlorides, metals, hydrocarbons and trace organics which cause toxic impacts. These may be either acute or chronic depending on the specific circumstances. The effect on the receiving water biota is to rapidly reduce species diversity and abundance leading to complete elimination at excessive concentrations. Downstream recovery is generally slower than the loss rate, with tolerant species returning, often at population densities greater than initial levels, due to lack of competition.

The difficulties in assessing toxic impacts based on single parameter values have led to the development of toxicity-based consents (TBCs). Typically, these are derived from laboratory and field ecotoxicological studies on fish and invertebrates. Results are expressed as LC50 values that indicate short-term lethal concentrations of a particular pollutant resulting in 50% mortality.

Public health

Relatively high concentrations of a variety of pathogens may be expected from both combined sewer overflow (CSO) and stormwater outfall (SWO) discharges. Bacterial contamination is a relatively short-term problem as die-off usually occurs within several days, although this is a longer period than the discharge itself. In addition, bacteria tend to adhere to suspended solids. As many of these particles will settle, bacteria can become established in the receiving waterbed, considerably extending their survival times.

The risk to public health depends on the degree of potential human exposure, and this will be greatest if the receiving water is used for contact recreational purposes. Swimmers are therefore at greatest risk. Studies have now demonstrated the relationship between gastro-enteritis and FS levels (Wyer *et al.*, 1995).

Aesthetics

In addition to chemical and biological impacts, public perception of water quality is also important. Research has shown that the public has a good idea of what might be considered a polluted river, but is less certain as to what might be considered a clean river. The public tends to misperceive as polluted even rivers of high chemical and biological quality. However, solids of obvious sanitary origin near to receiving waters are considered to be offensive.

3.6 Receiving water standards

3.6.1 Legislation and regulatory regime

Urban Waste Water Treatment Directive

The most important European legislation concerning discharges from sewerage systems is the Urban Waste Water Treatment Directive (CEC, 1991a). This sets out the standards required from WTPs in some detail, but is less prescriptive for collection systems. Compliance dates for various aspects of the directive range from 31 December 1998 to 31 December 2005.

The Directive requires Member States to ensure all urban areas with a population equivalent of 2000 or more are provided with collection systems. Further, the design, construction and maintenance of collecting systems needs be undertaken in accordance with best technical knowledge not entailing excessive costs with respect to:

- volume and characteristics of urban wastewater
- prevention of leaks
- limitation of pollution of receiving waters due to overflows.

The Directive acknowledges the difficulties that arise during periods of unusually heavy rainfall, and allows Member States to decide on their own measures to limit pollution from overflows. These could be based on one of the following:

- dilution rates
- collection system capacity in relation to dry weather flow
- a certain number of overflows per year.

Bathing Water Quality Directive

Also relevant is the Bathing Water Quality Directive (CEC, 1976) which regulates the bacteriological water quality of coastal bathing waters. The most relevant measure of wastewater contamination is the faecal coliform imperative standard set at 2000 FC/100 ml. However, in line with the epidemiological evidence cited in the previous section, the directive has been revised (CEC, 2002) to include two faecal indicator parameters: Intestinal Enterococci (IE) and Escherichia coli (EC). The 'good quality' or obligatory levels for these are set at 200 IE cfu/100 ml and 500 EC cfu/100 ml, both at 95 percentile compliance.

There are approximately 500 designated bathing waters in the UK. Failure to comply with the standards usually comes about by a combination of continuous wastewater discharges and intermittent CSO discharges. Hence, control of overflow discharges becomes important where they discharge directly into bathing waters or to estuarine or coastal waters that lead to bathing waters.

Asset management planning

Since the 1989 privatisation of the water industry in England and Wales, investment by water companies, and the impact on customer bills, has been regulated by OFWAT. Investment proposals or *asset management plans* are reviewed on a 5-year cycle – the Periodic Review. Government and the regulators define the targets that must be achieved in line with UK and EU legislation and other priorities. OFWAT sets price limits for each water company based on the implications of the targets.

As part of the planning process for the second Periodic review 1995–1999 (AMP2), the National Rivers Authority (now the Environment Agency), in consultation with government, the water companies and others, prepared a guideline document (NRA, 1993) representing policy on continuous and intermittent discharges of wastewater to the environment, and this remains important guidance, discussed in later chapters. The third and fourth reviews (AMP3: 2000–2005, AMP4: 2005–2010) take this process further, building on earlier experience.

Water Framework Directive

The Water Framework Directive (WFD) incorporates the main requirements for water management in Europe into one single, holistic system
based on river basins (CEC, 2000). New or re-organised river basin district
authorities are to be formed, each with a management plan aimed at
achieving the goals of the Directive. The key guiding goal is to achieve
'good status' of ground and surface waters; 'good' meaning that water
meets the standards established in existing water directives and, in addition, new ecological quality standards. A 'surface water' is defined as of
good ecological quality if there is only a slight departure from the biological community that would be expected in conditions of minimal anthropogenic impact. 'Good chemical status' is defined in terms of compliance
with all the quality standards established for chemical substances at the
European level. A new mechanism for controlling the discharge of dangerous substances is provided. A combined approach to setting standards is to
be used where both emission limit values and river quality standards are to
be legally binding. Derogations from good status are allowed in unforeseen
or exceptional circumstances (e.g. droughts, floods).

The WFD is designed to provide an 'umbrella' to existing directives as
well as incorporating new standards. River basin authorities will designate
specific protection zones within their area (i.e. bathing, drinking water or
protected natural areas) where the standards in the respective existing EU
directives apply, but zones with higher objectives may also be established
where more stringent standards must be met. Good ecological and chemical status is the minimum for all waters. The Urban Waste Water Treatment Directive (see above, p. 49) and the Nitrates Directive (CEC, 1991b)
are considered as tools to achieve the objectives of the river basin management plans and will be retained. The directives on the quality of surface
waters intended for drinking and the fish and shellfish directives will be
repealed, as all surface waters will now need to meet 'good status'.

The directive also sets new rules for groundwater. All direct discharges
to ground water are prohibited and a requirement is introduced to monitor
ground water bodies so as to detect changes in composition due to nonpoint pollution and take measures to reverse them.

The initial period of implementation of the directive is 15 years (9 years
to prepare plans and 6 years more to achieve specific targets) plus an additional deferment of up to two periods of 6 years if justified (Butler *et al.*,
2000; Kallis and Butler, 2001).

3.6.2 *UK standards*

The approach favoured in the UK is to link the standards required to be
met by discharges to the receiving water's ability to assimilate discharged
pollutants without detriment to legitimate uses of the water – so-called

environmental quality standards (EQSs) based on environmental quality objectives (EQOs). The 'uses' most influenced by urban drainage intermittent discharges are (NRA, 1993):

- aquatic life
- bathing
- amenity.

An outline discussion on these standards is given below, but the *UPM Manual* (FWR, 1998) should be consulted for further detail.

Intermittent standards

Existing environmental quality standards have been developed for continuous discharges. These are based on statistically checking the compliance of routine samples against quality criteria (usually 90 or 95 percentile). Discharges from drainage systems tend to be infrequent and of short duration, although they can be of high pollutant concentration, resulting in a disproportionately high impact on river life. In addition, routine sampling is not possible for intermittent discharges.

In response to this, intermittent standards have been derived which take into account the particular characteristics of the discharges and their impact (FWR, 1998). They provide acceptable concentrations of river quality determinands for short and long term exposure and the recovery period in between.

Aquatic life standards

Intermittent standards to protect aquatic life, based on the LC50 values mentioned earlier, are given in the *UPM Manual*. The standards consist of a relationship between three variables:

- pollutant concentration
- return period of an event in which that concentration is exceeded
- duration of the event.

Table 3.5 shows this three-way relationship for dissolved oxygen and unionised ammonia based on sustaining cyprinid fisheries. Thus, minimum river DO levels of 3.0–5.5 mg/l are allowed, depending on the duration and frequency of the storm event. Fish kills due to ammonia poisoning can be avoided if NH_3–N levels are limited to 0.03–0.25 mg/l.

The manual also describes how higher percentile water quality criteria (e.g. 99 percentile) can be set as an alternative means of protecting receiving water ecosystems.

Table 3.5 Intermittent standards for dissolved oxygen and ammonia concentration/duration thresholds for sustaining cyprinid fisheries (after FWR, 1998)

Return period (months)	DO concentration (mg/l)*		
	1 h	*6 h*	*24 h*
1	4.0	5.0	5.5
3	3.5	4.5	5.0
12	3.0	4.0	4.5
	NH₃–N concentration (mg/l)**		
1	0.150	0.075	0.030
3	0.225	0.125	0.050
12	0.250	0.150	0.065

* Applicable when NH_3–N < 0.02 mg/l ** Applicable when DO > 5 mg/l, pH > 7 and T > 5 °C

Bathing standards

Bathing water standards are based on limiting the concentration of faecal and total coliforms to 2000 and 10 000 no./100 ml respectively for at least 98.2% of the bathing season (May to September). Exceedance for up to 1.8% of the time is acceptable, as judged over an average period of about 10 years (NRA, 1993). A surrogate emission standard is also proposed (refer to Chapter 12).

Amenity standards

Receiving waters are classified, in terms of their amenity value, by the amount of public contact, as below.

- High amenity – water used for bathing and water-contact sports, watercourses through parks and picnic sites, shellfish waters
- Moderate amenity – water used for boating, watercourses near popular footpaths or through housing developments or town centres
- Low amenity – limited public interest.

Amenity guidelines are currently based on emission standards or provision of good engineering design. These are discussed further in Chapter 12.

Problems

3.1 Sediment transported in a sewer as bed-load has been measured at a concentration of 20 ppm (by volume). What is the concentration in mg/l if the specific gravity of the sediment is 2.65?   [53 mg/l]

3.2   Plot the following hydrograph and pollutograph (concentration and load-rate).

Time (hrs)	Flow (l/s)	COD (mg/l)
0.5	80	50
1	170	160
1.5	320	380
2	610	400
2.5	670	230
3	590	130
3.5	380	70
4	220	40
4.5	100	20
5	50	0

Compute the average and flow weighted (event mean) concentration of COD.   [148 mg/l, 208 mg/l]

3.3   A wastewater sample has an organic nitrogen content of 15 mg org.N-N/l and an ammonium concentration of 35 mg NH_4^+/l. If the nitrite and nitrate concentrations are negligible, what is the total nitrogen concentration of the sample?   [42 mg N/l].

3.4   Define the main types of solids found in urban drainage systems and discuss their importance.

3.5   Compare and contrast the main methods for determining the organic content of a wastewater sample.

3.6   Describe the main forms of nitrogen found in wastewater. Why are they of interest?

3.7   What type of emissions can be expected from an urban drainage system? Explain how their impact may be acute, delayed or chronic.

3.8   Discuss the main types of receiving water impact caused by intermittent discharges.

3.9   What are the implications of the Water Framework Directive for urban drainage discharges?

3.10 Explain the difference between Environmental Quality Objectives and Environmental Quality Standards.

3.11 How do intermittent standards differ from ordinary water quality standards? How are they derived?

3.12 The table below shows the number of times that dissolved oxygen fell below 4 mg/l for 6 hours or more in a river. Is this in compliance with the 4 mg/l–6 hour–1 year standard?   [No]

Year	1	2	3	4	5	6	7	8	9	10
Number	0	1	1	3	0	0	2	1	0	3

Key sources

Ellis, J.B. (ed.) (1985) *Urban Drainage and Receiving Water Impacts*, Pergamon Press, Oxford.

FWR (1998) *Urban Pollution Management Manual*, 2nd edn, Foundation for Water Research, FR/CL0009.

Hvitved-Jacobsen, T. (1986) Conventional pollutant impacts on receiving waters, in *Urban Runoff Pollution* (eds H.C. Torno, J. Marsalek and M. Desbordes), Springer-Verlag, 345–378.

Kallis, G. and Butler, D. (2001) The EU water framework directive: measures and implications. *Water Policy*, 3, 125–142.

Marsalek, J. (1998) Challenges in urban drainage – environmental impacts, impact mitigation, methods of analysis and institutional issues, in *Hydrodynamic Tools for Planning, Design, Operation and Rehabilitation of Sewer systems* (eds J. Marsalek, C. Maksimovic, E. Zeman and R. Price), NATO ASI Series 2. Environment, Vol 44, Kluwer Academic, 1–23.

References

Aalderink, R.H. and Lijklema, L. (1985) Water quality effects in surface waters receiving stormwater discharges, in *Water in Urban Areas*, TNO Committee on Hydrological Research, Proceedings and Information No. 33, 143–159.

AWWA (1999) *Standard Methods for the Examination of Water and Wastewater*, 20th edn, American Water Works Association.

DoE (various) *Methods for the Examination of Waters and Associated Materials* (Separate booklets for individual determinations), HMSO.

Butler, D., Kallis, G. and Mills, K. (2000) Implications of the new EU Water Framework Directive, *CIWEM Millennium Conference*, Leeds, UK.

CEC (1976) *Directive concerning Quality of Bathing Water*, 76/160/EEC.

CEC (1978) *Directive concerning Quality of Freshwater Supporting Fish*, 78/659/EEC.

CEC (1991a) *Directive concerning Urban Waste Water Treatment*, 91/271/EEC.

CEC (1991b) *Directive concerning protection of water against pollution by nitrates from agriculture*, 91/276/EEC.

CEC (2000) *Directive establishing a Framework for Community Action in the field of Water Policy*, 2000/60/EC.

CEC (2002) *Proposal for a Directive concerning Quality of Bathing Water*, COM(2002) 581 Final.

Droste, R.L. (1997) *Theory and Practice of Water and Wastewater Treatment*, John Wiley & Sons.

Gray, N.F. (1999) *Water Technology. An Introduction for Scientists and Engineers*. Arnold, London.

Henze, M. (1992) Characterisation of wastewater for modelling of activated sludge processes. *Water Science and Technology*, 25(6), 1–15.

House, M.A., Ellis, J.B., Herricks, E.E., Hvitved-Jacobsen, T., Seager, J., Lijklema, L., Aalderink, H. and Clifforde, I.T. (1993) Urban drainage – impacts on receiving water quality. *Water Science and Technology*, 27(12), 117–158.

Inman, J.B. (1979) Sewage and its pretreatment in sewers. Chap. 7 in *Developments in Sewerage – 1* (ed. R.E. Bartlett), Applied Science.

NRA (1993) *General Guidance Note for Preparatory Work for AMP2 (Version 2),* Oct.

Painter, H.A. (1958) Some characteristics of a domestic sewage. *Journal of Biochemical and Microbiological Technology and Engineering,* 1(2), 143–162.

Wyer, M.D., Fleisher, J.M., Gough, J., Kay, D. and Merrett, H. (1995) An investigation into parametric relationships between enterovirus and faecal indicator organisms in the coastal waters of England and Wales. *Water Research,* 29(8), 1863–1868.

4   Wastewater

4.1 Introduction

Wastewater, or sewage, is one of the two major urban water-based flows that form the basis of concern for the drainage engineer. The other, stormwater, is described in Chapter 6. Wastewater is the main liquid waste of the community. Safe and efficient drainage of wastewater is particularly important to maintain public health (because of the high levels of potentially disease-forming micro-organisms in wastewater) and to protect the receiving water environment (due to large amounts of oxygen-consuming organic material and other pollutants in wastewater). This chapter provides background information and summary data on wastewater. The quantification for design purposes is dealt with in Chapter 9.

The basic sources of wastewater are summarised in Fig. 4.1 and consist of:

- domestic
- non-domestic (commercial and industrial)
- infiltration/inflow.

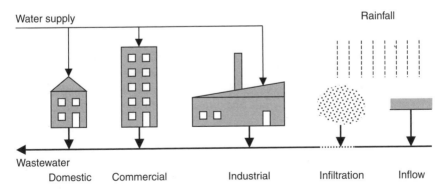

Fig. 4.1 Sources of wastewater

In practice, the relative importance of the components will vary with a number of factors, including:

- location (climatic conditions, the availability of water and its characteristics, and individual domestic water consumption)
- diet of the population
- presence of industrial and trade effluents
- the type of collection system (i.e. separate or combined)
- condition of the collection system.

This chapter is concerned with the generation and characteristics of wastewater. It collates quantity and quality information on the various sources of wastewater and discusses their relative importance.

4.2 Domestic

In many networks, the domestic component of wastewater is the most important. Domestic wastewater is generated primarily from residential properties but also includes contributions from institutions (for example, schools, hospitals) and recreational facilities (such as leisure centres). In terms of flow quantity, the defining variable is domestic water *consumption*, which is linked to human behaviour and habits. In fact, very little water is actually consumed, or lost from the system. Instead, it is used intermittently (degrading its quality) and then discharged as wastewater. Hence, in this section we shall look at the links between water usage and wastewater discharge and, in particular, how these vary with time.

4.2.1 Water use

Important factors affecting the magnitude of *per capita* water demand include the following.

Climate

Climatic effects such as temperature and rainfall can significantly affect water demand. Water use tends to be greatest when it is hot and dry, due largely to increased garden watering/sprinkling and landscape irrigation. The impact on wastewater is less pronounced, as this additional water will probably not find its way into the sewer.

Demography

It has been demonstrated that household occupancy levels are important, with larger families tending to have lower *per capita* demand. While, at the other end of the scale, retired people have been shown to use more water than the rest of the population (Russac *et al.*, 1991).

Socio-economic factors

The greater the affluence or economic capabilities of a community, the greater the water use tends to be. Work in the UK (Russac *et al.*, 1991) has confirmed the link established by Thackray *et al.* (1978) between water demand and economic indicators such as dwelling type or dwelling rateable value. This is probably due to greater ownership and use of water-using domestic appliances such as washing machines, dishwashers and power showers.

Development type

Dwelling type is important. In particular, dwellings with gardens may use more water than flats or apartments.

Extent of metering and water conservation measures

Water undertakers with metered supplies usually charge their customers based on the quantity of water used in a given period. Systems with unmetered services charge a flat rate for unlimited water use. In theory at least, metered supplies should prevent waste of water by users, reduce actual water use and therefore reduce wastewater flows. Water is not widely metered in the UK (about 20% of houses) but it has been shown to affect the amount of water consumed per household. An estimated reduction of 10% has been noted in metering trials in the Isle of Wight.

Water conservation measures such as low-flow taps/showers, low-flush toilets and recycling/re-use systems reduce water demand.

Quantification

Water consumption per head of population is extremely varied, as shown in Fig. 4.2. However, average domestic water usage in England has been estimated as 145 l/hd.d (Russac *et al.*, 1991; Edwards and Martin, 1995).

Water is used in three main areas in the home. Approximately one third of the water is used for WC flushing, one third for personal washing via the wash basin, bath and shower, and the final third for other uses such as washing-up, laundry and food/drink preparation (see Table 4.1). It is notable that only a very small percentage of this potable standard water is actually drunk (1–2 l/hd.d).

Table 4.1 Percentage of water consumed for various purposes (after DoE, 1992)

Component	Water consumed (%)		
	Household	Commercial	Industrial and agricultural
WC flushing	31	35	5
Washing/bathing	26	26	1
Urinal flushing	–	15	2
Food preparation/drinking	15	9	13
Laundry	12	8	–
Washing-up	10	2	–
Car washing/garden use	5	4	17
Other	1	1	62

4.2.2 Water–wastewater relationship

As mentioned earlier, there is a strong link between water usage and wastewater disposal, with relatively little supplied water being 'consumed' or taken out of the system. On a daily basis we can simply say:

$$G' = xG \qquad\qquad (4.1)$$

G water consumption per person (l/hd.d)
G' wastewater generated per person (l/hd.d)
x return factor, given in Table 4.2 (–)

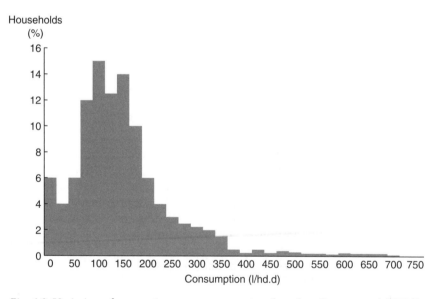

Fig. 4.2 Variation of *per capita* water consumption (based on Russac *et al.* [1991] with permission of the Chartered Institution of Water and Environmental Management, London)

Table 4.2 Percentage of water
discharged as wastewater

Country	x (%)
UK	95
Middle East	
Poor housing	85
Good housing	75
USA	60

It is estimated that, in the UK, about 95% of water used is returned to the sewer network (DoE, 1992). The other 5% is made up of water used externally (watering the garden and washing the car, for example) and to miscellaneous losses within the household. In hotter climates with low rainfall, this proportion can be up to 40%.

Fig. 4.3 shows a comparison made throughout the day between water use and wastewater flow in a catchment. In general, water use exceeds wastewater flow, especially in the early evening when gardens are being watered. At night this situation is reversed due to sewer infiltration flows.

4.2.3 Temporal variability

It is emphasised that both wastewater quantity and quality vary widely from the very long-term to the short-term. Hence, any particular reported value should be related to the timescale over which it was measured.

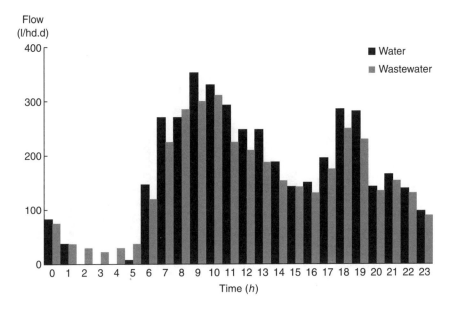

Fig. 4.3 Typical diurnal plot of water consumption and wastewater flow

Long-term

The major long-term trend is a steady increase in *per capita* consumption on an annual basis, reflecting a number of factors such as increased ownership of water-using appliances. The UK rate of increase in the 1990s was approximately 1% per annum.

Annual

Variations within the year due to seasonal effects can be observed in water demand. Evidence (Thackray *et al.*, 1978) suggests WC flushing decreases in summer (probably due to increased rate of body evaporation) and that bathing/showering increases. Outside water use increases significantly from gardening, and this can dominate the demand during summer months. For example, during the dry summer of 1995, increases in average demand of 50% or higher were observed in some areas. Fig. 4.4 shows the monthly trends in the Anglian region for 3 years where the average consumption in July was up to 25% greater than in one of the winter months. The effect on wastewater flows is less clearly defined but, typically, summer dry weather flow discharges normally exceed winter flows by 10–20%.

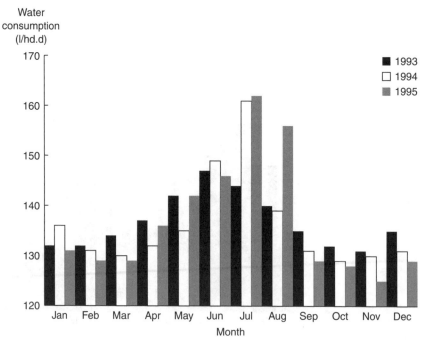

Fig. 4.4 Comparison of the effects on water consumption in a dry summer (Anglian region)

Weekly

Variations in water demand and wastewater production can occur within the week, from day-to-day. Thackray *et al.* (1978) and Butler (1991b) both found increased water consumption at weekends, probably due to increased WC flushing and bathing. In the UK, there is now little evidence of a specific 'washing day'.

Diurnal

A basic diurnal pattern showing variation from hour-to-hour of wastewater is given in Fig. 4.3. Minimum flows occur during the early morning hours when activity is at its lowest. The first peak generally occurs during the morning, the exact timing of which is dependent on the social activities of the community, but in this example, it is between 09:00–10:00. A second flow peak occurs in the early evening between 18:00 and 19:00, and then a third can also be distinguished between 21:00 and 22:00, but this is less clearly defined in magnitude and timing. Detailed timing within the diurnal cycle is also affected by the day of the week, with some differences noted at weekends (Butler, 1991b).

4.2.4 Appliances

Wastewater production is strongly linked to the widespread ownership and use of a wide range of domestic appliances, such as those in Table 4.3. The contribution of each individual appliance depends on both the volume of flow discharged after each operation and the frequency with which it is used.

Table 4.3 shows typical discharge volumes of six different domestic appliances. Particularly large volumes are discharged by washing machines and during bathing, whilst relatively little is used during each use of the wash basin.

Fig. 4.5 illustrates how the discharges from the individual appliances go to make up the general wastewater diurnal pattern. The most important contributor overall is the WC, which although only of modest volume is

Table 4.3 Domestic appliance discharge volumes (after Butler, 1991a)

Appliance	Volume (l/use)
WC	8.8
Bath	74
Shower	36
Wash basin	3.7
Kitchen sink	6.5
Washing machine	116

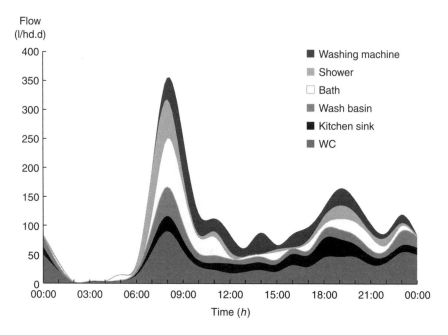

Fig. 4.5 Appliance diurnal discharge patterns

used very frequently throughout the day, and particularly at peak periods. Further discussion of the implications of the diurnal wastewater pattern is given in Chapter 9.

4.3 Non-domestic

4.3.1 Commercial

This category includes businesses such as shops, offices and light industrial units, and commercial establishments such as restaurants, laundries, public houses and hotels.

Demand is generated by drinking, washing and sanitary facilities, but patterns of use are inevitably different to those generated by domestic usage. For example, Table 4.1 shows how toilet/urinal usage is an even more dominant component of water use (50%) than in the domestic environment. Much less detailed information is available on commercial usage than on domestic usage.

4.3.2 Industrial

The component of wastewater generated by industrial processes can be important in specific situations, but is more difficult to characterise in general because of the large variety of industries. Table 4.1 shows that many of the most important components of usage found in domestic and commercial premises are much less important in industry and agriculture.

In most cases, effluents result from the following water uses:

- sanitary (e.g. washing, drinking, personal hygiene)
- processing (e.g. manufacture, waste and by-product removal, transportation)
- cleaning
- cooling.

The detailed rate of discharge will vary from industry to industry and will depend significantly on the actual processes used. Water consumption is often expressed in terms of volume used per mass of product. So, for example, papermaking consumes 50–150 m³/t and dairy products 3–35 m³/t. A detailed survey of industrial water usage has been carried out by Thackray and Archibald (1981).

Industrial effluents can be highly variable (in both quantity and quality) as a consequence of batch discharges, operation start-ups and shut-downs, working hours and other factors. These may change significantly at weekends. Depending on the relative magnitude of the flows, industrial discharges can completely alter the normal diurnal patterns of flow. There may also be significant seasonal changes in demand, for example due to agro-industrial practices responding to the needs of food production.

Other important factors include the size of organisation, the availability and cost of water, and the extent of process water recycling.

4.4 Infiltration and inflow

Unlike the other sources of wastewater, infiltration and inflow are not deliberate discharges, but occur as a consequence of the existence of a piped network. Infiltration and inflow have already been introduced in Chapter 2 and are defined as water that enters the sewer system through indirect and direct means respectively. Infiltration is extraneous groundwater or water from other leaking pipes that enters the sewer system through defective drains and sewers (cracks and fissures), pipe joints, couplings and manholes. Inflow is stormwater that enters separate foul sewers from illegal or misconnected yard gullies, roof downpipes or through manhole covers.

4.4.1 Problems

The presence of excessive amounts of infiltration may cause one or more of the following problems (Fiddes and Simmonds, 1981):

- reduced effective sewer capacity leading to possible surcharging and/or flooding
- overloading of pumping stations and wastewater treatment works
- higher frequency of CSO operation, possibly in dry weather during periods of high ground water levels
- increased entry of sediment (soil), resulting in higher maintenance requirements and possible surface subsidence.

4.4.2 Quantification

The extent of infiltration is site-specific but, when excessive, is usually a result of poor design and construction and will generally deteriorate as the system physically degrades. Influencing factors include (Bishop *et al.*, 1998; Martin *et al.*, 1982):

- age of the system
- standard of materials and methods
- standard of workmanship in laying pipes
- settlement due to ground movement
- height of ground water level (varies seasonally)
- type of soil
- aggressive chemicals in the ground
- extent of the network – total length of sewer (including house connections); type of pipe joint, number of joints and pipe size; number and size of manholes and inspection chambers
- frequency of surcharge.

The amount of infiltration may range widely from 0.01 to 1.0 m³ per day per mm diameter per km length (Metcalf and Eddy, 1991). Infiltration can reach serious proportions in old systems. In the UK, Stanley (1975) found rates in those existing sewers subject to infiltration ranging from 15% to 50% of average dry weather flow. Fuller details of the causes, costs and control of infiltration can be found in White *et al.* (1997).

4.4.3 Exfiltration

Exfiltration is the opposite of infiltration. Under certain circumstances, wastewater (or stormwater) is able to leak out of the sewer into the surrounding soil and groundwater. This creates the potential for groundwater contamination. Indeed, Environment Agency policy is to oppose the construction of new sewer systems within its groundwater Source Protection Zone I (NRA, 1992).

Factors affecting the likelihood of exfiltration are similar to those discussed for infiltration. Fuller details of the causes, costs, control and implications of exfiltration can be found in Anderson *et al.* (1996) and Reynolds and Barrett (2003).

4.5 Wastewater quality

Wastewater contains a complex mixture of natural organic and inorganic material present in various forms, from coarse grits, through fine suspended solids to colloidal and soluble matter. Much is in the form of highly putrescible compounds. In addition, a small proportion of man-made substances, derived from commercial and industrial practices, will be present.

In fact, wastewater is 99.9% water although the remaining 0.1% is very significant, particularly if it is allowed to enter the environment. Fresh domestic wastewater is cloudy-grey in colour with some recognisable solids and has a musty/soapy odour. With time (2–6 hours depending on ambient conditions), the waste 'ages' and gradually changes in character as a result of physical and biochemical processes. Stale wastewater is dark grey/black with smaller and fewer recognisable solids, and 'older' flows can have a pungent 'rotten eggs' odour due to the presence of hydrogen sulphide.

Wastewater quality is very variable in respect to both location and time. In addition, the techniques commonly used for sampling and analysis are subject to error (see Chapter 3). Therefore, caution is needed in interpreting standard or typical values. Such data should never be assumed to accurately represent the wastewater from a particular community – this can only be properly confirmed by a (possibly extensive) testing programme or access to historic data.

4.5.1 Pollutant sources

Wastewater quality is influenced by the contaminants discharged into it derived mainly from human, household and industrial activities. The quality of the carriage water (the original drinking water) or infiltrating groundwater can also be influential.

Human excreta

Human excreta are responsible for a large proportion of the pollutants in wastewater. Adults produce 200–300 g of faeces and 1–3 kg of urine per day. Faeces account for 25–30 g/hd.d of BOD and urine 10 g BOD/hd.d, which is 60% of the organic compounds found in wastewater (Feachem *et al.*, 1981). Excreta, however, contribute only a small proportion of wastewater fats.

Excreta are also an important source of nutrients. The bulk (94%) of the organic nitrogen in wastewater is derived from excreta. Of this percentage, 50% derives from urine (urea) which is most abundant in fresh wastewater as it is rapidly converted to ammonia under both aerobic and anaerobic conditions (see Chapter 3). Approximately 50% of the phosphorus discharged to sewer (1.5 g/hd.d) is derived from excreta. Excreta also contain about 1 g/hd.d of sulphur (Inman, 1979).

The bulk of the micro-organisms in wastewater originate in faeces; urine is relatively microbe-free.

Toilet

Toilet paper is used in large quantities. Although this disintegrates quickly in the turbulent flow in sewers, it is only slowly biodegradable due to the presence of the cellulose fibres. Approximately 7 g/hd.d is disposed of, most of which will become suspended solids (Friedler *et al.*, 1996). Tests have shown that coloured papers contribute some 15% of the wastewater COD.

A wide range of gross solids is discharged, either deliberately or accidentally, via the toilet. Meeds and Balmforth (1995) suggest the following categories:

- condoms
- sanitary towels
- tampons
- disposable nappies
- toilet tissue paper
- paper towels
- miscellaneous (paper origin)
- miscellaneous (fat origin).

In total, some 0.15 sanitary items/hd are disposed of each day (Friedler *et al.*, 1996).

A number of cleaning, disinfecting and descaling chemicals are also routinely discharged into the system via the toilet.

Food

Of course, digested food is the source of many of the excreta-related pollutants mentioned above. However, undigested food is a major contributor of fats including butter, margarine, vegetable fats, meats, cereals and nuts. Food residues are also a source of some organic nitrogen and phosphorus and of salt (NaCl).

Washing/laundry

Washing and laundry activities add soaps and detergents to the sewer. The polyphosphate builders used in synthetic detergents contribute approximately 50% of the phosphorus load. Phosphorus concentrations have diminished significantly in countries where legislation has imposed

significant reductions in the amounts of phosphorus used by manufacturers of detergents (Morse *et al.*, 1993).

Industry

The characteristics of industrial wastewaters, or trade effluents as they are
often called, are similar to those of domestic wastewater in that they are
likely to contain a very high proportion of water, and the impurities may
be present as suspended, colloidal or dissolved matter. But in addition, a
very large variety of pollutant types can be generated and industrial wastewater may contain:

- extremes of organic content
- a deficiency of nutrients
- inhibiting chemicals (acids, toxins, bactericides)
- resistant organic compounds
- heavy metals and accumulative persistent organics.

Processing liquors from the main industrial processes tend to be relatively
strong whilst wastewaters from rinsing, washing and condensing are comparatively weak. Discharges may be seasonal and vary considerably from
day to day both in volume and strength.

Carriage water and groundwater

The sulphate present in wastewater is derived principally from the mineral
content of the municipal water supply or from saline ground water infiltration (see Chapter 3).

In hard water areas, the use of softeners can result in significant
increases in the wastewater chloride concentrations. Infiltration of salt-
water (if present) can contribute similarly.

4.5.2 Pollutant levels

Typical values and ranges of pollutant levels in UK wastewater are given in
Table 4.4.

Problems

4.1 Classify the major sources of wastewater and discuss the factors
affecting their prevalence in practice.

4.2 Explain the quantitative link between water demand and wastewater
generation. What are the major factors influencing domestic water use?

4.3 What are the main differences between domestic, commercial and
industrial water demand?

Table 4.4 Pollutant concentrations and unit loads for wastewater (adapted from Ainger *et al.*, 1997)

Parameter type	Parameter	Unit load (g/hd.d)	Concentration (mg/l) mean (range)
Physical	Suspended solids		
	volatile	48	240
	fixed	12	60
	Total	60	300 (180–450)
	Gross (sanitary) solids		
	sanitary refuse	0.15*	
	toilet paper	7	
	Temperature		18 (15–20) °C: summer
			10 °C: winter
Chemical	BOD$_5$		
	soluble	20	100
	particulate	40	200
	Total	60	300 (200–400)
	COD		
	soluble	35	175
	particulate	75	375
	Total	110	550 (350–750)
	TOC	40	200 (100–300)
	Nitrogen		
	organic N	4	20
	ammonia	8	40
	nitrites		0
	nitrates		<1
	Total	12	60 (30–85)
	Phosphorus		
	organic	1	5
	inorganic	2	10
	Total	3	15
	pH		7.2 (6.7–7.5): hard water
			7.8 (7.6–8.2): soft water
	Sulphates	20	100: dependent on water supply
	FOG		100
Microbiological	Total coliforms		10^7–10^8 MPN/100 ml
	Faecal coliforms		10^6–10^7 MPN/100 ml
	Viruses		10^2–10^3 infectious units/100 ml

*items/hd.d

4.4 Describe how wastewater varies at various timescales and explain the significance of this.

4.5 Compare and contrast the mechanisms, amounts and implications of infiltration and exfiltration.

4.6 What are the main sources of pollutants in wastewater and what is their importance?

Key sources

Ainger, C.M., Armstrong, R.J. and Butler, D. (1997) *Dry Weather Flow in Sewers*, Report R177, CIRIA, London.

Butler, D. (1991a) A small scale case-study of wastewater discharges from domestic appliances. *Journal of the Institution of Water and Environmental Management*, 5(2), April, 178–185.

Males, D.B. and Turton, P.S. (1979) *Design Flow Criteria in Sewers and Water Mains*. Technical Note No. 32, Central Water Planning Unit.

References

Anderson, G., Bishop, B., Misstear, B. and White, M. (1996) *Reliability of Sewers in Environmentally Sensitive Areas*, Report PR44, CIRIA, London.

Bishop, P.K., Misstear, B.D., White, M. and Harding, N.J. (1998) Impacts of sewers on groundwater quality. *Journal of the Chartered Institution of Water and Environmental Management*, 12, June, 216–223.

Butler, D. (1991b) The influence of dwelling occupancy and day of the week on domestic appliance wastewater discharge. *Building and Environment*, 28(1), 73–79.

Butler, D., Friedler, E. and Gatt, K. (1995) Characterising the quantity and quality of domestic wastewater inflows. *Water Science and Technology*, 31(7), 13–24.

Department of Environment (1992) *Using Water Wisely. A Consultation Paper*, HMSO.

Edwards, K. and Martin, L. (1995) A methodology of surveying domestic water consumption. *Journal of the Chartered Institution of Water and Environmental Management*, 9, October, 477–488.

Feachem, R.G., Bradley, D.J., Garelick, H. and Mara, D.D. (1981) *Health Aspects of Excreta and Sullage Management – A State-of-the-Art Review*, The World Bank.

Fiddes, D. and Simmonds, N. (1981) Infiltration – do we have to live with it? *The Public Health Engineer*, 9(1), 11–13.

Friedler, E., Brown, D.M. and Butler, D. (1996) A study of WC derived sewer solids. *Water Science and Technology*, 33(9), 17–24.

Inman, J.B. (1979) Sewage and its pretreatment in sewers. Chap. 7 in *Developments in Sewerage – 1* (ed. R.E. Bartlett), Applied Science.

Martin, C., King, D., Quick, N.J. and Nott, N.A. (1982) Infiltration investigation, analysis and cost/benefit of remedial action. Paper 17 in *ICE Conference on Restoration of Sewerage Systems*, Thomas Telford, 175–185.

Meeds, B. and Balmforth, D.J. (1995) Full-scale testing of mechanically raked bar screens. *Journal of the Chartered Institution of Water and Environmental Management*, 9, December, 614–620.

Metcalf and Eddy, Inc. (1991) *Wastewater Engineering. Treatment, Disposal and Re-use*, 3rd Ed, McGraw-Hill.

Morse, G.K., Lester, J.N. and Perry, R. (1993) *The Economic and Environmental Impact of Phosphorus Removal from Wastewater in the European Community*, Selper Publications.

National Rivers Authority (1992) *Policy and Practice for the Protection of Groundwater*.

Nichol, E.H. (1988) *Small Water Pollution Control Works: Design and Practice*, Ellis Horwood.

Reynolds, J.H. and Barrett, M.H. (2003) A review of the effects of sewer leakage on groundwater quality. *Journal of the Chartered Institution of Water and Environmental Management*, 17(1), March, 34–39.

Russac, D.A.V., Rushton, K.R. and Simpson, R.J. (1991). Insights into domestic demand from a metering trial. *Journal of the Institution of Water and Environmental Management*, 5(3), June, 342–351.

Stanley, G.D. (1975) *Design Flows in Foul Sewerage Systems*, DOE. Project Report No. 2.

Thackray, J.E. and Archibald, G. (1981) The Severn–Trent studies of industrial water use. *Proceedings of the Institution of Civil Engineers*, Part 1, 70, August, 403–432.

Thackray, J.E., Cocker, V. and Archibald, G. (1978) The Malvern and Mansfield studies of water usage. *Proceedings of the Institution of Civil Engineers*, Part 1, 64, February, 37–61.

White, M., Johnson, H., Anderson, G. and Misstear, B. (1997) *Control of Infiltration to Sewers*, Report R175, CIRIA, London.

5 Rainfall

5.1 Introduction

As already noted, urban drainage systems deal with both wastewater and stormwater. Most stormwater is the result of rainfall. Other forms of precipitation – snow for example – are contributors too, but rainfall is by far the most significant in most places. Methods of representing and predicting rainfall are therefore crucial in the design, analysis and operation of drainage systems.

The detailed study of rainfall is the work of hydrologists, and their work primarily entails interpreting and predicting nature – always a difficult task. They work primarily using observation, which is the origin of all our knowledge about rainfall. The more we observe, the more we learn.

Observation provides historical records and allows derivation of relationships between rainfall event properties (particularly intensity, duration and frequency). Often, urban drainage engineers and modellers require long periods of rainfall (including different storms and the dry periods between) to input to models. These may be real historical records, though it is hard to know exactly how representative a particular portion of history actually is. For this reason, it may be more appropriate to use specially-created synthetic sets of data that represent the properties of actual rainfall.

This chapter describes the main methods of rainfall measurement, and considers rainfall data requirements for different applications. The representation of historical rainfall data and the generation of synthetic rainfall are discussed, and alternative forms of rainfall data are presented. Data and techniques presented in this chapter will be used in subsequent chapters on design and analysis.

5.2 Measurement

5.2.1 Rain gauges

Rain gauges are the most common device for measuring rainfall. A standard non-recording gauge (Fig. 5.1(a)) collects rain falling on a standard area (in the UK, a 127 mm diameter funnel with the rim placed 300 mm above ground level) over a known period of time. The volume of the stored rainfall is measured manually and, if necessary, converted to rainfall intensity (depth/time) by dividing by the collection area. Collection periods range from 6 hours to one month, but one day is typical. Since urban drainage systems can respond in less than 6 hours, the data from non-recording gauges is of limited value in this application.

Recording gauges are able to provide a continuous record of rainfall. The tipping-bucket rain gauge collects rainfall over short periods of time in a balanced reservoir consisting of two miniature compartments. Rainwater enters the first compartment until the weight of the water makes it tilt. Water begins to enter the second compartment while the first empties (Fig. 5.1(b)). Thus, the gauge produces a series of tips with a changing frequency depending on the rainfall intensity. The number of tips per unit time is therefore related to the rainfall intensity. A record is made either of the number of tips in a set time interval or the time of each tip. Typically, this is recorded electronically and stored in the memory of a data logger on site or transmitted over telephone lines to a central station. The data can be downloaded to a computer at convenient time intervals for processing. The range of rainfall depth resolution is 0.1 to 0.5 mm/tip.

Fig. 5.1 Rain gauges: (a) standard; (b) tipping bucket (courtesy of Hydrokit, Poole)

Siting

The siting of gauges must be carefully planned in order to obtain representative data for the catchment. There are general rules (Meteorological Office, 1982; WRc, 1987) about the distances from obstacles, the level of the gauge orifice above ground level and so on, but these rules cannot always be fully adhered to in the urban environment. In addition, in urban areas, some rain falls on roofs so the siting of gauges on roofs is acceptable if the number of gauges on the different levels corresponds approximately to the proportions of the different types of urban surface.

If more than one gauge is in use, careful synchronisation of records is crucial. This can be achieved by accurate clock setting at each gauge on the same quartz watch. Settings should be regularly checked. As a rule, rain gauges should be visited and checked at least once a week.

5.2.2 Other forms of measurement

An emerging technique for rainfall measurement is ground-based radar. In principle, radar can provide a continuous and almost instantaneous picture of rainfall when it is still in the atmosphere. The method works by directing a radar beam at falling raindrops, collecting and measuring the intensity of the reflected radiation and relating this to the rainfall intensity (Collier, 1996).

However, practical factors such as distortion due to hills and tall buildings ('ground clutter'), wind, raindrop size and ice crystals mean the relationship is not simple or straightforward. As all of these effects vary during the passage of even a single storm, frequent calibration by conventional gauges is essential.

For urban catchments, relatively fine spatial resolution is preferable. This is particularly relevant to urban flood forecasting and real time control of sewer systems (see Chapter 22).

Satellite imagery has also been used, usually by indirectly relating cloud-top temperature to rainfall (Rosenfeld and Collier, 1998). Petrovic and Elgy (1994) used satellite infra-red data to estimate the areal distribution of rain.

5.2.3 Data requirements

The appropriate level of detail in rainfall measurement depends on how the data will be used. Three broad categories can be identified.

In *design and planning*, the task is to produce the overall dimensions of the system. Examples include determining the peak flow rate in storm sewers (see Chapter 11) or the total volume of storm detention tanks (Chapters 9 and 12).

In *checking and evaluation*, the performance of the designed system is assessed under extreme or onerous conditions. This usually requires more effort than design and, consequently, more detailed rain data.

The third task, *analysis and operation*, is concerned with the evaluation of systems that already exist and examples include the verification of a flow simulation model with real flow data (Chapter 19) or operation of a system in real-time (Chapter 22). This latter task has the most stringent requirements for rainfall data.

Table 5.1 lists the rainfall data requirements for examples of each of the three main engineering tasks. The *rainfall record duration* is the length of historical data available for analysis, measured in years. This should be significantly longer than the return period (defined in the next section) of the storm event used in system design. The *gauge location* is ideally within the catchment, but this is less important in design than analysis. *Temporal resolution* is the desired time period between rainfall measurements and *spatial resolution* indicates the desired distance between rain gauges. It is preferable to have several gauges in all catchments to provide data checks and detect spatial variations, including storm movement. Minimising *synchronisation errors* becomes important when multiple gauges are used.

5.3 Analysis

5.3.1 Basics

Rain data measured at an individual rain gauge is most commonly expressed either as *depth* in mm or *intensity* in mm/h. This type of *point* rainfall data is therefore representative of one particular location on the

Table 5.1 Requirements for rainfall data in urban drainage applications (adapted from Schilling, 1991)

Engineering task	Rainfall record duration (yrs)	Rain gauge location (relative to catchment)	Temporal resolution (min)	Spatial resolution (km²/gauge)	Synchronisation error (min)
Design/planning					
Sewers	>10	near vicinity	block rain	homogeneous	≤30
CSO volumes	>5	near vicinity	≤15	homogeneous	≤30
Checking/evaluation					
Sewers	>20	adjacent	1	homogeneous	≤10
CSO volumes	>10	adjacent	5	≤5	≤5
Analysis/operation					
Calibration/ verification	several events	within	2	2*	0.25
Real-time control	on-line	within	2	2*	0.25

* No less than 3 in total

catchment. Such data is of greater value if it can be related statistically to two other important rainfall variables: duration and frequency.

The rainfall *duration* refers to the time period D minutes over which the rainfall falls. However, duration is not necessarily the time period for the whole storm, as any event can be subdivided and analysed for a range of durations. It is common to represent the *frequency* of the rainfall as a *return period*. An annual maximum rainfall event has a return period of T years if it is equalled or exceeded in magnitude once, on average, every T years. Thus a rainfall event that occurs on average twenty times in 100 years has a return period of 5 years. Annual maximum storm events are normally used to determine return period because it is assumed that the largest event in one year is statistically independent of the largest event in any other year.

5.3.2 IDF relationships

Definition

A convenient form of rainfall information is the intensity-duration-frequency (IDF) relationship. A typical set of IDF curves is given in Fig. 5.2 where it can be seen that (for an event with a particular return period) rainfall intensity and duration are inversely related. As the duration increases, the intensity reduces. This confirms the common-sense observation

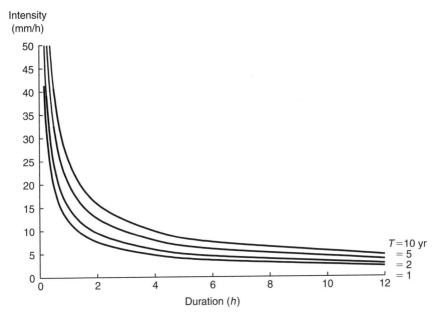

Fig. 5.2 Typical intensity-duration-frequency curves

Example 5.1

Using the data presented in Fig. 5.2, determine the intensity of a 1 yr return period 2 h duration rainfall event. For a similar duration event of 10 yr return period, find the appropriate rainfall depth.

Solution

For $T = 1$ yr, $D = 2$ h $\therefore i = 7.5$ mm/h
For $T = 10$ yr, $D = 2$ h $\therefore i = 16$ mm/h $\therefore d = 16 \times 2 = 32$ mm

---

that heavy storms only last a short time, but drizzle can go on for long periods. Also, frequency and intensity are related, as rarer events (greater return periods) tend to have higher intensities (for a given duration).

Derivation

IDF relationships can be derived, for a particular location, by a procedure known as rainfall frequency analysis. Rainfall depths monitored during individual storms are abstracted from recording gauges and the annual maximum values ranked from 1 to n (the number of years of record) in decreasing order of magnitude. The relevant return period (T) in years is then estimated using, for example, Weibull's plotting position formula:

$$T = \frac{n + 1}{m} \tag{5.1}$$

where m is the event rank number $(1,2,.......n)$.

The data set of depths and their associated return periods can be fitted to a statistical distribution (e.g. log-normal, Gumbel) using methods based on the moments of the data or maximum likelihood data. This is a manual procedure based on plotting on probability paper, or can be accomplished with appropriate frequency analysis software. It is possible to interpolate or extrapolate (although with increasingly uncertain predictions) the intensity of any return period rainfall. Shaw (1994) gives fuller details of this approach.

Prediction

In most situations, however, it is not necessary to derive such a set of curves, but rather to use previously-derived ones. Several mathematically-similar expressions may be fitted to describe IDF relationships. The simplest relate the average rainfall intensity i (mm/h) and duration D (min) for a fixed return period T (yr):

$$i = \frac{a}{D + b} \tag{5.2}$$

where a, b are constants. An early UK example of this was the so-called 'Ministry of Health' formula in which $a = 750$ and $b = 10$ for $5 \leq D \leq 20$ min, and $a = 1000$, $b = 20$ for $20 \leq D \leq 100$ min (Ministry of Health, 1930). Norris (1948) later showed this formula corresponded to a $T = 1$ year storm event.

Bilham (1936) improved this basic approach by proposing a formula, based on 10 years of continuous rain gauge data, that relates intensity and duration to storm frequency of occurrence:

$$N = 1.25 \, D \, (I/25.4 + 0.1)^{-3.55} \tag{5.3}$$

N number of times in 10 years during which rainfall occurs
I rainfall depth (mm)
D duration (h)

If $N = 2$, the storm return period is (approximately) 5 years. The equation is valid for rainfall durations of 5 min to 2 h, but has been extrapolated to longer durations. Later, Holland (1967) simplified and updated the formula to give:

$$N = D \, (I/25.4)^{-3.14} \tag{5.4}$$

valid up to rainfall durations of 25 h. Bilham's formula still gives good results, but tends to overestimate the probability of higher-intensity storms.

The *Flood Studies Report* (NERC, 1975) gave point rainfall depth-duration-frequency data for the whole of the United Kingdom for durations from 1 min to 48 h. The procedure used to analyse and present the data is explained in Volumes 1 and 2 of the Report. Options available to obtain this data are:

- contacting the Meteorological Office with details of the National Grid reference for the location of interest
- following the procedures in Volume 2 of the *Flood Studies Report* used in conjunction with the maps in Volume 5 of the report
- using one of the urban drainage models (see Chapters 19 and 20) in conjunction with the maps in Volume 3 of the *Wallingford Procedure* (DoE/NWC, 1981)
- following a manual method in Volume 4 of the *Wallingford Procedure*.

5.3.3 Wallingford Procedure manual method

Rainfall information for any location in the UK may be abstracted from maps associated with the *Wallingford Procedure*. The method itself will be described in more detail in Chapter 11 but the rainfall estimation approach will be explained here.

The *Flood Studies Report* and the *Wallingford Procedure* both use a standard notation when specifying rainfall information. Thus *MT-D* represents the depth of rainfall (in mm) occurring for duration D with a return period T years. Durations specified in minutes start at any minute in the hour, those in hours start 'on the hour' and those in days begin at 9 a.m. GMT.

The method is based on working from standard M5–60 min rainfall and the ratio (r) of M5–60 min/M5–2 day rainfall depth, both of which are mapped for the UK and given in Figs 5.3 and 5.4 respectively. The M5–60 min rainfall effectively denotes the quantity of rainfall in an area and the ratio r reflects the 'type' of rainfall. Low values of r (<0.2) represent rain that mostly falls as drizzle, whereas values >0.4 indicate the prevalence of much higher intensity storms.

By means of coefficients (or *growth factors*) Z_1 (Fig. 5.5) and Z_2 (Table 5.2), the standard M5–60 min rainfall can be related to:

- the 5 year rain depth for the required duration (M5–D) and
- the depth of rain for the required duration and return period (MT-D).

Example 5.2 shows how this approach can be used to produce an IDF relationship.

IDF relationships of point rainfall are widely used in urban drainage applications. In particular, they are essential in the application of the Rational Method (Chapter 11).

Table 5.2 Ratio Z_2 – Relationship between rainfall of return period T(MT) and M5 for England and Wales (after DoE/NWC 1981)

M5 (mm)	M1	M2	M3	M4	M5	M10	M20	M50
5	0.62	0.79	0.89	0.97	1.02	1.19	1.36	1.56
10	0.61	0.79	0.90	0.97	1.03	1.22	1.41	1.65
15	0.62	0.80	0.90	0.97	1.03	1.24	1.44	1.70
20	0.64	0.81	0.90	0.97	1.03	1.24	1.45	1.73
25	0.66	0.82	0.91	0.97	1.03	1.24	1.44	1.72
30	0.68	0.83	0.91	0.97	1.03	1.22	1.42	1.70
40	0.70	0.84	0.92	0.97	1.02	1.19	1.38	1.64
50	0.72	0.85	0.93	0.98	1.02	1.17	1.34	1.58
75	0.76	0.87	0.93	0.98	1.02	1.14	1.28	1.47
100	0.78	0.88	0.94	0.98	1.02	1.13	1.25	1.40

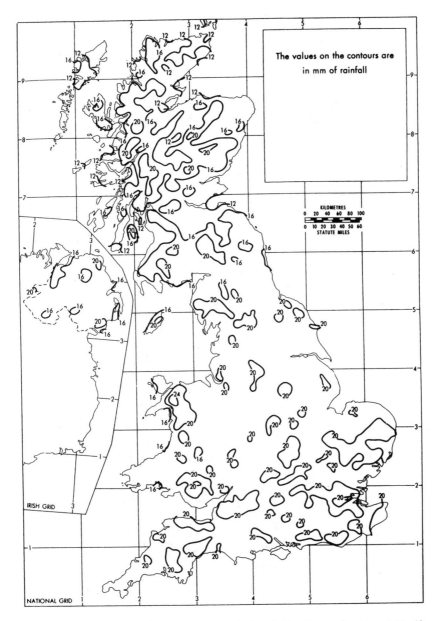

The values on the contours are in mm of rainfall

Fig. 5.3 Rainfall depths of 5 year return period and 60 minutes duration: M5–60 min (reproduced from 'The Wallingford Procedure' with permission of HR Wallingford Ltd)

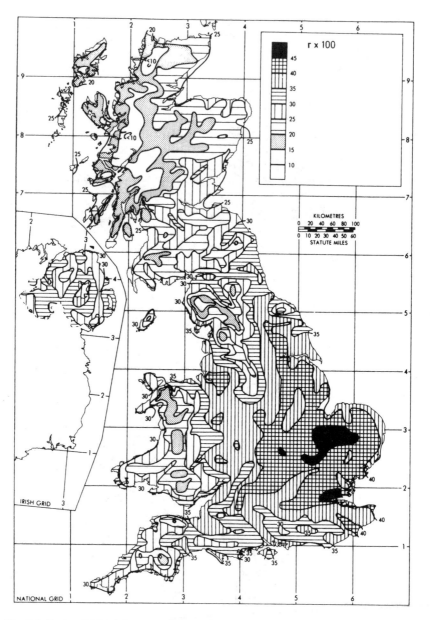

Fig. 5.4 Ratio of 60 minute to 2 day rainfalls of 5 year return period: *r* (reproduced from 'The Wallingford Procedure' with permission of HR Wallingford Ltd)

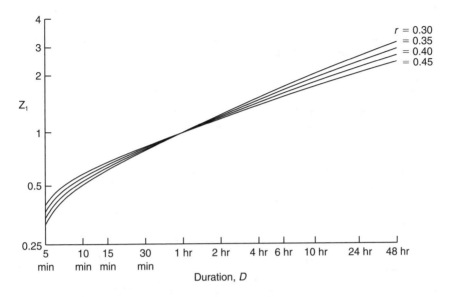

Fig. 5.5 Relationship between Z_1 and D for different values of r ($0.30 \leq r \leq 0.45$) (based on 'The Wallingford Procedure' with permission of HR Wallingford Ltd)

5.3.4 Areal extent

Point rainfall is not necessarily representative of rainfall over a larger area because average rainfall intensity decreases with increasing area. In order to deal with this problem, and avoid overestimating flows from larger catchments, *areal reduction factors* (ARF) have been developed, based on the comparison of point and areal data from areas where several gauges exist.

In the *Wallingford Procedure*, the ARF is calculated from:

$$ARF = 1 - f_1 D^{-f_2} \tag{5.5}$$

where

$$f_1 = 0.0394 \, A^{0.354}$$
$$f_2 = 0.040 - 0.0208 \ln (4.6 - \ln A)$$

and A is catchment area (km^2).

The expression is valid for UK catchment areas <20 km^2 and storm durations of 5 mins to 48 h (see Example 5.3). For most urban situations, the ARF will exceed 0.9.

Example 5.2

Determine the relationship between intensity and duration for 10-year return period storms in the London area.

Solution

Calculate the range of M10–D rainfall intensities.
Read from Fig. 5.3: M5–60 min = 20 mm
Read from Fig. 5.4: M5–60 min/M5–2 day, $r = 0.45$
Read from Fig. 5.5: Z_1 values for various Ds
Read from Table 5.2: Z_2 values for various Ms

(1) Storm duration (D)		(2) M5–60 min rainfall total (mm)	(3) Z_1	(4) M5–D rainfall total (mm) (2) × (3)	(5) Z_2	(6) M10–D rainfall total (mm) (4) × (5)	(7) Intensity (mm/h) (6) ÷ (1)
h	min	(mm)					
	5	20	0.39	7.8	1.21	9.4	112.8
	10	20	0.55	11.0	1.22	13.4	80.4
	15	20	0.63	12.6	1.23	15.5	62.0
	30	20	0.77	15.4	1.24	19.1	38.2
1		20	1.0	20	1.24	24.8	24.8
2		20	1.2	24	1.24	29.8	14.9
4		20	1.4	28	1.22	34.2	8.6
6		20	1.5	30	1.22	36.6	6.1
10		20	1.65	33	1.21	39.9	4.0
24		20	2.0	40	1.19	47.6	2.0
48		20	2.3	46	1.17	53.8	1.1

Example 5.3

Adjust the point rainfall intensity of 25 mm/h for a 15 minute storm falling over a 200 ha urban catchment.

Solution

$D = 0.25$ h, $A = 2$ km^2 ∴ equation 5.5 is valid

$f_1 = 0.0394 \times 2^{0.354} = 0.050$
$f_2 = 0.040 - 0.0208 \ln (4.6 - \ln 2) = 0.012$

$\text{ARF} = 1 - 0.050 \times 0.25^{-0.012} = 0.95$

Areal intensity = 24 mm/h

5.3.5 Flood Estimation Handbook

The *Flood Estimation Handbook (FEH)* was published by the Institute of Hydrology (1999) as a major revision to supersede the Flood Studies Report and its supplementary reports. It is a consolidation of research on rainfall and flood frequency estimation and new procedures for application in river and flood defence. It is supported by software packages and digital catchment descriptors.

Experience had shown that rainfall frequency estimates using the FSR approach did not allow for all regional and local variations. This led to under- or over-design in certain areas, a problem the FEH was designed to overcome. For example, Allitt (2001) showed in a comparison of 7 UK sites that for the 30-year return period event the FSR always underestimated the 24-hour rainfall depth and sometimes underestimated shorter duration events. A design based on the 30-year FSR storm may therefore only give a 20-year or lower protection.

The difficulty for the urban drainage engineer lies in the limitations of the *Handbook*:

- it does not apply to catchments of area less than 50 ha
- it is not recommended for use on heavily urbanised catchments
- storm duration data ranges from 1 hour to 8 days (although shorter durations may be extrapolated).

It is clear that the focus of the manual, in its present form, is at river catchment scale and not urban drainage schemes. So we might provisionally accept the *FEH* as a better estimate of present day rainfall, but await an update of the *Wallingford Procedure* to allow its interpretation and use in urban areas.

5.4 Single events

So far we have considered rainfall to consist of just a fixed rainfall depth for a given duration. Clearly, this is unrealistic as rainfall intensity varies with time throughout the storm. This is represented as a plot of rainfall intensity against time called an *hyetograph* (or 'storm profile').

5.4.1 Synthetic design storms

A design storm is an idealised storm profile to which a statistically-based return period has been attached. The defined pattern in time is designed to reproduce (albeit imperfectly) the 'shape' of observed storms. The shape depends mainly on the type of event: a frontal storm usually has the highest intensities near the middle, and in a convective storm intensities are highest near the beginning.

The simplest (and least realistic) form of design storm is *block rainfall*, which may be simply derived from an IDF curve. Block rainfall has the same intensity over its duration and therefore has a rectangular time distribution. It is widely used in the Rational Method (given in Chapter 11) and has the advantage of being simple, quick to use and easily understandable.

In order to facilitate more accurate design solutions, profiles that better represent observed rainfall profiles have been produced. This has become more important with the advent of more sophisticated surface routing methods and flow simulation models. In the UK, a number of shapes have been proposed over the years, based successively on more comprehensive data sets.

Information on storm profiles can be found in the *Flood Studies Report* (NERC, 1975), based on the analysis of a wide range of storm events. A family of standard, symmetrical profiles was produced, with maximum rainfall intensity at the centre of the storm and varied in amplitude. The *peakedness* of a profile is defined as the ratio of maximum to mean intensity and the *percentile peakedness* is the percentage of storms that are equally or less peaked. The profile shape was not found to vary significantly with storm duration, return period or geographical region. However, on average, summer storms were found to be more peaked than winter ones.

The *Wallingford Procedure* recommends the 50 percentile summer profile (i.e. the storm that is more peaked than 50% of all summer storms) for design of drainage systems (see Fig. 5.6). A rainfall profile can be estimated by distributing the mean intensity over the storm duration as shown

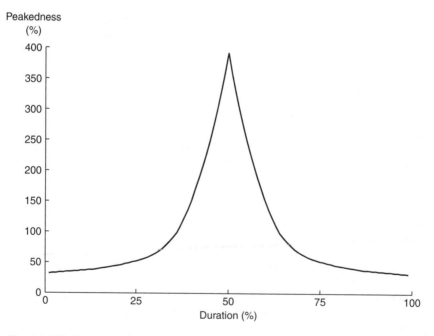

Fig. 5.6 FSR 50 percentile summer storm profile

Example 5.4

Determine the intensity of a 50 percentile summer storm of mean intensity 25 mm/h and duration 15 min at its $\frac{1}{3}$ and $\frac{1}{2}$ points.

Solution

$\frac{1}{3}$ point = 33.3% duration. From Fig. 5.6, % of mean intensity = 80

$$i_{33} = 0.8 \times 25 = 21 \text{ mm/h} @ t = 5 \text{ min}$$

$\frac{1}{2}$ point = 50% duration. From Fig. 5.6:

$$i_{50} = 3.9 \times 25 = 98 \text{ mm/h} @ t = 7.5 \text{ min}$$

---

in Example 5.4. The *Procedure* also recommends a method of smoothing the point rainfall profile to allow for areal extent.

5.4.2 Historical single events

Historical single events are hyetographs of point rainfall constructed from measured data. Unlike design storms they are not idealised and do not have an attached return period. Recorded data should be at intervals of 5 minutes, and preferably 1 minute. Their main use is in the verification of flow simulation models with measured hyetographs and simultaneous observations of flow. A model so verified is then assumed to give an accurate picture of catchment response (as considered in Chapter 19).

If the model allows consideration of spatially varying rainfall, the hyetographs used should adequately reflect the patterns of rainfall in time *and* space. The direction and movement (tracking) of a storm can be important in certain catchments (especially large ones), and can be a source of error if neglected (Ngirane-Katashaya and Wheater, 1985). Storms moving longitudinally through the catchment (relative to the drainage system) have the greatest influence.

5.5 Multiple events

5.5.1 Historical time–series

An historical series of rainfall events is the full set of all measured point rainfall for a particular time period (which would include all the single historical events and the intervening dry periods) at a particular location. These are used in conjunction with pre-calibrated, continuous simulation models, for long-term analyses. In effect, the use of time–series rainfall shifts the frequency analysis from the rainfall stage to the simulated runoff stage, since the complete series of historical events is run through the model. The return period of any characteristic of interest (e.g. peak flow, CSO operation) is obtained by the conventional ranking procedure and the use of a plotting formula, as described in Section 5.3.2. A typical time–series is shown in Fig. 5.7.

The advantage of time–series is that they are almost certain to contain the conditions that are critical for the catchment being studied. Their main disadvantage is that large amounts of data from recording rain gauges are required (which also needs extensive data analysis), and it is unlikely that this will be available for the particular site under consideration. This objection can be overcome to a limited extent by use of regional annual time–series.

Annual time–series

An annual rainfall time–series is a sequence of historic rainfall events that is statistically representative of the annual pattern at a given location. Three time–series have so far been derived in the UK by selecting typical months from a 40-year rainfall record and assembling them to form a typical year (Henderson, 1986). Their regions of applicability and basic hydrological properties are given in Table 5.3.

The time–series recommended for use is the one with the closest rainfall characteristics and not necessarily the closest geographically. Nevertheless, regionalisation procedures are still required to account for differences in hydrological factors between the catchment being studied and the region. Garside (1991) describes the procedures in detail.

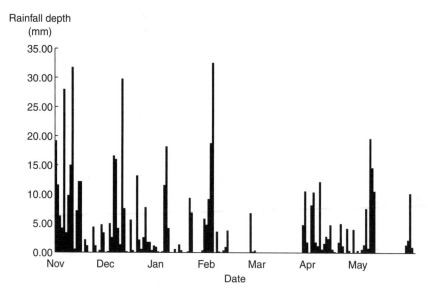

Fig. 5.7 Time–series rainfall (6 months of daily data)

Table 5.3 Hydrological characteristics of WRc annual time series

Region	M5–60 (mm)	r	SAAR** (mm)
South-east England	20	0.40	630
Central eastern England	19	0.40	590
South-west England*	19	0.33	930

* Can also be used for north-west England and Wales
** Standard annual average rainfall

Annual time–series may be used either in chronological order or ranked with the highest intensity/depth event first and the lowest last. The three series mentioned above are available as 99-event sequences, with each event associated with a particular catchment wetness (UCWI) and duration of antecedent dry weather period. The series can be sampled (for example, by selecting the first five storms and every fifth thereafter) to reduce computing time requirements.

Their main use is to represent the significant events in a typical year for situations where only short return periods are significant, such as day-to-day hydraulic performance of existing and rehabilitated systems including overflow spill events. The results of most interest generally are the total annual spill volume and the spill frequency from an overflow in particular, and the sewer system in general.

Unfortunately, the annual time–series suffers from several faults that make it difficult to use, including its limitation to just one year, the assumptions in its derivation and the need to rely on regionalisation. These concerns have led to the development of synthetic time–series rainfall.

5.5.2 Synthetic time–series

Synthetic series

A synthetic series simply overcomes the problem of using large numbers of different events in the annual time–series by representing their pattern with just a small number of synthetic storms. The synthetic storms can be based on conventional rainfall parameters such as storm depth, catchment wetness and storm peakedness.

Rainey and Osborne (1991) have derived two synthetic series based on the annual time–series described previously. Their regions of applicability and basic hydrological characteristics are given in Table 5.4.

Table 5.4 HR synthetic rainfall series

Region	r	SAAR (mm)
East	0.40	610
West	0.33	930

Stochastic rainfall generation

An alternative approach is the development of statistically-based rainfall models. Output from these models is a continuous rainfall time–series, which is statistically similar to the historical data for the catchment.

Cowperthwaite *et al.* (1991) report on a stochastic rainfall generator based on the Neyman-Scott Rectangular Pulses cluster model. The model has a plausible physical basis that assumes any rainfall event is triggered by arriving *storm origins* from which *rain cells* are generated. It is assumed that:

- the storm origins arrive based on a statistical Poisson process
- each storm origin generates a random number of rain cells
- the intensity and duration of each cell follow statistical exponential distributions
- the intensity of each cell is constant throughout the duration
- the total intensity at any point in time is the sum of the intensities of the rain cells.

The model's five parameters have been successfully fitted to hourly rainfall records for a number of UK sites. Moreover, a 5-minute interval time–series can be obtained by disaggregating the hourly series. Accuracy is greatly increased (to ±10% observed) if results are calibrated with monthly rainfall totals for several years of record. A commercial version of the procedure STORMPAC is available which in version 3 has some enhancements over the original method (Potter *et al.*, 2001).

Other models of British rainfall are also available (e.g. Onof and Wheater, 1994) and are being extended to include variability in space and time (e.g. Northop, 1998).

5.6 Climate change

5.6.1 Causes

An emerging challenge in the field of urban drainage is global warming, potentially leading to climate change. The evidence for global warming is compelling, with records showing that global-average surface air temperature has risen by around 0.6 °C since the beginning of the twentieth century, with about 0.4 °C of this warming occurring since the 1970s (Fig. 5.8). The year 1998 was the warmest on record, and 2001 was the third warmest. In the UK, the 1990s was the warmest decade in central England since records began in the 1660s (Hulme *et al.*, 2002).

The main clue to the cause of these rises is the significant increase in the concentration of greenhouse gases (e.g. carbon dioxide, methane) observed over the past 200 years. Pre-industrial carbon dioxide concentration was approximately 270 ppm whereas today it is more than 360 ppm (Bridge-

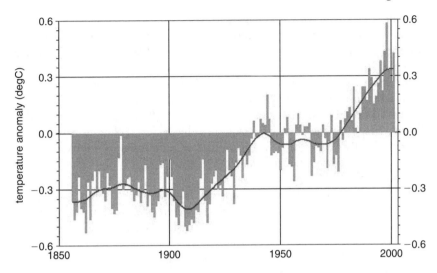

Fig. 5.8 Change in global-average surface air temperature anomalies relative to 1961–1990 average (UKCIP02 Climate Change Scenarios, funded by DEFRA, produced by Tyndall and Hadley Centres for UKCIP)

man and Gregory, 1999). These changes are thought to be due to a combination of natural and human causes. Currently, humankind emits some 6.5 billion tonnes of carbon into the atmosphere each year, mostly through the combustion of coal, oil and gas for energy. Land use results in further emissions of up to 2 billion tonnes of carbon. This intensifies the natural greenhouse effect, trapping more energy in the lower atmosphere. On the other hand, other human activity generates pollutants that actually cool the climate (sulphur dioxide which transforms into aerosols). All this is overlain on the natural background variability in the climate (Hulme *et al.*, 2002).

Unravelling these phenomena and making predictions into the future requires a significant modelling exercise. For example, the UK's Hadley Centre Global Climate Model (GCM) has been used to simulate global climate from 1860 to 2000. Only by incorporating natural *and* human factors could the model fit the data adequately, especially the warming since the 1970s. This and other evidence has led the Intergovernmental Panel on Climate Change to conclude that 'most of the warming observed over the last 50 years is likely to have been due to increasing concentrations of greenhouse gases' (IPPC, 2001).

5.6.2 *Future trends*

Prediction of future trends in climate is based on work carried out by the IPPC (2001).

Four scenarios have been developed based on different views of how the world might develop over the coming years. No attempt is made to prioritise these or attach a probability of occurrence. These have been used as a basis for the UK predictions and are summarised in Table 5.5.

To enable more spatially detailed predictions to be made, a number of regional climate models (RCMs) have been developed, such as the Hadley Centre's Europe model that has a resolution of 50 km². Application of this model, based on the four scenarios in Table 5.5 and three time horizons (2020s, 2050s and 2080s) has produced a wide range of climate predictions. Those most relevant to urban drainage are highlighted below (Hulme *et al.*, 2002).

Generally, the climate will become warmer, so by the 2080s the average annual temperature is predicted to rise by between 2 °C (for the low emissions scenario) and 3.5 °C (for the high emissions scenario). The south-east

Table 5.5 UKCIP02 Climate Change and IPCC Scenarios

Scenario		
UKCIP02	IPCC	Description
High emissions	A1F1	Rapid economic growth, global population that peaks in mid-century and declines thereafter, and the rapid introduction of new and more efficient technologies. Convergence among regions, capacity building and increased cultural and social interactions, with a substantial reduction in regional differences in per capita income.
Medium–High	A2	Heterogeneous world with great self-reliance and preservation of local identities. Fertility patterns across regions converge very slowly, which results in continuously increasing global population. Economic development is regionally oriented and per capita economic growth and technological change are more fragmented and slower than in other scenarios.
Medium–Low	B2	Emphasis is on local solutions to economic, social and environmental sustainability. Continuously increasing global population at a rate lower than in A2, intermediate levels of economic development, and less rapid and more diverse technological change than in B1 and A1F1. Environmental protection and social equity is mainly at local and regional levels.
Low	B1	Convergent world with the same global population as in A1F1, but with rapid changes in economic structures towards a service and information economy, with reductions in material intensity, and the introduction of clean and resource-efficient technologies. Emphasis on global solutions to economic, social and environmental sustainability, including improved equity, but without additional climate initiatives.

of England is expected to warm more than the north-west, with more frequent high summer temperatures and fewer very cold winters.

Temperature rises will lead to increases in annual precipitation of up to 10% by the end of the century, with increases of up to 35% occurring in winter under the high emissions scenario. Almost the whole of the UK is expected to be drier in the summer, with the greatest decreases in rainfall (up to 50% in the high emissions scenario) in the south-east. Summer soil moisture is also reduced by 40% or more over much of England under the high emissions scenario.

Heavy winter rainfall will become more frequent, with intensities that are currently experienced around once every 2 years becoming between 5% (low emissions) and 20% (high emissions) heavier by the 2080s. Storm events in the summer will become more intense and more frequent.

5.6.3 Implications

The main findings of climate change studies, of relevance to urban drainage, are an increase in total precipitation (and hence runoff) and increased storm intensities. The potential implications are as follows:

- increased flows that may exceed the capacity of existing sewer systems leading to more frequent surcharging and surface flooding
- greater deterioration of sewers due to more frequent surcharging
- more frequent CSO spills
- greater build-up and mobilisation of surface pollutants in summer
- poorer water quality in rivers due to extra SWO and CSO spills and reduced base flows in summer
- increased flows of dilute wastewater at WTPs due to higher rainfall and infiltration, potentially leading to poorer treatment by biological processes.

5.6.4 Solutions

It is unlikely that major upgrades of the existing sewer network will be carried out in response to *potential* climate change. Increases in runoff will need to be dealt with in a number of ways, such as:

- increased application of SUDS: infiltration devices, above-ground storage (see Chapter 21)
- more widespread capture and reuse of rainwater (see Chapter 24)
- planned urban overland flow routing (major–minor systems) (see Chapter 11)
- increased application of real-time control (see Chapter 22).

Problems

5.1 Describe the main types of rain gauge and assess their relative merits in urban drainage applications.

5.2 Calculate the average rainfall intensity for a 30 min, 1 yr return period storm using the Ministry of Health, Bilham and Holland equations.   [20, 18.2, 19.6 mm/h]

5.3 Derive a series of IDF curves (1, 2 & 5 yr) for a location in Coventry.

5.4 Construct the 50 percentile rainfall intensity profile for a 20 mm/h, 20 minute duration summer storm.

5.5 What is a synthetic design storm and how can it be represented? What are the differences from an historical event?

5.6 Explain what you understand by an historical series of rainfall. What is an annual time–series?

5.7 What are the benefits of using rainfall synthetic time–series?

5.8 What are the main impacts of climate change and how might urban drainage systems be adapted to take account of them?

Key sources

DoE/NWC (1981) *Design and Analysis of Urban Storm Drainage. The Wallingford Procedure. Volume 1: Principles, Methods and Practice*, Department of the Environment, Standing Technical Committee Report No. 28.

Einfalt, T., Krejci, V. and Schilling, W. (1998) Rainfall data in urban hydrology, in *Hydroinformatic Tools for Planning, Operation, Design, Operation and Rehabilitation of Sewer Systems* (eds J. Marsalek, C. Maksimovic, E. Zeman and R. Price), NATO ASI Series 2: Environment – Vol. 44, Kluwer Academic Press.

Hall, M.J. (1984) *Urban Hydrology*, Elsevier Applied Science.

Hulme, M., Jenkins, G.J., Lu, X., Turnpenny, J.R., Mitchell, T.D., Jones, R.G., Lowe, J., Murphy, J.M., Hassell, D., Boorman, P., McDonald, R. and Hill, S. (2002) *Climate Change Scenarios for the United Kingdom: The UKCIP02 Scientific Report*. Tyndall Centre for Climate Change Research, University of East Anglia, Norwich.

IPPC (2001) *Climate Change 2001: Synthesis Report*. Contribution of Working Groups I, II & III to the Third Assessment Report. www.grida.no/climate.

Natural Environment Research Council (1975) *Flood Studies Report*, 5 volumes, Institute of Hydrology, Wallingford.

Niemczynowicz, J. (1996) Rainfall measurement, processing and application in urban hydrology, in *Rain and Floods in Our Cities. Gauging the Problem* (ed. C. Maksimovic), WMO Technical Report in Hydrology and Water Resources No. 53, 9–78.

References

Allitt, R. (2001) Modelling FEH Storms. *WaPUG Spring Meeting*.

Ashfaq, A. and Webster, P. (2002) Evaluation of the FEH rainfall-runoff method for catchments in the UK. *Journal of Chartered Institution of Water and Environmental Management*, **16**, August, 223–228.

Bilham, E.G. (1936) Classification of heavy falls in short periods. *British Rainfall 1935*, 262–280.

Bridgeman, J.M. and Gregory, J.M. (1999) The potential impact of climate change upon sewerage. *WaPUG Autumn Meeting*. www.wapug.org.uk.

Collier, C.G. (1996) *Applications of Weather Radar Systems: A Guide to Uses of Radar Data in Meteorology and Hydrology*, 2nd edition, Praxis Publishers, John Wiley & Sons.

Cowperthwaite, P.S.P., Metcalfe, A.V., O'Connell, P.E., Mawdsley, J.A. and Threlfall, J.L. (1991) *Stochastic Rainfall Generation of Rainfall Time Series*, FWR Report No. FR0217.

Garside, I.G. (1991) *Using Annual Rainfall Time Series*, WaPUG User Note No. 23.

Henderson, R.J. (1986) *Rainfall Time Series for Sewer System Modelling*, WRc Report No. ER195E.

Holland, D.J. (1967) Rain intensity frequency relationships in Britain. *British Rainfall 1961*, Part III, 43–51.

Hurcombe, P. (2001) Climate change and potential effects on sewerage systems. *WaPUG Spring Meeting*. www.wapug.org.uk.

Institute of Hydrology (1999) *Flood Estimation Handbook*, 5 volumes, Institute of Hydrology, Wallingford.

Meteorological Office (1982) *Rules for Rainfall Observers*, Meteorological Office Leaflet No. 6.

Ministry of Health, Departmental Committee on Rainfall and Runoff (1930) Rainfall and runoff. *Journal of Institution of Municipal County Engineers*, 56, 1172–1176.

Ngirane-Katashaya, G. and Wheater, H.S. (1985) Hydrograph sensitivity to storm kinematics. *Water Resources Research*, 2(3), 337–345.

Norris, W.H. (1948) Sewer design and frequency of heavy rain. *Proceedings of Institution of Municipal Engineers*, 75(6), 349–364.

Northop, P. (1998) A clustered spatial-temporal model of rainfall. *Proceedings of Royal Society of London*, A, **454**, 1875–1888.

Onof, C. and Wheater, H.S. (1994) Improvements of the modelling of British rainfall using a modified random parameter Bartlett-Lewis model. *Journal of Hydrology*, **157**, 177–195.

Petrovic, J. and Elgy, J. (1994) The use of Meteostat infrared imagery for enhancement of areal rainfall estimates. *Remote Sensing and GIS in Urban Waters, UDT '94* (eds C. Maksimovic, J. Elgy and V. Dragalov), Moscow, 29–40.

Potter, R., Lane, A. and Cowpertwaite, P. (2001) Testing of the new rainfall generator model of STORMPAC3 against STORMPAC2 and historic data. *WaPUG Spring Meeting*. www.wapug.org.uk.

Rainey, C.M.L. and Osborne, M.P. (1991) Design for storage using synthetic rainfall. *WaPUG Autumn Meeting*. www.wapug.org.uk.

Rosenfeld, D. and Collier, C.G. (1998) Estimating surface precipitation. Chapter 14.1 in *Global Energy and Water Cycles* (eds K.A. Browning and R. Gurney), Cambridge University Press, 124–133.

Schilling, W. (1991) Rainfall data for urban hydrology: what do we need? *Atmospheric Research*, 27, 5–21.

Shaw, E.M. (1994) *Hydrology in Practice*, 3rd Edition, Van Nostrand Reinhold.

WRc (1987) *A Guide to Short Term Flow Surveys of Sewer Systems*, Water Authorities Association.

6 Stormwater

6.1 Introduction

Stormwater (surface runoff) is the second major urban flow of concern to the drainage engineer. Safe and efficient drainage of stormwater is particularly important to maintain public health and safety (due to the potential impact of flooding on life and property) and to protect the receiving water environment. Reliable data on the quantity and quality of existing and projected stormwater flows is a prerequisite for cost-effective urban drainage design and analysis.

Stormwater is generated by rainfall, and consists of that proportion of rainfall that runs off from urban surfaces (see Fig. 6.1). Hence, the properties of stormwater, in terms of quantity and quality, are intrinsically linked to the nature and characteristics of both the rainfall and the catchment.

This chapter is concerned with the generation and characteristics of stormwater. Details concerning rainfall have been presented in Chapter 5. Many of the concepts presented here underlie the design and analysis techniques described in later chapters.

6.2 Runoff generation

The transformation of a rainfall hyetograph into a surface runoff hydrograph involves two principal parts. Firstly, *losses* due to interception, depression storage, infiltration and evapo-transpiration are deducted from the rainfall. Secondly, the resulting *effective rainfall* is transformed by *surface routing* into an *overland flow* hydrograph.

Conventionally, little attention has been given to the description of the overland flow phase. Yet, for most urban drainage applications, the runoff processes are at least as important as the pipe flow processes (discussed later in Chapter 8) and of equal importance to the rainfall processes.

Much of the rainfall that reaches the ground does not, in fact, run off. It is 'lost' immediately or as it runs overland. The water may be completely lost from the catchment surface by processes such as by evapo-transpiration, it may be temporarily retained in depression storage or it may eventually find its way to the drainage system via groundwater.

6.2.1 Initial losses

Interception and wetting losses

Interception consists of the collection and retention of rainfall by vegetation cover. There is an initial retention period, after which excess rain falls through the foliage or flows to the soil over the stems. The interception rate then rapidly approaches zero. The interception loss for impervious areas is small in magnitude (<1 mm) and is normally neglected or combined with depression storage.

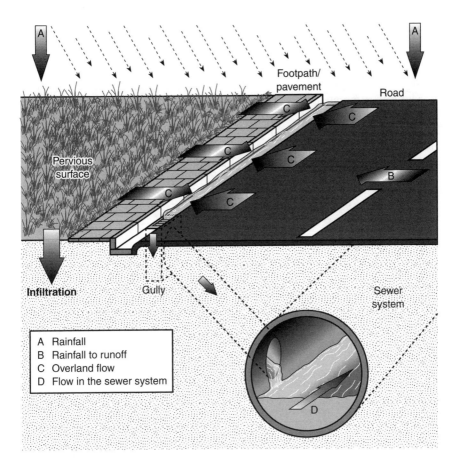

Fig. 6.1 Stormwater runoff generation processes

Depression storage

Depression storage accounts for rainwater that has become trapped in small depressions on the catchment surface, preventing the water from running off. Infiltration, evaporation or leakage will eventually remove the water that has been retained. Factors affecting the magnitude of depression storage are surface type, slope and rainfall return period (Kidd and Lowring, 1979). Depression storage d (mm) can be represented as:

$$d = \frac{k_1}{\sqrt{s}} \tag{6.1}$$

k_1 coefficient depending on surface type (0.07 for impervious surfaces and 0.28 for pervious surfaces) (mm)

s ground slope (–)

Typical values for d are 0.5–2 mm for impervious areas, 2.5–7.5 mm for flat roofs, and up to 10 mm for gardens.

Representation

For intense summer storms in urban areas, the initial losses are not important, but for less severe storms or for less urbanised catchments they should not be neglected. For modelling purposes, the combined initial losses are usually subtracted from the rainfall at the beginning of the storm to leave the *net* rainfall. This is illustrated in Example 6.1.

6.2.2 Continuing losses

Evapo-transpiration

Evapo-transpiration is the vaporisation of water from plants and open water bodies and therefore its removal from surface runoff. Although it is a continuing, constant loss, its effect during short duration rainfall events is negligible. For example, the average daily value of potential evaporation in the UK during the summer months is 2 to 3 mm. Consequently, it is normally neglected in most models or considered to be lumped into the initial losses.

Infiltration

Infiltration represents the process of rainfall passing through the ground surface into the pores of the soil. The infiltration capacity of a soil is defined as the rate at which water infiltrates into it. The magnitude depends on factors including soil type, structure and compaction, initial

---

Example 6.1

For an urban catchment of average slope 1% with an estimated interception loss of 0.5 mm, calculate the net rainfall profile (based on initial losses only) of the following storm:

Time (min)	0–10	10–20	20–30	30–40
Rainfall intensity (mm/h)	6	12	18	6

Take $k_1 = 0.1$ mm.

Solution

Interception loss = 0.5 mm
Depression storage loss from equation 6.1: $d = 0.1/\sqrt{0.01} = 1$ mm

Time (min)	0–10	10–20	20–30	30–40
Rainfall intensity (mm/h)	6	12	18	6
Rainfall depth (mm)	1	2	3	1
Net rainfall depth (mm)	0	1.5	3	1
Net rainfall intensity (mm/h)	0	9	18	6

Initial losses are deducted from rainfall *depth* at the beginning of the storm.

---

moisture content, surface cover and the depth of water on the soil. The infiltration rate tends to be high initially but decreases exponentially to a final quasi-steady rate when the upper soil zone becomes saturated.

A common empirical relationship used to represent infiltration is Horton's (1940) equation:

$$f_t = f_c + (f_o - f_c)e^{-k_2 t} \tag{6.2}$$

f_t	infiltration rate at time t (mm/h)
f_c	final (steady state) infiltration rate or capacity (mm/h)
f_o	initial rate (mm/h)
k_2	decay constant (h^{-1})

The equation is valid when $i > f_c$. These parameters depend primarily on soil/surface type and initial moisture content of the soil. The range of values encountered for f_c, f_o and k are given in Table 6.1. Careful adaptation of the equation is required to render it suitable for application in continuous simulation models.

Other, more physically-based approaches, have been formulated such as Green and Ampt's (1911) equation and Richard's (1933) equation. These are not widely implemented in urban drainage models.

Table 6.1 Typical Horton parameters for various surface types

Surface type	f_o (mm/h)	f_c (mm/h)	k_2 (h^{-1})
Coarse textured soils	250	25	2
Medium textured soils	200	12	2
Fine textured soils	125	6	2
Clays/paved areas	75	3	2

Representation

Continuing losses are always important in urban catchments, but are of most prominence in areas with relatively large open spaces. A simplified, but common, approach to representing them is by a constant proportional loss model applied after initial losses have been deducted to produce the *effective* rainfall:

$$i_e = C\, i_n \tag{6.3}$$

i_e effective rainfall intensity (mm/h)
C dimensionless runoff coefficient (–)
i_n net rainfall intensity (mm/h)

The runoff coefficient C depends primarily on land use, soil and vegetation type and slope. It is also influenced by rainfall characteristics (e.g. intensity, duration) and antecedent conditions. Values of C range from 0.70 to 0.95 for impervious surfaces such as pavements and roofs, and from 0.05 to 0.35 for pervious surfaces. A more comprehensive listing of coefficients is given in Table 11.3. This model also forms the basis for the Rational Method used for estimating stormwater peak flow rates. This important method is described in more detail in Chapter 11.

6.2.3 Percentage runoff equation

For urban catchments in the UK, the dimensionless runoff coefficient can be estimated from the so-called PR (percentage runoff) equation ($C = PR/100$) prepared as part of the *Wallingford Procedure* (DoE/NWC, 1983). This is a regression equation derived from data obtained from 17 catchments and 510 (summer) events:

$$PR = 0.829\ \text{PIMP} + 25.0\ \text{SOIL} + 0.078\ \text{UCWI} - 20.7$$
$$[PR > 0.4\ \text{PIMP}] \tag{6.4}$$

$$PR = 0.4\ \text{PIMP} \qquad\qquad [PR \le 0.4\ \text{PIMP}]$$

PIMP   percentage impervious area of the catchment (25–100)
SOIL   a soil index for the UK (0.15–0.50)
UCWI   urban catchment (antecedent) wetness index (30–300)

This equation is reasonably reliable provided it is used with variables that are within the range of those upon which it is based (shown in brackets). Since its development, it has been used successfully to represent many hundreds of catchments throughout the UK (see Example 6.2). The principal variables are described in further detail below.

PIMP

The percentage imperviousness represents the degree of urban development of the catchment and is defined as:

$$\text{PIMP} = \frac{A_i}{A} \times 100 \tag{6.5}$$

A_i impervious (roofs and paved areas) area (ha)
A total catchment area (ha)

SOIL

The *SOIL* index is based on the winter rain acceptance parameter in the *Flood Studies Report* (NERC, 1975) and is a measure of infiltration potential of the soil. It can be obtained from maps in the *Flood Studies Report* or the *Wallingford Procedure* (DoE/NEC, 1981).

UCWI

The urban catchment wetness index *(UCWI)* represents the degree of wetness of the catchment at the start of a storm event. As *UCWI* increases, so does the *PR* value reflecting the increased runoff expected from a wetter catchment. It can be estimated for design purposes from its relationship with the standard average annual rainfall *(SAAR)* given in Fig. 6.2. A map of average annual rainfall is given in the *Wallingford Procedure*.

When simulating historical events:

$$UCWI = 125 + 8API5 - SMD \tag{6.6}$$

API5   5-day antecedent precipitation index
SMD   soil moisture deficit

API5 is calculated according to a methodology described in the Wallingford Procedure based on rainfall depths in the 5 days prior to the event. *SMD* is a measure of the amount of water that can be retained within the soil matrix, values of which are available for UK locations from the Meteorological Office.

Limitations

In specific circumstances, the PR equation has been found to have limitations.

- For catchments with relatively low PIMP (that is, with a large proportion of pervious surface), particularly those with light soils in dry conditions, the equation tends to under-predict the runoff volume. This has led to various 'work around' strategies being developed, but these in turn have their difficulties and can be complicated to apply in practice (Osborne, 2000).
- During long-duration storms, catchment surfaces can be significantly wetted, increasing the proportion of runoff. This expected increase in runoff is not properly represented.
- The equation was developed for use with discrete rainfall events and is not directly applicable for continuous simulation using rainfall time series (see Section 5.5).

6.2.4 New runoff equation

The so-called NR (new runoff) equation has been developed to try to overcome some of the limitations mentioned in the previous section. The model has two major differences. The first is that runoff is calculated separately for impervious and pervious areas (not combined, as in the PR equation). The second difference lies in the way API is allowed to vary during the storm rather than being a fixed value.

The model has three components: initial losses (see Section 6.2.1), runoff from impervious areas and runoff from pervious areas.

Impervious area runoff

The model deals with continuous losses following deduction of the initial losses. Impervious areas are dealt with, simply, in two parts:

- a proportion of the surface is assumed to be directly connected to the drainage network and to generate 100% runoff. This can be estimated from Table 6.2.
- the rest of the surface is assumed to be less effectively connected and to behave hydrologically as if it were pervious. This part is therefore added to the pervious area total.

Table 6.2 Percentage connectedness (after Osborne, 2001)

Surface type	Percentage connected
Normal urban paved surfaces	60
Roof surfaces	80
Well-drained roads	80
Very high-quality roads	100

Pervious area runoff

The pervious area and the less effectively connected impervious area are taken together, and the runoff is calculated using a soil moisture storage model, applied progressively throughout a storm (rather than once, before the storm):

$$R_t = I_t \, API_t \, / \, S_t \tag{6.7}$$

R_t	Runoff depth at time t (mm)
I_t	Rainfall depth at time t (mm)
API_t	Antecedent precipitation index at time t (mm)
S_t	Soil storage depth at time t (mm)

The value of API_t used in equation 6.7 gives a better definition of the wetness of the soil than conventional $API5$ introduced earlier. It takes account of evaporation, better represents the rate of drying out of different soil types, and can be continuously updated during the storm. Details of its calculation are given by Osborne (2001). The default value for the soil storage depth S is 0.2 m, which notionally represents the soil depth that is wetting and drying.

6.3 Overland flow

Once the losses from the catchment have been accounted for, the effective rainfall hyetograph can be transformed into a surface runoff hydrograph – a process known as *overland flow* or *surface routing*. In this process, the runoff moves across the surface of the sub-catchment to the nearest entry point to the sewerage system.

There are two general approaches currently used for routing overland flow. The most common utilises the *unit hydrograph* method, although this is actually implemented in a number of different ways. The second, more physically-based approach, usually utilises a *kinematic wave* model.

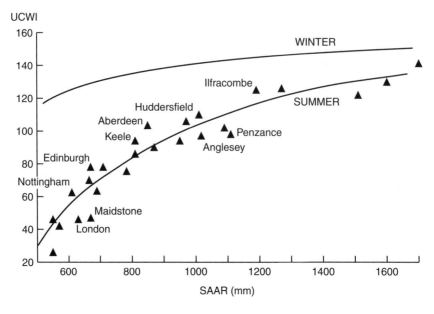

Fig. 6.2 Relationship between *UCWI* and *SAAR* (based on Packman [1986] with permission of the Chartered Institution of Water and Environmental Management, London)

6.3.1 Unit hydrographs

The unit hydrograph is a widely used concept in hydrology that has also found application in urban hydrology. It is based on the premise that a unique and time-invariant hydrograph results from effective rain falling over a particular catchment. Formally, it represents the outflow hydrograph resulting from a unit depth (generally 10 mm) of effective rain falling uniformly over a catchment at a constant rate for a unit duration D: the D-h unit hydrograph is shown in Fig. 6.3. The ordinates of the D-h unit hydrograph are given as $u(D,t)$, at any time t. D is typically 1 h for natural catchments but could, in principle, be any time period.

Once derived, the unit hydrograph can be used to construct the hydrograph response to any rainfall event based on three guiding principles:

- constancy: the time base of the unit hydrograph is constant, regardless of the intensity of the rain
- proportionality: the ordinates of the runoff hydrograph are directly proportional to the volume of effective rain – doubling the rainfall intensity doubles the runoff flow-rates

Example 6.2

Calculate the effective rainfall profile for the storm specified in Example 6.1. The rain falls on a catchment that is 78% impervious, has a soil type index of 0.25 and a *SAAR* of 540 mm.

Solution

$SAAR = 540$ mm
Read from Fig. 6.2: $UCWI = 40$
Equation 6.4:

$$PR = 0.829 \times 78 + 25.0 \times 0.25 + 0.078 \times 40 - 20.7 = 53\%$$

Equation is valid as $[PR = 53] > [0.4 \times 78 = 31]$
Total net rainfall depth $= 5.5$ mm (From Example 6.1)
Runoff rainfall depth $= 0.53 \times 5.5 = 2.9$ mm
Runoff loss $= 5.5 - 2.9 = 2.6$ mm
Continuing loss $= 2.6/0.5 = 5.2$ mm/h (over 30 minutes)
The profile is therefore:

Time (min)	0–10	10–20	20–30	30–40
Net rainfall profile (mm/h)	0	9	18	6
Effective rainfall profile (mm/h)	0	3.8	12.8	0.8

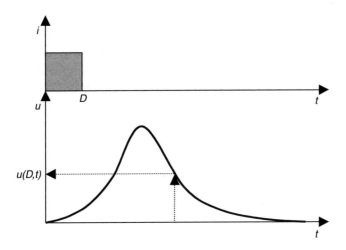

Fig. 6.3 The unit hydrograph

- superposition: the response to successive blocks of effective rainfall, each starting at particular times, may be obtained by summing the individual runoff hydrographs starting at the corresponding times.

This approximate, linear approach (known as 'convolution') is stated succinctly in equation 6.8. If a rainfall event has n blocks of rainfall of duration D, the runoff $Q(t)$ at time t is:

$$Q(t) = \sum_{\omega=1}^{N} u(D,j)\, I_\omega$$

(6.8)

$$Q(t) = u(D,t)I_1 + u(D,t-D)I_2 + \ldots + (D,t-(N-1)D)I_n$$

$Q(t)$ runoff hydrograph ordinate at time t (m^3/s)
$u(D,j)$ D-h unit hydrograph ordinate at time j (m^3/s)
I_ω is the rainfall depth in the ωth of N blocks of duration D (m)
j $t-(\omega-1)D$ (s)

Further detail and examples on using unit hydrographs are given in Shaw (1994).

In order to use this concept in ungauged urban catchments for design purposes, some way of predicting unit hydrographs is required, based on catchment characteristics. Three methods of doing this are in current use: synthetic unit hydrographs, the time–area method or reservoir models.

6.3.2 *Synthetic unit hydrographs*

The detailed shape of the unit hydrograph reflects the characteristics of the catchment from which it has been derived. When converted into dimensionless form, it is found that very similar shapes are observed in catchments in the same region.

Harms and Verworn (1984) have derived a dimensionless unit hydrograph suitable for urban areas, as shown in Fig. 6.4. This has a linear rise up to the peak flow, an exponential recession and an end point at 1% of peak flow:

$$Q = \frac{t}{t_p} Q_p \qquad\qquad 0 < t < t_p$$

(6.9a)

$$Q = Q_p e^{-\frac{t-t_p}{k_3}} \qquad\qquad t \geq t_p$$

(6.9b)

Q flow-rate (m^3/s)
Q_p peak flow rate (m^3/s)
t time (s)
t_p time to peak (s)
k_3 exponential decay constant (s^{-1})

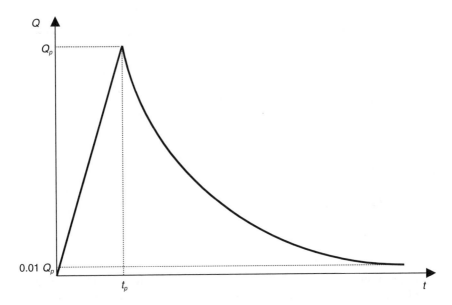

Fig. 6.4 Synthetic unit hydrograph

The three parameters, Q_p, t_p and k can be related to catchment characteristics.

This is the most direct application of the unit hydrograph approach, but is the least common in practice.

6.3.3 Time–area diagrams

An alternative approach is to derive a time–area diagram, which can be shown (Hall, 1984) to be a special case of a unit hydrograph. To do this, lines of equal flow 'travel time' to the catchment outfall are delineated, so-called *isochrones*. The maximum travel time represents the time of concentration (t_c) of the catchment (considered in Chapter 11). The time–area diagram that is constructed by summing the areas between the isochrones (see Fig. 6.5) defines the response of the catchment.

When combined with rainfall in depth increments of I_1, I_2, ... I_N, flow at any time $Q(t)$ is:

$$Q(t) = \sum_{\omega=1}^{N} \frac{dA(j)}{dt} I_\omega \tag{6.10}$$

where $dA(j)/dt$ is the slope of the time–area diagram at time j.

The time–area diagram can be used in design as an extension to the Rational Method and in the TRRL method (Chapter 11 has a description and

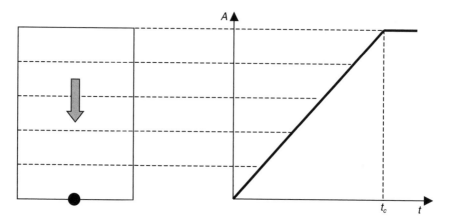

Fig. 6.5 Linear time–area diagram

examples) and is also implemented in some simulation models using standard time–area diagram profiles and empirical equations to estimate t_c.

6.3.4 Reservoir models

The third approach is to propose the analogy that the catchment surface acts on the flow generated by an effective rainfall profile as one or more reservoirs connected in series. Each reservoir then experiences inflows of rainfall (and/or inflows from upstream reservoirs) and outflows of runoff. The model is based on the two equations of continuity and storage:

$$\frac{dS}{dt} = I - O \tag{6.11}$$

$$S = KO^m \tag{6.12}$$

I	inflow rate (m^3/s)
O	outflow rate (m^3/s)
S	storage volume (m^3)
K	reservoir time constant (s)
m	exponent (–)

These equations will be returned to in Chapter 9, in the context of reservoir routing.

If m is taken to be equal to unity (a physical impossibility, but a conceptual convenience), then the reservoir is referred to as 'linear'. Nash (1957) proposed that the overland flow process could be represented as a series of identical linear reservoirs, where the output from one reservoir is considered as the input to a second, and so on. Assuming an instantaneous inflow of unit volume, the unit hydrograph time to peak t_p and peak flow Q_p are:

$$t_p = (\alpha - 1)K \tag{6.13a}$$

and

$$Q_p = \frac{1}{K(\alpha - 1)!} \left(\frac{\alpha - 1}{e} \right)^{n-1} \tag{6.13b}$$

This approach can be used by specifying the number of reservoirs (α) and the time constant K, where α and K can be related to catchment characteristics. Alternatively, α and K can be used as calibration constants in models.

Again, it can be shown mathematically that a linear reservoir cascade is a special case of the unit hydrograph approach.

The reservoirs are termed linear because, in equation 6.12, the storage S is linearly related to outflow O, so from equation 6.11 outflow must be linearly related to inflow. However, other non-unity values of m (for example, 0.67) can be used, resulting in a non-linear response from a *single* conceptual reservoir. The parameters K and m become calibration constants.

6.3.5 *Kinematic wave*

A more physically-based approach is to simplify and solve the equations of motion to give:

$$\frac{\partial d}{\partial t} + \frac{\partial q}{\partial x} = i_e \tag{6.14}$$

$$q = \frac{1}{n} d^{5/3} s^{1/2} \tag{6.15}$$

d depth of flow (m)
q flow per unit width $= Q/b$ (m²/s)
i_e effective rainfall intensity (m/s)
t time (s)
x longitudinal distance (m)
n Manning's roughness coefficient (m$^{-1/3}$/s)
s catchment slope (–)

Equation 6.14 is a continuity equation and 6.15 is a simplified momentum equation based on Manning's equation. The kinematic wave approximation is explained further in Chapter 19 and Manning's equation in Chapter 8.

Manning's *n* for urban surfaces is typically 0.05–0.15, an order of magnitude greater than pipe roughness.

6.4 Stormwater quality

It is not tenable to assume that rainwater, and certainly not stormwater runoff, is 'pure'. Numerous studies over the last twenty years have shown that urban stormwater can be heavily contaminated with a range of polluting substances. Stormwater contains a complex mixture of natural organic and inorganic materials, with a small proportion of man-made substances derived from transport, commercial and industrial practices. These materials find their way into the drainage system from atmospheric sources and as a result of being washed off or eroded from urban surfaces. In certain respects, stormwater can be as polluting as wastewater.

It was stressed in Chapter 4 that wastewater is variable in character. The quality of stormwater is even more variable from place to place and from time to time. As with wastewater, care should be taken in interpreting 'standard' or 'typical' values.

6.4.1 Pollutant sources

As mentioned earlier, stormwater quality is influenced by rainfall and, especially, by the catchment. The major catchment sources include vehicle emissions, corrosion and abrasion; building and road corrosion and erosion; bird and animal faeces; street litter deposition, fallen leaves and grass residues; and spills.

Atmospheric pollution

Pollutants in the urban atmosphere result mainly from human activities: heating, vehicular traffic, industry or waste incineration, for example. They may either be absorbed and dissolved by precipitation (known as wet fallout), to be carried directly into the drainage system with the stormwater; or they may settle on land surfaces (as dry fallout), and subsequently be washed off. Dry fallout particles can be transported by winds over long distances.

Although atmospheric sources are accepted as a major contributor to stormwater pollution, the importance of dry and wet fallout appears to be dependent on the site and the pollutant. In Gothenburg, for example, wet fallout has been identified as the dominant form of atmospheric pollutant (at 60%) for nitrogen, phosphorus, lead, zinc and cadmium. Even higher percentages have been noted in some locations.

Dry fallout is thought to be of more importance in urban areas or areas with significant sources of solids. In Sweden, it was estimated that 20% of the organic matter, 25% of phosphorus and 70% of the total nitrogen in stormwater can be attributed to atmospheric fallout (Malmqvist, 1979). Granier *et al.* (1990) found that approximately half the total loads of lead and chromium come from the atmosphere, with a significantly lower proportion of zinc.

Vehicles

Vehicle emissions include volatile solids and PAHs derived from unburned fuel, exhaust gases and vapours, lead compounds (from petrol additives), and hydrocarbon losses from fuels, lubrication and hydraulic systems.

Pollutants are generated by the everyday passage of traffic. Tyre wear releases zinc and hydrocarbons. Vehicle corrosion releases pollutants such as iron, chromium, lead and zinc. Other pollutants include metal particles, especially copper and nickel, released by wear of clutch and brake linings. Most metals are predominantly associated with the particulate phase.

Wear of the paved surface will release various substances: bitumen and aromatic hydrocarbons, tar and emulsifiers, carbonates, metals and fine sediments, depending on the road construction technique and materials used.

Buildings and roads

Urban erosion produces particles of brick, concrete, asphalt and glass. These particles can form a significant constituent of sediment in stormwater. The extent of pollution will depend on the condition of the buildings/roads. Roofs, gutters and exterior paint can release varying amounts of particles, again depending on condition. Metallic structures, such as street furniture (e.g. fences, benches) corrode, releasing toxic substances such as chromium. Roads and pavements degrade over time, releasing particles of various sizes.

Animals

Urine and faeces deposited on roads and pavements by animals (pets or wild) are a source of bacterial pollution in the form of faecal coliforms and faecal streptococci. They are also a source of high oxygen demand.

De-icing

The most commonly used de-icing agent is salt (sodium chloride). Salt applications to roads cause the annual chloride loads in stormwater to be (on average) 50–500 times higher than would occur naturally (Stotz, 1987). Rock salt contains other impurities, including an insoluble fraction shown to contribute 25% of the winter suspended solids load in a motorway study (Colwill *et al.*, 1984). The presence of salt also accelerates corrosion of vehicles and metal structures.

Urban debris

Urban surfaces can contain large amounts of street debris, litter and organic materials such as dead and decaying vegetation. Litter will generally result in elevated levels of solids and greater consequential oxygen demand. Fallen leaves and grass cuttings may lie on urban surfaces, particularly in road gutters, and decompose, or may be washed into gullies.

Spills/leaks

Household cleaners and motor fluids/lubricants are sometimes illegally discarded or spilled into gutters and gullies. The range and amounts of these pollutants vary considerably, depending on land use and public behaviour. However, domestic sources of chemical pollutants are usually minor when compared with industrial spills or illicit toxic waste disposal.

6.4.2 Surface pollutants

The bulk of the pollutants, derived from the sources mentioned, are attached to particles of sediment that deposit temporarily on the catchment surface. Analysis of this particulate material shows a large range of sizes from below 1 μm to above 10 mm. Larger, denser sediment causes particular problems in the drainage system itself and this is addressed in detail in Chapter 15. Despite typically comprising less than 5% of the material present, particulates less than 50 μm in size have most of the pollutants associated with them: 25% of the COD, up to 50% of the nutrients and 15% of total coliforms (Ellis, 1986). These, of course, are the particles most readily washed off by the stormwater.

6.4.3 Pollutant levels

Pollutants in stormwater include solids, oxygen-consuming materials, nutrients, hydrocarbons, heavy metals and trace organics, and bacteria. Typical values and ranges of pollutant discharges from stormwater systems in the UK are given in Table 6.2. The table demonstrates the inherent variability of runoff quality. The average quality of 'clean' and 'dirty' catchments can vary by a factor of ten, and the variation in quality between stormwater events for any single catchment can vary by a factor of three (Ellis, 1986). Runoff quality will depend upon a number of factors, including:

- geographical location
- road and traffic characteristics
- building and roofing types
- weather, particularly rainfall.

Table 6.2 Pollutant event mean concentrations and unit loads for stormwater (after Ellis, 1986)

Quality parameter	EMC (mg/l)	Unit load (kg/imp ha.yr)*
Suspended solids (SS)	21–2582 (190)	347–2340 (487)
BOD$_5$	7–22 (11)	35–172 (59)
COD	20–365 (85)	22–703 (358)
Ammoniacal nitrogen	0.2–4.6 (1.45)	1.2–25.1 (1.76)
Total nitrogen	0.4–20.0 (3.2)	0.9–24.2 (9.0)
Total phosphorus	0.02–4.30 (0.34)	0.5–4.9 (1.8)
Total lead	0.01–3.1 (0.21)	0.09–1.91 (0.83)
Total zinc	0.01–3.68 (0.30)	0.21–2.68 (1.15)
Hydrocarbons	0.09–2.8 (0.4)	–
Faecal coliforms	400–50 000 (6430) (MPN/100 ml)	0.9–3.8 (2.1) ($\times 10^9$ counts/ha)

*imp ha = impervious area measured in hectares

6.4.3 Representation

The most common methods used to quantify stormwater quality are event mean concentrations, regression equations/rating curves and build-up/washoff models. These are described below.

Event mean concentrations

In this simplest of methods, the assumption is made that stormwater has a constant concentration, such as those given in Table 6.2. Thus, the method lends itself to easy integration with standard flow (not quality) simulation models. The method cannot represent variations in quality within the storm and, therefore, is most suitable in situations where calculation of total pollutant load only is required.

Regression equations

In this approach, the quality of stormwater is statistically regressed against a number of describing variables, e.g. catchment characteristics or land use. This is the quality equivalent to the PR equation discussed earlier in the chapter. Regression equations can usually be relied on to give good representations on the catchment(s) on which they were based and perhaps similar ones. They will be less accurate on other catchments, but can often give a reasonable first approximation.

Build-up

The most common model-based approach to quality representation is by separately predicting pollutant build-up and washoff. In practice, the distinction between these two processes is not clearly defined. The factors affecting build-up of pollutants on impervious surfaces include:

- land use
- population
- traffic flow
- effectiveness of street cleaning
- season of the year
- meteorological conditions
- antecedent dry period
- street surface type and condition.

Build-up on the surface dM_s/dt can be assumed to be linear, so:

$$\frac{dM_s}{dt} = aA \qquad (6.16)$$

M_s mass of pollutant on surface (kg)
a surface accumulation rate constant (kg/ha.d)
A catchment area (ha)
t time since the last rainfall event or road sweeping (d)

Accumulation rate a values for solids in residential areas are up to 5 kg/imp ha.d.

Detailed observation of sites in the US (Sartor and Boyd, 1972) revealed that, despite there being no rainfall or street cleaning, the pollutant deposition often has a reducing rate of increase rather than a uniform linear increase. The first-order removal concept can be used to represent this, which implies that equilibrium is reached when the supply rate of pollutants matches their removal:

$$\frac{dM_s}{dt} = aA - bM_s \qquad (6.17)$$

where b is the removal constant (d^{-1}).

The equilibrium mass on the catchment is therefore $A\ (a/b)$. Novotny and Chesters (1981) reported values for b from a US medium density residential area of 0.2–0.4 d^{-1}. Studies in London (Ellis, 1986) suggest equilibrium is reached within 4–5 days where vehicle-induced re-suspension is dominant. The levelling-off phenomenon is most profound in areas where:

- adjacent pollution traps (pervious areas) are available
- vehicle-induced wind and vibration is high.

This will be the case on motorways and trunk roads, and in busy commercial/industrial areas.

Pollutants other than solids can be predicted using 'potency' factors (defined in Chapter 20).

Washoff

Washoff occurs during rainfall/runoff by raindrop impact, erosion or solution of the pollutants from the impervious surface. Important factors include:

- rainfall characteristics
- topography
- solid particle characteristics
- street surface type and condition.

The simplest approach is to assume there is effectively an infinite store of pollutants always available on the surface to be washed off, and hence no build-up. Experimental evidence suggests this assumption may be valid in British conditions (Mance and Harman, 1978). Washoff can then be modelled as a function of rainfall intensity:

$$W = z_1 i^{z_2} \tag{6.18}$$

W pollutant washoff rate (kg/h)
i rainfall intensity (mm/h)
z_1, z_2 pollutant-specific constants.

The exponent z_2 usually has values between 1.5 and 3.0 for particulate pollutants and <1.0 for those in solution (Delleur, 1998). Price and Mance (1978) found z_2 to be 1.5 and z_1 0.02 on some British catchments. This is a convenient form for ready addition to flow models.

Alternatively, a first-order relationship, where the rate at which pollutants washoff is assumed to be directly proportional to the amount of pollutant remaining on the surface, can be used:

$$W = -\frac{dM_s}{dt} = k_4 i M_s(t) \tag{6.19}$$

where k_4 is the washoff constant (mm^{-1}).

Integrating equation 6.19 gives:

$$M_s(t) = M_s(0)e^{-k_4 i t} \tag{6.20}$$

$M_s(0)$ initial amount of pollutant on surface (kg)
$M_s(t)$ amount of pollutant on surface after time t (kg)
$M_w(t)$ amount of pollutant washed off after time t (kg)

Since $M_s(t) = M_s(0) - M_w$:

$$M_w(t) = M_s(0)[1 - e^{-k_4 it}] \tag{6.21}$$

Typical values for k_4 are 0.1–0.2 mm^{-1}. As with the build-up parameters, this requires calibration for each individual catchment. Washoff concentration (c) can be obtained as:

$$c = \frac{W}{Q} = \frac{k_4 M_s}{A_i} \tag{6.22}$$

where A_i is the catchment impervious area (ha).

Example 6.3

A storm of duration 30 mins and intensity 10 mm/h falls on a 1.5 ha impervious area urban catchment. If the initial pollutant mass on the surface is 12 kg/ha, calculate:
a) the mass of pollutant washed off during the storm ($k_4 = 0.19$ mm^{-1})
b) the average pollutant concentration.

Solution

(a) $M_s(0) = 12 \times 1.5 = 18$ kg

From equation 6.21:

$$M_w(0.5) = 18[1 - e^{-0.19 \times 10 \times 0.5}] = 11.0 \text{ kg}$$

(b) $c = \dfrac{M_w(0.5)}{Q} = \dfrac{11.0 \text{ (kg)}}{0.01 \text{ (m/h)} \times 0.5 \text{ (h)} \times 15\,000 \text{ (m}^2)} \begin{array}{l} = 0.147 \text{ kg/m}^3 \\ = 147 \text{ mg/l} \end{array}$

A disadvantage of this formulation is that pollutant concentration will only decrease with time as M_s decreases. This can be remedied by introducing an exponent w for i in equation 6.19 where i is in the range 1.4–1.8.

$$W = k_5 i^w M_s \tag{6.23}$$

where k_5 is the amended washoff constant (mm^{-1}).

Problems

6.1 List and explain the initial rainfall losses. How are they represented mathematically?

6.2 List and explain the continuing rainfall losses. How are they represented mathematically?

6.3 What are the disadvantages of the PR equation? How are some of these overcome with the NR equation?

6.4 Reassess the effective rainfall profile calculated in Example 6.2 using Horton's equation as the continuing loss model where $f_0 = 10$ mm/h, $f_c = 1$ mm/h and $k_2 = 1$ h^{-1}.   [0, 0.4, 10.6, 0 mm/h]

6.5 A 1 mm, 10 min unit hydrograph for an urban catchment is given in the table. Derive the runoff hydrograph resulting from the effective rainfall hyetograph also given.   [Peak = 2500 l/s @ 40 mins]

Time (mins)	0–10	10–20	20–30	30–40	40–50	50–60	60–70
UH flow-rate (l/s)	0	250	500	375	250	125	0
Rainfall intensity (mm)	1	2	3	0	1	0	0

6.6 Compare and contrast the three main approaches to predicting unit hydrographs in ungauged urban catchments.

6.7 A three-reservoir Nash cascade ($K = 12$ min) is used to represent the runoff response of a 10 ha urban catchment. Calculate the 10 mm 1 hr unit hydrograph peak flow and time to peak.   [278 l/s, 24 min]

6.8 What are the main sources of pollutants in stormwater and what is their importance?

6.9 Explain the main ways in which stormwater quality is modelled. What are their relative merits?

6.10 For the conditions described in Example 6.4, re-evaluate the storm-water pollutant concentration at 10 minute time intervals.
[166, 122, 88 mg/l]

Key sources

Colyer, P.J. and Pethick, R.W. (1976) *Storm Drainage Design Methods. A Literature Review*, Report No. INT154, Hydraulics Research Station.

Hall, M.J. (1984) *Urban Hydrology*, Elsevier Applied Science.

Marsalek, J, Maksimovic, C., Zeman, E. and Price, R. (eds) (1998) *Hydroinformatic Tools for Planning, Operation, Design, Operation and Rehabilitation of Sewer Systems*, NATO ASI Series 2: Environment – Vol. 44, Kluwer Academic Press.

Luker, M. and Montague, K.N. (1994) *Control of Pollution from Highway Drainage Discharges*, Report R142, CIRIA, London.

Torno, H.C., Marsalek, J. and Desbordes, M. (eds) (1986) *Urban Runoff Pollution*, NATO ASI Series G: Ecological Sciences – Vol. 10, Springer-Verlag.

References

Colwill, D.M., Peters, C.J. and Perry, R. (1984) *Water quality in motorway runoff*, TRRL Supplementary Report 823, Transport and Road Research Laboratory.

Delleur, J.W. (1998) Modelling quality of urban runoff, in *Hydroinformatic Tools for Planning, Operation, Design, Operation and Rehabilitation of Sewer Systems* (eds J. Marsalek, C. Maksimovic, E. Zeman and R. Price), NATO ASI Series 2: Environment – Vol. 44, Kluwer Academic Press, 241–285.

DoE/NWC (1981) *Design and Analysis of Urban Storm Drainage. The Wallingford Procedure. Volume 1: Principles, Methods and Practice*, Department of the Enviroment, Standing Technical Committee Report No. 28.

Ellis, J.B. (1986) Pollutional aspects of urban runoff, in *Urban Runoff Pollution* (eds H.C. Torno, J. Marsalek and M. Desbordes), NATO ASI Series G: Ecological Sciences – Vol. 10, Springer-Verlag, 1–38.

Granier, L., Chevrevil, M., Carru, A. and Letolle, R. (1990) Urban runoff pollution by organochlorines (polychlorinated biphenyls and lindane) and heavy metals (lead, zinc and chromium). *Chemosphere*, **21**(9), 1101–1107.

Green, W.H. and Ampt, G.A. (1911) Studies of soil physics, 1: The flow of air and water through soils. *Journal of Agricultural Science*, **4**(1), 1–24.

Harms, R.W. and Verworn, H.-R. (1984) HYSTEM – ein hydrologisches Stadtentwaesser-ungsmodell. Teil I: Modellbeschreibung. *Korrespondenz Abwasser*, **31**(2).

Horton, R.E. (1940) An approach towards a physical interpretation of infiltration capacity. *Proceedings of Soil Science Society of America*, **5**, 399–417.

Kidd, C.H.R. and Lowring, M.J. (1979) *The Wallingford Urban Sub-catchment Model*. IoH Report No. 60, Institute of Hydrology.

Malmqvist, P-A. (1979) Atmospheric fallout and street cleaning – effect on urban storm water and snow. *Progress in Water Technology*, **10**, 417–431.

Mance, G. and Harman, M. (1978) The quality of urban stormwater runoff, in *Urban Storm Drainage* (ed. P.R. Helliwell), Pentech Press, 603–617.

Nash, J.E. (1957) The form of the instantaneous unit hydrograph. *International Association of Science Hydrology Publication*, **45**(3), 114–121.

Natural Environment Research Council (1975) *Flood Studies Report*, 5 volumes, Institute of Hydrology.

Novotny, V. and Chesters, G. (1981) *Handbook of Nonpoint Pollution, Sources and Management*, Van Nostrand Reinhold.

Osborne, M.P. (2000) Runoff models – lessons from study audits. WaPUG Spring Meeting, Blackpool.

Osborne, M.P. (2001) *A New Runoff Volume Model*, WaPUG User Note No. 28, version 3.

Packman, J.C. (1986) Runoff estimation in urbanising and mixed urban/rural catchments. *IPHE Seminar on Sewers for Adoption*, Imperial College, London.

Price, R, and Mance, G. (1978) A suspended solids model for storm water runoff, in *Urban Storm Drainage* (ed. P.R. Helliwell), Pentech Press, 546–555.

Richards, L.A. (1933) Capillary conduction of liquids through porous mediums. *Physics*, **1**, 318–333.

Sartor, J.D. and Boyd, G.B. (1972) *Water Pollution Aspects of Street Surface Contaminants*. EPA Report No. EPA/R2/72/081.

Shaw, E.M. (1994) *Hydrology in Practice*, 3rd edition, Van Nostrand Reinhold.

Stotz, G. (1987) Investigations of the properties of the surface water runoff from federal highways in the FRG. *The Science of the Total Environment*, **59**, 329–337.

7 System components and layout

7.1 Introduction

This chapter gives an overview of the elements that make up any urban drainage system, including building drainage and other main system components. The main stages in the design process are also described.

7.2 Building drainage

Even though urban drainage engineers are not normally involved directly in the planning, design and construction of building drainage, it is important that they are at least aware of the main components and layout of systems in and around buildings. This includes, in particular, an understanding of how building drains connect with the main sewer system.

Building drainage in the UK is subject to the 2000 Building Regulations, supported by Approved Document H (DTLR, 2001).

7.2.1 Soil and waste drainage

Inside

A common arrangement for the soil (WC) and waste (other appliances) drainage of modern domestic properties is shown in Fig. 7.1. This illustrates a two-storey dwelling with appliances on both floors connected to a single vertical stack. Each appliance is protected by a trap (U-bend or S-bend) and water seal to prevent odours reaching the house from the downstream drainage system.

The stack (typically 100–150 mm in diameter) has a top open to atmosphere that should be at least 900 mm from the top of any adjacent opening into the property. The flow regime in a vertical stack is quite different to that in sloping pipes. Flow tends to adhere to the perimeter of the pipe forming an annulus with a central air core. The pressure of the air in the core varies with height, depending on the appliances in use and can be both positive and negative. Design rules and details have been devised to avoid the risk of water being siphoned from the traps.

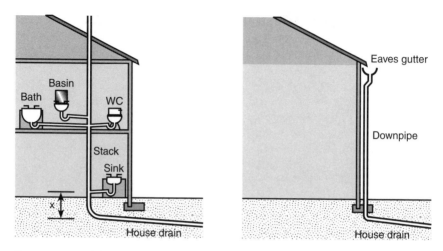

Fig. 7.1 Typical building drainage arrangement in a two-storey house

The distance *x* from the lowest branch to the invert of the building drain typically exceeds 450 mm for dwellings up to three storeys high.

Outside

The individual lengths of drains connecting each property to the public sewer tend to be short (usually <20 m) and small (diameter <150 mm). However, in terms of the total length of the whole piped system, they make up a surprisingly large fraction – perhaps as much as half.

Components

Outside building drainage systems have a number of common components, particularly associated with providing access for testing, inspection and blockage-clearance from the surface. A *rodding eye* permits rodding along the drain from the surface. It consists of a vertical or inclined riser pipe with a sealed, removable cover. Access can also be gained using an *access chamber* over a pipe fitting with a sealed, removable cover. *Inspection chambers* are also used and consist of shallow access points on the drain, and also have a sealed, removable cover. Woolley (1988) gives comprehensive coverage of building drainage details.

Building drains are typically designed using procedures similar to small foul sewers (described in Chapter 10). Gradients tend to be quite steep (>1:80 for 100 dia. pipes), although field evidence suggests that very flat drains are no more likely to block than steep ones (Lillywhite and Webster, 1979); good quality construction is more influential in reducing blockage potential. Drains and private sewers are relatively shallow with a minimum cover of 0.75 m under gardens and 1.25 m under roads and paths. The height x, plus the length and gradient of the drain, determines the minimum feasible depth of the public sewer.

Layout

The main aim of the layout of external building drainage is to minimise the length of pipework and associated components, whilst ensuring that adequate accessibility is maintained. Generally, changes of direction should be minimised and appropriate access points provided where necessary.

Building drains carrying soil and waste should discharge only to a public foul or combined sewer. Many existing installations still feature an interceptor trap with water seal in the last inspection chamber before the sewer. These were provided to reduce the risk of odour release into the building drainage and to discourage the entry of rodents. However, they have tended to fall into disuse and disrepair and can be a source of block-age and odour problems in their own right. Today, they are not normally specified.

7.2.2 *Roof drainage*

A conventional arrangement for the roof drainage of domestic properties is given in Fig. 7.1. This shows a two-storey property with a pitched roof, drained by an eaves gutter connected to a single, vertical downpipe, posi-tioned at one end. A typical eaves gutter is a 75 mm half-round channel with a nominal fall. Its capacity can be estimated using the theory of spatially varied flow, and also depends on the position and spacing of the outlets.

Flow in the rainwater downpipe is annular, just like in the soil and waste stack. The type of inlet dictates capacity. For single family dwellings, downpipes are 75–100 mm in diameter.

The downpipes can discharge directly to a separate storm sewer, but will need a water seal trap if connected to a combined sewer. Roof drainage should not be discharged to separate foul sewers.

The design of roof drainage systems is similar to that of small storm sewer networks (Chapter 11 gives details). In this situation, the catchments are very small (<60 m^2), the time of concentration is low (1–2 min) and so short duration, high intensity rainfall events are critical for pipe capacity estimation. Often a fixed rainfall intensity of 50 mm/h is used in design.

7.3 System components

The design of sewer systems is covered in Chapters 10 to 12, but this section introduces the main 'hardware' components of the sewerage.

7.3.1 Sewers

Most new sewers are circular in cross-section and range upwards in diameter from 150 mm. They can be made of vitrified clay, concrete, fibre cement, PVC-U and other polymers, pitch fibre or brick. Information on pipe materials and jointing methods is given in Chapter 15.

Vertical alignment

Fig. 7.2 illustrates how the vertical position of a sewer is defined by its invert level (IL). The invert of a pipe refers to the lowest point on the inside of the pipe. The invert level is the vertical distance of the invert above some fixed level or *datum* (for example, in the UK, above ordnance datum (AOD)). Other important levels shown in Fig. 7.2 are the *soffit level* which is the highest point on the inside of the pipe and the *crown* which is the highest point on the outside of the pipe. Using the nomenclature defined in Fig. 7.2:

$$b = a + D$$

and

$$c = b + t = a + D + t$$

D internal diameter of the pipe (mm)
t pipe wall thickness (mm)

The depth of the pipe (y_1) is therefore:

$$y_1 = d - a + t \tag{7.1}$$

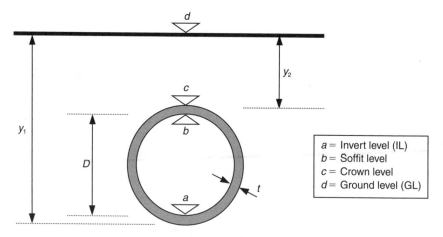

Fig. 7.2 Level definitions associated with sewers

Example 7.1

A 375 mm diameter pipe with 15 mm walls has an invert level of 52.665 m. If the ground level is 54.930, calculate the pipe: (a) soffit level, (b) depth and (c) cover.

Solution

(a) soffit level: $\quad b = a + D = 52.665 + 0.375 = 53.040$ m

(b) depth (equation 7.1): $\quad y_1 = d - a + t = 54.930 - 52.665 + 0.015$
$$= 2.280 \text{ m}$$

(c) cover (equation 7.2): $\quad y_2 = y_1 - D - 2t = 2.280 - 0.375 - 0.030$
$$= 1.875 \text{ m}$$

---

and the cover of the pipe (y_2) is:

$$y_2 = d - c = y_1 - D - 2t \tag{7.2}$$

Fig. 7.3 shows a typical sewer vertical alignment plotted on a *longitudinal profile*. The profile contains the main information required in the vertical plane to construct the pipeline.

Two invert levels are given at two of the manholes, since one refers to the exit and one to the entry level. At MH34, dissimilar diameter pipes meet and good practice recommends (as shown) that soffit not invert levels are matched. 'Chainage' refers to the plan (horizontal) distance along the pipe from a specific point. The top line of the profile box is fixed at a given level above datum (in this case 75 m). All other vertical levels can then be scaled from this line. The scale of the drawing is usually distorted to give more detail in the vertical plane.

Normal practice is to ensure individual pipes between manholes have a constant gradient. Sewers are usually constructed under the highway with the storm sewer being on the centre-line and the foul sewer being offset laterally and slightly lower. Should exfiltration occur, this will avoid pollution of the stormwater system. Sewers should be laid deep enough:

- to drain the lowest appliance in the premises served
- to withstand surface loads
- to prevent the contents from freezing.

Typically, minimum cover for rigid pipes is 0.9 m under gardens and fields and 1.2 m under roads. More detailed aspects of the structural design of sewers will be discussed in Chapter 15.

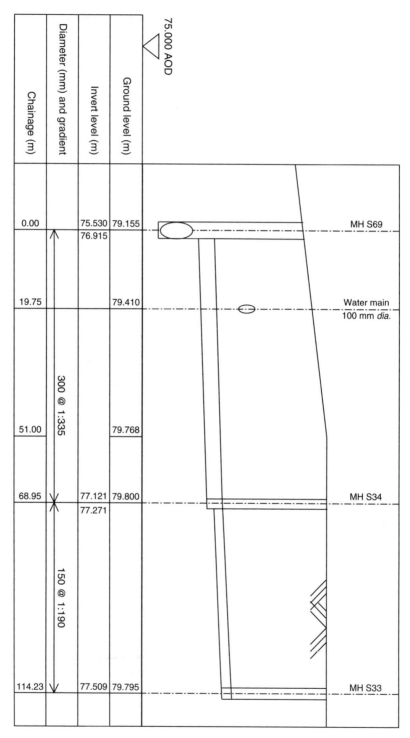

Fig. 7.3 Longitudinal profile of a sewer

Horizontal alignment

Fig. 7.4 shows a typical sewer horizontal alignment with two possible ways of numbering the system. Fig. 7.4 (a) numbers the *pipes*, and is based on the computer coding method originally recommended in *Road Note 35* (TRRL, 1976) and the *Wallingford Procedure* (DoE/NWC, 1981). It is suitable for both computer and manual methods. Sewers are numbered in the form (*x.y*) where *x* refers to the sewer branch and y refers to the individual pipe within the branch.

An alternative procedure is to attach numbers or other unique code to the manholes, as in Fig. 7.4 (b). Standard symbols are given in the key.

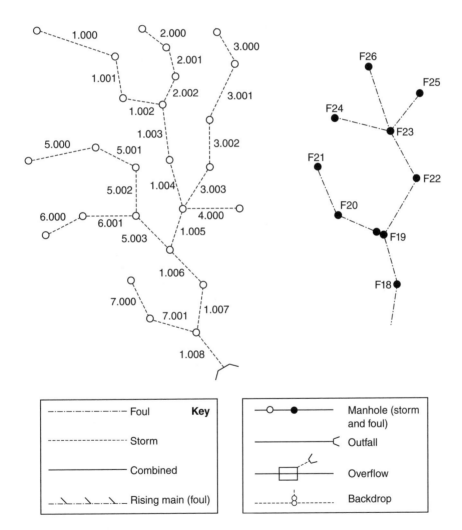

Fig. 7.4 Standard sewer symbols and numbering systems

This is more appropriate during the construction phase where manholes are usually numbered in sequence working away from the outfall. The horizontal position of manholes may be identified by their grid reference.

Good engineering practice is to ensure individual pipe runs (between manholes) are straight in plan. However, larger (man-entry) sewers can be built with slight curves if necessary. Further aspects of good practice in horizontal and vertical alignment are discussed later in the chapter under layout design.

7.3.2 Manholes

As with building drainage systems, access points are required for testing, inspection and cleaning. In sewer systems, access is usually by manholes that differ from inspection chambers in that they are deeper (>1 m) and can be entered if necessary. Manholes are provided at (BS EN 752–3: 1997):

- changes in direction
- heads of runs
- changes in gradient
- changes in size
- major junctions with other sewers
- every 90 m.

In larger pipes, where man-access is possible (although undesirable), the spacing of manholes may be increased up to 200 m. Manholes are commonly constructed of precast concrete rings as specified in BS 5911. Fig. 7.5 shows a detail of a precast concrete ring manhole. Smaller manholes may have precast benching. The diameter of the manhole will depend on the size of sewer and the orientation and number of inlets. Requirements for manhole covers and frames are given in BS EN 124: 1994.

In situations where a high level sewer is connected to one of significantly lower level, a backdrop manhole can be used. These are typically used to bring the flow from higher level laterals into a manhole rather than lowering the length of the last sewer lengths. Drops may be external or internal to the manhole, or sloping ramps may be used, depending on the drop height and the diameter of the pipe. Fig. 7.6 shows an externally-placed vertical backdrop manhole. Drop manholes can require additional maintenance.

More information on manhole details can be found in Woolley (1988) and WRc (1995).

7.3.3 Gully inlets

Surface runoff is admitted from roads and other paved areas via inlets known as 'gullies'. Gullies consist of a grating and usually an underlying

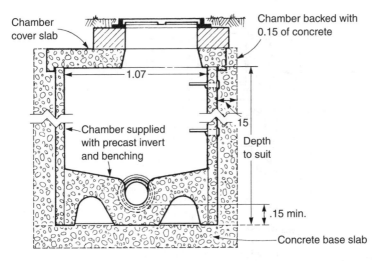

Fig. 7.5 Precast concrete ring manhole (reproduced from Woolley 1988 with permission of E & FN Spon)

sump (a 'gully pot') to collect heavy material in the flow. A water seal is incorporated to act as an odour trap for those gullies connected to combined sewers (see Fig. 7.7). The gully is connected to the sewer by a lateral pipe. The relevant standard for gully gratings and frames is BS EN 124: 1994 and for precast concrete and vitrified clay gully pots, BS 5911 or BS EN 295: 1991 respectively.

The size, number and spacing of gullies will determine the extent of surface ponding of runoff during storm events. Gullies are always placed at low points and, typically, are spaced along the road channel, adjacent to the kerb. The simplest approach is to specify a standard of 50 m spacing or to require one gully per 200 m² of impervious area. Alternatively, Mollinson (1958) proposed:

$$L = \frac{280\sqrt{s}}{W} \tag{7.3}$$

L gully spacing (m)
s longitudinal road gradient (%)
W width of drainage area (m²)

More accurate (but more involved) gully spacing methods are explained in Chapter 9.

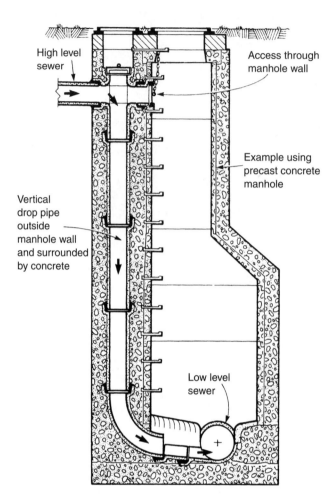

High level sewer

Access through manhole wall

Example using precast concrete manhole

Vertical drop pipe outside manhole wall and surrounded by concrete

Low level sewer

Fig. 7.6 Backdrop manhole (reproduced from Woolley 1988 with permission of E & FN Spon)

7.3.4 Ventilation

Ventilation is required in all urban drainage systems, but particularly in foul and combined sewers. It is needed to ensure that aerobic conditions are maintained within the pipe, and to avoid the possibility of build-up of toxic or explosive gases. The implications of anaerobic conditions and health and safety issues will be discussed in Chapter 17.

Nearly all sewer systems are ventilated passively, without air extraction equipment. Some major pumping stations and WTPs are mechanically ventilated. In larger and older schemes, above-ground ventilation shafts have been used to ensure good circulation of air. Care is needed in siting these

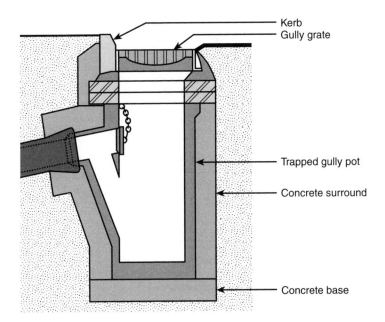

Fig. 7.7 Trapped road gully

structures to avoid odour nuisance. Some schemes use ventilated manhole covers. More modern practice is to utilise the ventilation provided by the soil stacks on individual buildings (Fig. 7.1). Air is drawn through the system by the low pressure induced by the flow of air over the top of the stacks, and by the fall of wastewater. The water seals on domestic appliances avoid backup of sewer gases into the building interior.

7.4 Design

7.4.1 Stages

A number of fundamental stages need to be followed to design a rational and cost-effective urban drainage system. These are illustrated in Fig. 7.8 and are valid for any type of system. Further details on sewer sizing are given in Chapters 10–12.

The first stage is to define the *contributing area* (catchment area and population) and mark it on a topographical map. In the UK, this will probably be a 1:1250 or 1:2500 Ordnance Survey (OS) map. The map should already include contours, but other pertinent natural (e.g. rivers) and man-made (e.g. buildings, roads, services) features should also be marked up. Possible outfall or overflow points should be identified and investigations made as to the capacity of the receiving water body.

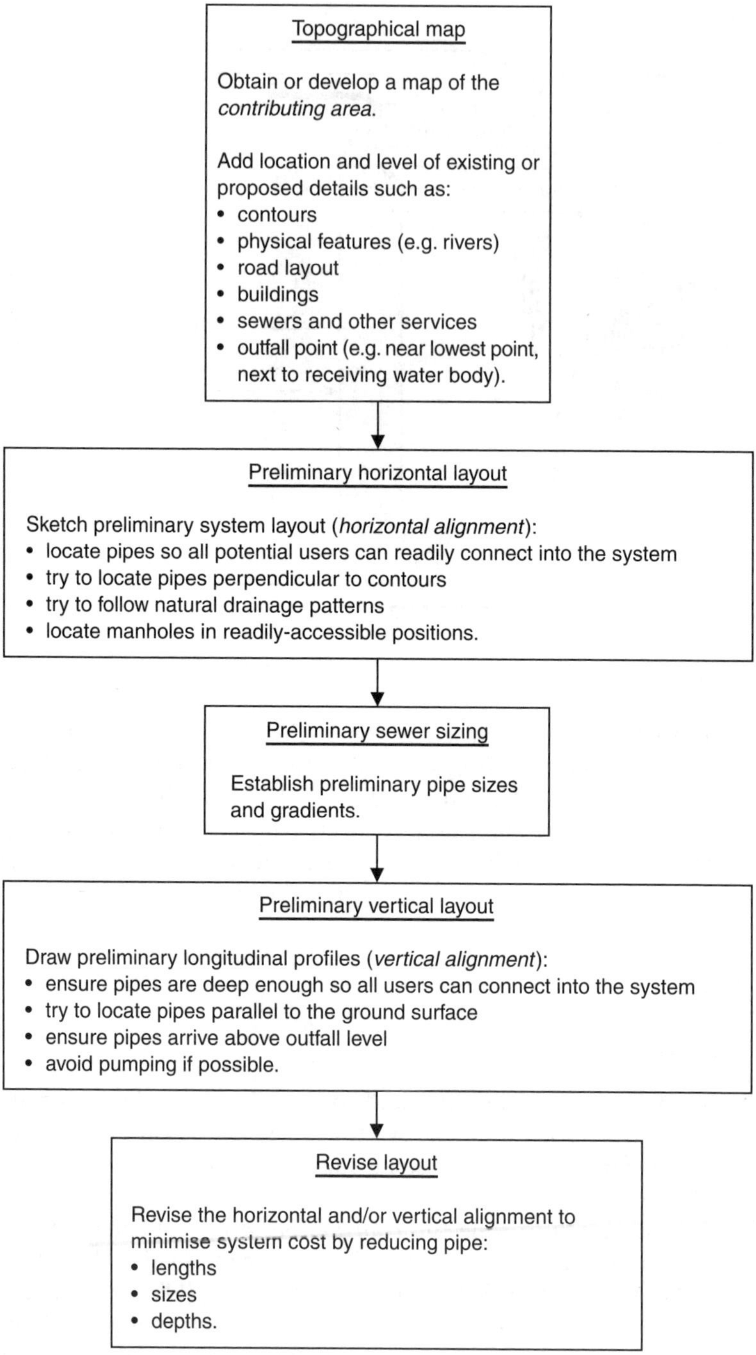

```
┌─────────────────────────────────────────────┐
│           Topographical map                 │
│                                             │
│  Obtain or develop a map of the             │
│  contributing area.                         │
│                                             │
│  Add location and level of existing or      │
│  proposed details such as:                  │
│  • contours                                 │
│  • physical features (e.g. rivers)          │
│  • road layout                              │
│  • buildings                                │
│  • sewers and other services                │
│  • outfall point (e.g. near lowest point,   │
│    next to receiving water body).           │
└─────────────────────────────────────────────┘
                      │
                      ▼
┌─────────────────────────────────────────────────────────────┐
│              Preliminary horizontal layout                   │
│                                                             │
│  Sketch preliminary system layout (horizontal alignment):    │
│  • locate pipes so all potential users can readily connect   │
│    into the system                                          │
│  • try to locate pipes perpendicular to contours            │
│  • try to follow natural drainage patterns                  │
│  • locate manholes in readily-accessible positions.         │
└─────────────────────────────────────────────────────────────┘
                      │
                      ▼
┌─────────────────────────────────────────────┐
│           Preliminary sewer sizing          │
│                                             │
│   Establish preliminary pipe sizes          │
│   and gradients.                            │
└─────────────────────────────────────────────┘
                      │
                      ▼
┌─────────────────────────────────────────────────────────────┐
│               Preliminary vertical layout                    │
│                                                             │
│  Draw preliminary longitudinal profiles (vertical alignment):│
│  • ensure pipes are deep enough so all users can connect     │
│    into the system                                          │
│  • try to locate pipes parallel to the ground surface        │
│  • ensure pipes arrive above outfall level                   │
│  • avoid pumping if possible.                               │
└─────────────────────────────────────────────────────────────┘
                      │
                      ▼
┌─────────────────────────────────────────────┐
│               Revise layout                 │
│                                             │
│  Revise the horizontal and/or vertical      │
│  alignment to minimise system cost by       │
│  reducing pipe:                             │
│     • lengths                               │
│     • sizes                                 │
│     • depths.                               │
└─────────────────────────────────────────────┘
```

Fig. 7.8 Urban drainage design procedure

The next stage is to produce a preliminary *horizontal alignment* aiming to achieve a balance between the requirement to drain the whole contributing area and the need to minimise pipe run lengths. Least-cost designs tend to result when the pipe network broadly follows the natural drainage patterns and is branched, converging to a single major outfall.

Having located the pipes horizontally, the pipe sizes and gradients can now be calculated based on estimated flows from the contributing area as described in Chapters 10–12. Generally, sewers should follow the slope of the ground as far as possible to minimise excavation. However, gradients flatter than 1:500 should be avoided as they are difficult to construct accurately. A preliminary *vertical alignment* can then be produced, again bearing in mind the balance between coverage of the area and depths of pipes. The alignment can be plotted on longitudinal profiles as shown in Fig. 7.3. Ground levels can initially be taken from the OS map contours, but eventually an on-site level survey will be required. Pumping should be avoided, particularly on storm sewer systems, but will be needed if excavations exceed about 10 m.

The final stage involves revising both the horizontal and vertical alignment to minimise cost by reducing pipe lengths, sizes and depths whilst meeting the hydraulic design criteria. Longer sewer runs may be cost-effective if shorter runs would require costlier excavation and/or pumping.

7.4.2 Sewers for Adoption

When new residential areas are built, it is common for developers to construct the building drainage and some of the 'communal' sewers. The aim of the developers and of the subsequent house owners is for the drainage network to be accepted by the local Sewerage Undertaker and 'adopted' as part of the network under its control. To aid this process, a set of design and construction guidelines, called 'Sewers for Adoption' has been produced, which is generally agreed and accepted (WRc, 2001).

The main areas covered by this document are:

- design and construction advice
- standard details
- model civil engineering specification
- form of agreement (Section 104 of the Water Industry Act 1991).

Appropriate reference to the recommendations in Sewers for Adoption is made in later chapters.

Further government advice is available on design and construction standards to ensure sewers are adoptable, with the aim of minimising the proliferation of private sewers (DEFRA, 2002).

Problems

7.1 Describe the main components of building drainage. Compare and contrast them with those used in main sewer systems.

7.2 A 375 mm internal diameter, 1:258 gradient sewer connects two manholes A and B. The upstream manhole A has co-ordinates E 274.698, N 842.393, and the soffit level of the sewer leaving it is 16.438 m. Assuming negligible fall across manhole B (E 342.812, N 864.844), what is the invert level of the 450 mm diameter exiting pipe? If the cover level at manhole B is 18.590 m, what is its depth?

[15.710 m, 2.880 m]

7.3 Explain how manholes differ from inspection chambers. Why and where would you locate manholes? Where are backdrop manholes used and why?

7.4 Explain the function of road gullies. Why, where and how would you locate them?

7.5 How are sewer systems ventilated?

7.6 List the main stages in urban drainage design. Assess the main factors affecting the horizontal and vertical alignment of the system.

Key sources

Wise, A.F.E. and Swaffield, J.A. (2002) *Water, Sanitary and Waste Services for Buildings*, 5th edn, Butterworth-Heinemann.

Woolley, L. (1988) *Drainage Details*, 2nd edn, E & FN Spon.

WRc (1993) *Materials Selection Manual for Sewers, Pumping Mains and Manholes*, Water Services Association/Foundation for Water Research.

WRc (2001) *Sewers for Adoption. A Design and Construction Guide for Developers*, 5th edn, Water UK.

References

BS 5911 *Precast Concrete Pipes and Fittings for Drainage and Sewerage. Part 200: 1989 Unreinforced and Reinforced Manholes and Soakaways of Circular Cross-section.*

BS EN 124: 1994 *Gully Tops and Manhole Tops for Vehicular and Pedestrian Areas: Design Requirements, Type Testing, Marking.*

BS EN 295: 1991 *Vitrified clay pipes and fittings and pipe joints for drains and sewers. Parts 1–3.*

BS EN 752–3: 1997 *Drain and Sewer Systems Outside Buildings. Part 3; Planning.*

The Building Regulations 2000, *Approved Document H, Drainage and Waste Disposal*, The Stationery Office.

DoE/NWC (1981) *Design and Analysis of Urban Storm Drainage. The Wallingford Procedure. Volume 1: Principles, Methods and Practice*, Department of the Environment, Standing Technical Committee Report No. 28.

DEFRA (2002) *Protocol on Design, Construction and Adoption of Sewers in England and Wales*, DEFRA Publications, London. www.defra.gov.uk/environment/water/industry/sewers

Lillywhite, M.S.T. and Webster, C.J.D. (1979) Investigations of drain blockages and their implications for design. *The Public Health Engineer*, **7**(2), 53–60.

Mollinson, A.R. (1958) Road surface water drainage. *Journal of the Institution of Highway Engineers*, **59**(16), 889.

Standing Technical Committee on Sewers and Water Mains (1980) *Sewer and Water Main Records*, STC Report No. 25, National Water Council.

Transport and Road Research Laboratory (1976) *A Guide for Engineers to the Design of Storm Sewer Systems*, Road Note 35, Department of the Environment, HMSO.

8   Hydraulics

8.1 Introduction

An understanding of hydraulics is needed in the design of new drainage systems in order to specify the appropriate size of system components, especially pipes, channels and tanks. It is also needed in the analysis and modelling of existing systems in order to predict the relationship between flow-rate and depth for varying inflows and conditions.

Study of civil engineering hydraulics tends to concentrate on two main types of flow. The first is *pipe flow* in which a liquid flows in a pipe under pressure. The liquid always fills the whole cross-section, and the pipe may be horizontal, or inclined up or down in the direction of flow. The second is *open-channel flow*, in which a liquid flows in a channel by gravity, with a free surface at atmospheric pressure. The liquid only fills the channel when the flow-rate equals or exceeds the designed capacity, and the bed of the channel slopes down in the direction of flow.

The most common type of flow in sewer systems is a hybrid of these two: *part-full pipe flow*, in which a liquid flows in a pipe by gravity, with a free surface. The liquid only fills the pipe area when the flow-rate equals or exceeds the designed capacity, and the bed of the pipe slopes down in the direction of flow. Traditionally the theories used are most closely related to those for full pipes, though actually part-full pipe flow is a special case of open-channel flow.

In this chapter we will deal with the types of flow in the following order: pipe flow, part-full pipe flow, open-channel flow. Aspects of hydraulics are developed where appropriate in other chapters, for example those relating to special features in Chapter 9, to the design of sewers in Chapters 10, 11 and 12, storage in Chapter 13, pumped systems in Chapter 14, and flow models in Chapter 19.

This chapter presents the main principles of hydraulics relevant to urban drainage. It is intended as an introduction to the subject for those who have not studied it before, or as a refresher course for those who have. For more information, a number of sources are referred to in the text. For general reference, Chadwick and Morfett (1998) give a practical engineering treatment of all aspects. Hamill (2001) provides helpful explanations, and Kay (1999) uses a more descriptive approach.

8.2 Basic principles

8.2.1 Pressure

Pressure is defined as force per unit area. The common units for pressure are kN/m² or bars (1 bar = 100 kN/m²).

Absolute pressure is pressure relative to a vacuum, and *gauge pressure* is pressure relative to atmospheric pressure. Gauge pressure is used in most hydraulic calculations. Atmospheric pressure varies but is approximately 1 bar.

In a still liquid, pressure increases with vertical depth:

$$\Delta p = \rho g \Delta y \tag{8.1}$$

Δp increase in pressure (N/m²)
ρ density of liquid (for water, 1000 kg/m³)
g gravitational acceleration (9.81 m/s²)
Δy increase in depth (m)

Pressure at a point is equal in all directions.

8.2.2 Continuity of flow

In a section of pipe with constant diameter and no side connections (Fig. 8.1), in any period of time the mass of liquid entering (at 1) must equal the mass leaving (at 2). Assuming that the liquid has a constant density (mass per unit volume), the volume entering (at 1) must equal the volume leaving (at 2).

In terms of flow-rate (volume per unit time, Q):

$$Q_1 = Q_2 \tag{8.2}$$

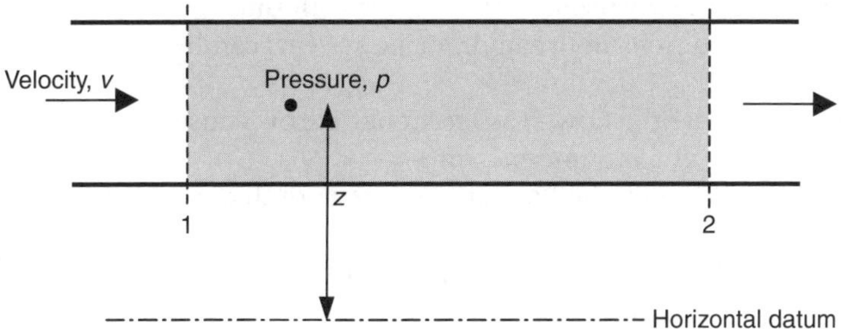

Fig. 8.1 Continuity and definition of symbols for pipe-flow

The common units for flow-rate are m³/s or l/s (1 l (litre) = 10^{-3} m³).

The velocity of the liquid varies across the flow cross-section, with the maximum for a full pipe in the centre. 'Mean velocity' (v) is defined as flow-rate per unit area (A) through which the flow passes:

$$v = \frac{Q}{A} \tag{8.3}$$

The common units for velocity are m/s (see Example 8.1).

Equation (8.2) can be rewritten as:

$$v_1 A_1 = v_2 A_2$$

Example 8.1

The diameter of a pipe flowing full increases from 150 mm to 200 mm. The flow-rate is 20 l/s (0.020 m³/s). Determine the mean velocity upstream and downstream of the expansion.

Solution

Upstream (1): $A_1 = \dfrac{\pi 0.15^2}{4}$ so $v_1 = \dfrac{Q}{A_1} = \dfrac{0.02 \times 4}{\pi 0.15^2} = 1.13$ m/s

Downstream (2): $A_2 = \dfrac{\pi 0.2^2}{4}$ so $v_2 = \dfrac{Q}{A_2} = \dfrac{0.02 \times 4}{\pi 0.2^2} = 0.64$ m/s

---

8.2.3 Flow classification

In hydraulics there are two terms for 'constant': *uniform* and *steady*. Uniform means constant with distance, and steady means constant with time. The words have negative forms: *nonuniform* means not constant with distance, *unsteady* means not constant with time.

Hydraulic conditions in urban drainage systems can be:

- uniform steady: the flow cross-sectional area is constant with distance, and flow-rate is constant with time
- nonuniform steady: the flow area varies with distance, but flow-rate is constant with time
- uniform unsteady: the flow area is constant with distance, but flow-rate varies with time
- nonuniform unsteady: the flow area varies with distance, and flow-rate varies with time.

Flow in sewers is generally unsteady to some extent: wastewater varies with the time of day, and storm flow varies during a storm. However, in many hydraulic calculations, it is not necessary to take this into account, and conditions are treated as steady for the sake of simplicity. Most of the rest of this chapter considers only steady flow. In some cases, unsteady effects are significant and must be considered: for example, storage effects (considered in Chapter 13), sudden changes in pumping systems (referred to in Section 14.4.3), and storm waves in sewers (see Sections 11.7 and 19.4).

8.2.4 Laminar and turbulent flow

The property of a fluid that opposes its motion is called viscosity. Viscosity is caused by the interaction of the fluid molecules creating friction forces between the layers of fluid travelling at different velocities. Where velocities are low, fluid particles move in straight, parallel trajectories – *laminar* flow. Where velocities are high, fluid particles follow more chaotic paths and the flow is *turbulent*.

Flow can be identified as laminar or turbulent using a dimensionless number, the Reynolds number (R_e), defined for a pipe flowing full as:

$$R_e = \frac{vD}{v} \qquad (8.4)$$

v mean velocity (m/s)
D pipe diameter (m)
v kinematic viscosity of liquid (for water, typically 1.1×10^{-6} m²/s, dependent on temperature)

Example 8.2

For the pipe in Example 8.1, determine the Reynolds number for both sides of the change in diameter. Is the flow laminar or turbulent? Assume that the pipe carries water, kinematic viscosity 1.1×10^{-6} m²/s.

Solution

Reynolds number upstream (1) $= \dfrac{v_1 D_1}{v} = \dfrac{1.13 \times 0.15}{1.1 \times 10^{-6}} = 154\,100$

Reynolds number downstream (2) $= \dfrac{v_2 D_2}{v} = \dfrac{0.64 \times 0.2}{1.1 \times 10^{-6}} = 116\,400$

Both are well into the turbulent zone.

When $R_e < 2000$, the flow in the pipe is laminar; when $R_e > 4000$, flow is turbulent. In most urban drainage applications, flow is firmly in the turbulent region (see Example 8.2).

8.2.5 Energy and head

A flowing liquid has three main types of energy: pressure, velocity and potential. In hydraulics, the most common way of expressing energy is in terms of *head*, energy per unit weight (common units, m).

The three types of energy expressed as head are:

pressure head $\dfrac{p}{\rho g}$ velocity head $\dfrac{v^2}{2g}$ potential head z

The symbols p, v and z are illustrated on Fig. 8.1.

Total head (H) is the sum of the three types of head, given by 'the Bernoulli equation':

$$H = \frac{p_1}{\rho g} + \frac{v^2}{2g} + z \tag{8.5}$$

When a liquid flows in a pipe or a channel, some head (h_L) is 'lost' from the liquid. So, for water flowing between sections 1 and 2 in the full pipe on Fig. 8.1:

$$H_1 - h_L = H_2 \tag{8.6}$$

or

$$\frac{p_1}{\rho g} + \frac{v_1^2}{2g} + z_1 - h_L = \frac{p_2}{\rho g} + \frac{v_2^2}{2g} + z_2 \tag{8.7}$$

If flow in Fig. 8.1 is uniform, and the pipe is horizontal, then:

$$v_1 = v_2 \quad \text{and} \quad z_1 = z_2$$

Therefore equation (8.7) becomes:

$$h_L = \frac{p_1}{\rho g} - \frac{p_2}{\rho g}$$

The head loss is equal to the difference in pressure head.

8.3 Pipe flow

8.3.1 Head (energy) losses

The head or energy losses in flow in a pipe are made up of *friction losses* and *local losses*. Friction losses are caused by forces between the liquid and the solid boundary (distributed along the length of the pipe), and local losses are caused by disruptions to the flow at local features like bends and changes in cross-section. Total head loss h_L is the sum of the two components.

The distribution of losses, and the other components in equation 8.7 can be shown by two imaginary lines.

The *energy grade line* (EGL) is drawn a vertical distance from the datum equal to the total head.

The *hydraulic grade line* (HGL) is drawn a vertical distance below the energy grade line equal to the velocity head.

The two lines are drawn for a pipe flowing full on Fig. 8.2. The lines allow all the terms in equation 8.7 to be identified.

8.3.2 Friction losses

A fundamental requirement in the hydraulic design and analysis of urban drainage systems is the estimation of friction loss. The basic representation

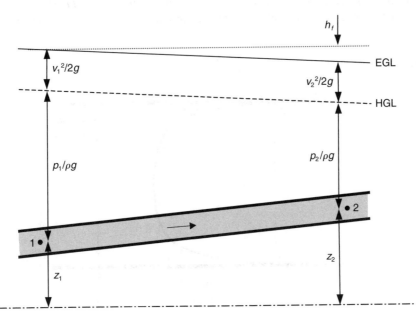

Fig. 8.2 EGL and HGL for a pipe flowing full

of friction losses, valid for both laminar and turbulent flow, is the Darcy-Weisbach equation:

$$h_f = \frac{\lambda L}{D} \cdot \frac{v^2}{2g} \qquad (8.8)$$

h_f head loss due to friction (m) (Fig. 8.2)
λ friction factor (no units)
L pipe length (m)
D pipe diameter (m)

The term h_f/L is the gradient of the energy grade line, and (for uniform flow) of the hydraulic grade line, and is often referred to as the 'hydraulic gradient' or 'friction slope'.

8.3.3 Friction factor

The friction factor λ is one of the most interesting aspects of pipe hydraulics. Is it constant for a particular pipe or does it vary? Does it depend on pipe roughness or not?

As mentioned, velocity varies over the flow area. In turbulent flow (the type of flow of most significance in urban drainage), velocity levels are quite similar across most of the cross-section, but fall rapidly near the pipe wall (Fig. 8.3). Very near to the wall, a boundary layer exists, where the velocity is low and laminar conditions occur: a *laminar sub-layer*.

Frictional losses are affected by the thickness of the laminar sub-layer

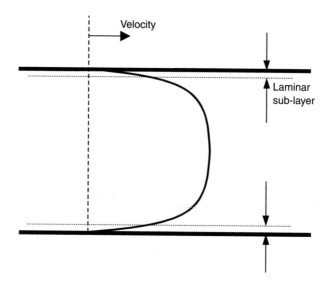

Fig. 8.3 Velocity profile in turbulent flow, with laminar sub-layer

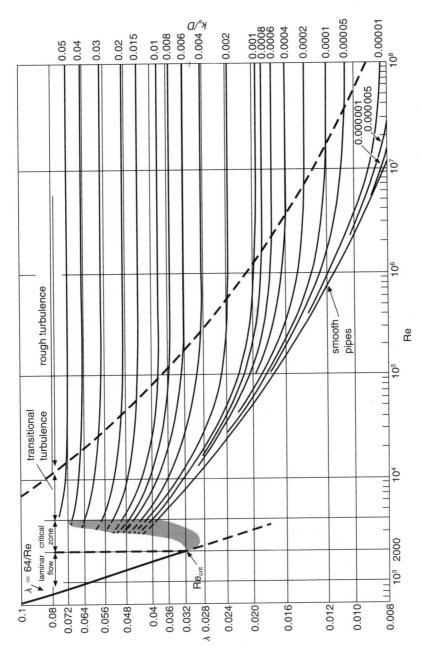

Fig. 8.4 Moody diagram (reproduced from Chadwick and Morfett [1998] with permission of E & FN Spon)

relative to the 'size' of the roughness of the pipe wall. In commercial pipes, the wall roughness is measured in terms of an equivalent sand roughness size (k_s), and can be thought of as the mean projection height of the roughness from the pipe wall.

Fig. 8.4 presents the Moody diagram. This is a plot of the friction factor λ, against Reynolds number R_e for a range of values of relative roughness k_s/D. The Moody diagram demonstrates the relative effects of the thickness of the laminar sub-layer and the size of the roughness.

When the roughness projections are small compared with the thickness of the laminar sub-layer, the friction losses are independent of pipe roughness and dependent on Reynolds number. Flow is said to be 'smooth turbulent'. λ is a function of R_e, but not of k_s/D.

When the roughness projections are much greater in height, losses are linked to pipe roughness only (not Reynolds number) and conditions are known as 'rough turbulent'. λ is a function of k_s/D, but not of R_e.

Between these two conditions lies a transitional turbulent region where conditions are dependent on roughness height and Reynolds number. Most urban drainage flows are rough or transitionally turbulent flows.

Whichever is the greater, the thickness of the laminar sub-layer or the size of the roughness, the friction factor can be determined mathematically from the Colebrook-White equation:

$$\frac{1}{\sqrt{\lambda}} = -2\log_{10}\left(\frac{k_s}{3.7D} + \frac{2.51}{R_e\sqrt{\lambda}}\right) \tag{8.9}$$

The Moody diagram illustrates the relationship between the basic hydraulic variables and their relative importance, but it does not directly represent the variables routinely used in engineering practice: flow-rate, velocity, pipe diameter, roughness and gradient.

The Colebrook-White equation can provide an explicit expression for velocity by substitution of λ using equation 8.8 and R_e using equation 8.4, giving:

$$v = -2\sqrt{2gS_fD}\log_{10}\left(\frac{k_s}{3.7D} + \frac{2.51v}{D\sqrt{2gS_fD}}\right) \tag{8.10}$$

k_s pipe roughness (m)
S_f hydraulic gradient or friction slope, h_f/L (–)
v kinematic viscosity (m²/s)

As illustrated by Example 8.3, this equation can be easily solved for v. However, it is rather intractable if the variable of concern is the hydraulic gradient (S_f) or pipe diameter (D). Three possible options are available: to formulate an iterative solution using a computer or programmable calculator, to use design charts or tables, or to rely on approximate equations.

8.3.4 Wallingford charts and tables

Charts

A popular graphical method consists of a series of charts produced by Hydraulics Research, Wallingford (Hydraulics Research, 1990), an example of which is given as Fig. 8.5.

In these charts, the three dependent variables (D, Q and S_f) are graphically related to velocity v, with each chart being valid for a particular roughness height k_s. Using the chart for a particular pipe roughness, it is possible to determine any two of the variables if the other two are known (Examples 8.3 and 8.4). The charts are drawn for water at 15 °C, for which kinematic viscosity, $v = 1.141 \times 10^{-6}$ m²/s.

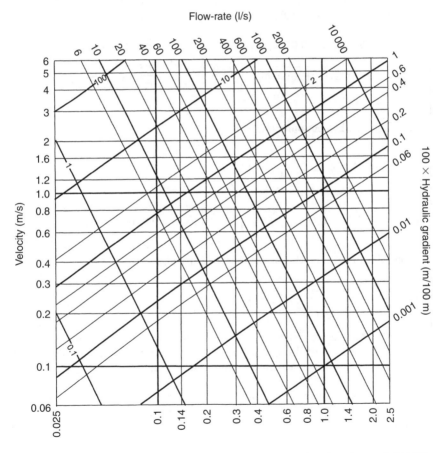

Fig. 8.5 Wallingford chart for $k_s = 0.6$ mm (based on Hydraulics Research [1990] with permission of HR Wallingford Ltd)

Example 8.3

i) A 600 mm diameter circular pipe is used to pump stormwater. Calculate the mean velocity and flow-rate in the pipe if the hydraulic gradient is 0.01 by solving the Colebrook-White equation. Assume the roughness height k_s is 0.6 mm and kinematic viscosity is 1.14×10^{-6} m²/s.

ii) What value of hydraulic gradient is suggested by the Moody diagram in conjunction with the Darcy-Weisbach equation? (Use the value of v already determined, together with the remaining data given.) Are conditions smooth, transitional or rough turbulent?

iii) Check v and Q using the appropriate Wallingford chart.

Solution

i) Velocity can be calculated by direct substitution into the Colebrook-White equation (8.10) given:

$D = 0.6$ m, $S_f = 0.01$, $k_s = 0.6$ mm and $v = 1.14 \times 10^{-6}$ m²/s

$$v = -2\sqrt{2g0.01 \times 0.6} \, \log_{10}\left(\frac{0.06 \times 10^{-3}}{3.7 \times 0.6} + \frac{2.51 \times 1.14 \times 10^{-6}}{0.6\sqrt{2g0.01 \times 0.6}}\right)$$

$$= 2.43 \text{ m/s}$$

and flow-rate by continuity (8.3):

$$Q = vA = 2.43 \times \frac{\pi 0.6^2}{4} = 688 \text{ l/s}$$

ii) $\dfrac{k_s}{D} = \dfrac{0.6}{600} = 0.001$ $R_e = \dfrac{2.43 \times 0.6}{1.14 \times 10^{-6}} = 1\,279\,000$

from Moody diagram (Fig. 8.4), $\lambda = 0.019$ (rough), so from (8.8):

$$\frac{h_f}{L} = \frac{\lambda}{D}\frac{v^2}{2g} = \frac{0.019}{0.6}\frac{2.43^2}{2g} = 0.01$$

iii) The Wallingford chart (Fig. 8.5) can be used by finding the point of intersection of the hydraulic gradient line (read downwards sloping right to left) with the diameter line (read vertically upwards).

Hydraulic gradient = 0.01, or 1 in 100, $D = 0.6$ m

This gives:

$v \simeq 2.4$ m/s $Q \simeq 700$ l/s

Example 8.4

A circular storm sewer is to be designed to run just full when conveying a flow-rate of 0.07 m³/s. If the mean velocity is specified at 1 m/s, calculate the required pipe diameter and gradient using an appropriate method. Assume the roughness height is 0.6 mm and kinematic viscosity 1.141×10^{-6} m²/s.

Solution

In principle, this problem could be solved using the Colebrook-White equation. However, this would require an iterative solution, best suited to a computational method. A direct solution can be achieved by reading from the Wallingford chart for $k_s = 0.6$ mm – Fig. 8.5.

$Q = 70$ l/s, $v = 1.0$ m/s

This gives:

$D = 0.3$ m, and gradient $= 0.42$ in 100, or 1 in 240

Tables

Output from the Colebrook-White equation can also be presented in tables. Table 8.1 is an example, giving Q and v for a variety of values of D and S_f, for k_s of 1.5 mm and kinematic viscosity of 1.14×10^{-6} m²/s. Comprehensive sets of tables that fulfil the same function as the Wallingford charts are also commercially available (HR Wallingford and Barr, 1998).

8.3.5 Approximate equations

Barr (1975) has developed a series of explicit approximate equations which enable determination of D and S_f with only minor loss of accuracy. For determination of S_f the following equation may be used (in terms of v, D and k_s):

$$S_f \approx \frac{v^2}{8gD\left[\log_{10}\left\{\dfrac{k_s}{3.7D} + \left(\dfrac{6.28v}{vD}\right)^{0.89}\right\}\right]^2} \tag{8.11}$$

Table 8.1 Table of output from the Colebrook-White equation, for $k_s = 1.5$ mm Q (l/s) in **bold**; v (m/s) in *italic*

Dia (m)	$S_f = 0.001$		$S_f = 0.002$		$S_f = 0.005$		$S_f = 0.01$		$S_f = 0.02$		$S_f = 0.05$	
0.2	**10**	*0.33*	**15**	*0.47*	**23**	*0.75*	**33**	*1.06*	**47**	*1.50*	**75**	*2.38*
0.3	**31**	*0.43*	**43**	*0.62*	**69**	*0.98*	**98**	*1.38*	**139**	*1.96*	**220**	*3.11*
0.375	**55**	*0.50*	**79**	*0.71*	**125**	*1.13*	**177**	*1.60*	**250**	*2.27*	**396**	*3.59*
0.75	**347**	*0.79*	**492**	*1.11*	**780**	*1.76*	**1104**	*2.50*	**1562**	*3.54*	**2472**	*5.60*

and for determination of D:

$$D \approx \frac{v^2}{8gS_f\left[\log_{10}\left\{\dfrac{1.558}{(v^2/2gS_fk_s)^{0.8}} + \dfrac{15.045}{(v^3/2gS_f v)^{0.73}}\right\}\right]^2} \qquad (8.12)$$

These have the advantage of being tractable, but the disadvantage of being rather lengthy to solve.

8.3.6 Roughness

Typical values of roughness k_s for use in the Colebrook-White equation, related to sewer type and age, are given in Table 8.2. The values described as 'new' are appropriate for new, clean and well-aligned pipes. They are appropriate in the design of stormwater pipes (where excessive sediment content is not expected) or in establishing the initial flow conditions in newly-laid foul or combined sewers. 'Old' values are generally preferred in the design or analysis of foul and combined sewers, where roughness is related more to the effect of biological slime than the pipe material.

For preliminary design purposes, or where existing pipe conditions are unknown, a value of $k_s = 0.6$ mm is suggested for storm sewers and $k_s = 1.5$ mm for foul sewers (irrespective of pipe material). Sewers subject to sediment deposition can have k_s values in the order of 30–60 mm (see Chapter 15 for further discussion). As an example of the effect of roughness height, for a 150 mm diameter pipe conveying 10 l/s, a change in the value of k_s from 0.6 mm to 1.5 mm would result in an increase in flow depth and a decrease in flow velocity of about 10%.

For rising mains, roughness can be empirically related to velocity (Flaxman and Dawes, 1983):

$$k_s \approx 0.3v^{-0.93} \qquad (8.13)$$

The Wallingford tables (HR Wallingford and Barr, 1998) recommend that for rising mains in 'normal' condition, k_s should be taken as 0.3 mm for a typical mean velocity of 1 m/s, 0.15 mm for 1.5 m/s, and 0.06 mm for 2 m/s.

Table 8.2 Typical values of roughness (k_s)

Pipe material	k_s range (mm)	
	new	old
Clay	0.03–0.15	0.3–3.0
PVC–U (and other polymers)	0.03–0.06	0.15–1.50
Concrete	0.06–1.50	1.5–6.0
Fibre cement	0.015–0.030	0.6–6.0
Brickwork – good condition	0.6–6.0	3.0–15
Brickwork – poor condition	–	15–30
Rising mains	0.03–0.60	

8.3.7 Local losses

Local losses occur at points where the flow is disrupted, such as bends, valves and changes of area. In certain circumstances (for example, where there are many fittings in a short length of pipe) these can be equal to or greater than the friction losses. Local losses are usually expressed in terms of velocity head as follows:

$$h_{local} = k_L \frac{v^2}{2g}$$

(8.14)

h_{local} local head loss (m)
k_L a constant for the particular type of fitting (–)

In gravity sewers, local losses occur at manholes, but these are only usually significant when the system is surcharged.

Examples of the value of k_L used in design practice are given in Table 8.3.

For design cases when velocity is unknown and calculations are based on the Wallingford charts, a useful alternative to equation 8.14 is to express local losses as an *equivalent length* of pipe. The sum of this equivalent length and the actual pipe length is then multiplied by the hydraulic gradient (taken from the chart) to give the total (friction + local) losses.

It is helpful to relate this to the friction factor λ. Combining equations 8.8 and 8.14 gives:

$$\text{total losses} = \frac{\lambda L}{D} \frac{v^2}{2g} + k_L \frac{v^2}{2g}$$

So, if an equivalent length L_E is to be added to L in order to replace the separate term for local losses,

$$\frac{L_E}{D} = \frac{k_L}{\lambda}$$

(8.15)

Table 8.3 Local head loss constants

Fitting	k_L
Pipe entry (sharp-edged)	0.50
Pipe entry (slightly rounded)	0.25
Pipe entry (bell-mouthed)	0.05
Pipe exit (sudden)	1.0
90° pipe bend ('elbow' – sharp bend)	1.0
90° pipe bend (long)	0.2
Straight manhole on gravity sewer (part-full)	<0.1
Straight manhole on gravity sewer (surcharged)	0.15
Manhole with 30° bend (surcharged)	0.5
Manhole with 60° bend (surcharged)	1.0

In rough turbulence (Fig. 8.4) L_E is independent of v, but in transitional or smooth turbulence it is not (because λ is affected by R_e). As well as being available from the Moody diagram, the value of λ can be inferred from the Wallingford chart, since

$$\lambda = \frac{S_f D 2g}{v^2}$$

we have $L_E = \dfrac{k_L}{S_f} \dfrac{v^2}{2g}$

Determination of equivalent length is shown in Example 8.5. Its use is demonstrated in Example 9.2.

Example 8.5

A surcharged manhole with a bend has local loss constant $k_L = 1.0$. Determine $\dfrac{L_E}{D}$ (assuming that it is independent of velocity) if this feature occurs

in a pipe with a diameter of (i) 300 mm or (ii) 600 mm (k_s 1.5 mm for both). For both cases determine the conditions for which the equivalent length is independent of velocity. (Assume kinematic viscosity $= 1.14 \times 10^{-6}$ m²/s.)

Solution

i) $\dfrac{k_s}{D} = \dfrac{1.5}{300} = 0.005$

Equivalent length is independent of velocity in rough turbulence. From Moody diagram (Fig. 8.4) $\lambda = 0.03$, so (equation 8.15)

$$\frac{L_E}{D} = \frac{k_L}{\lambda} = \frac{1.0}{0.03} = 33$$

This is valid if R_e is greater than 200 000, i.e. velocity greater than 0.76 m/s.

ii) $\dfrac{k_s}{D} = \dfrac{1.5}{600} = 0.0025$

Equivalent length is independent of velocity in rough turbulence. From

Moody diagram (Fig. 8.4) $\lambda = 0.025$, so $\dfrac{L_E}{D} = \dfrac{k_L}{\lambda} = \dfrac{1.0}{0.025} = 40$

This is valid if R_e is greater than 500 000, i.e. velocity greater than 0.95 m/s.

8.4 Part-full pipe flow

The common flow condition in urban drainage pipes is part-full pipe flow. The presence of the free surface must be taken into account in hydraulic computations.

8.4.1 Normal depth

In uniform steady gravity flow, an equilibrium exists along a part-full pipe or channel. The energy consumed by friction between the liquid and the pipe wall is in balance with the fall along the pipe length. If pipe slope could be increased for the same flow-rate, additional energy would be available to the flow, resulting in higher velocity and lower depth. The equilibrium depth is referred to as the *normal depth*.

Since depth of flow and velocity are constant when conditions are uniform, and pressure at the surface is atmospheric, the EGL and HGL are parallel to the bed, and the HGL coincides with the water surface (Fig. 8.6).

8.4.2 Geometric and hydraulic elements

When water flows along a part-full pipe, a number of properties, shown in Fig. 8.7, can be defined as in Table 8.4.

Hydraulic radius can be related to geometrical properties, and for a circular pipe running full (for example):

$$R = \frac{A}{P} = \frac{\pi D^2/4}{\pi D} = \frac{D}{4} \tag{8.16}$$

where D is the pipe diameter (m) (see Example 8.6).

A circular cross-sectional shape is most common for sewers and drains. The need to understand the hydraulic conditions at a range of depths results from the wide variations in flow experienced by sewers during their working life.

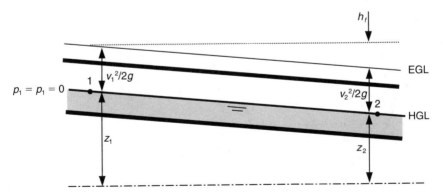

Fig. 8.6 EGL and HGL for a pipe flowing part-full

Table 8.4 Geometric elements

Property	Symbol	Definition	Common units
Depth	d	Height of water above the channel invert	m
Area	A	Cross-sectional area of flow	m²
Wetted perimeter	P	Portion of the flow area's perimeter that is in contact with the channel	m
Hydraulic radius	R	A per unit P	m
Top width	B	Flow width at the water surface	m
Hydraulic mean depth	d_m	A per unit B	m

Fig. 8.7 shows the cross-section of a pipe of diameter D with flow of depth d. The angle subtended at the pipe centre by the free surface is θ. From geometrical considerations, θ is related to proportional depth of flow d/D as follows:

$$\theta = 2\cos^{-1}\left[1 - \frac{2d}{D}\right]$$
(8.17)

Expressions for area (A), wetted perimeter (P), hydraulic radius (R), top width (B) and hydraulic mean depth (d_m), based on D and θ are given in Table 8.5.

Using these relationships, Fig. 8.8 gives dimensionless relationships for part-full flow depth, cross-sectional area, wetted perimeter and hydraulic radius as a proportion of the full-depth value (that is d/D, A/A_f, P/P_f and R/R_f). Also shown are lines indicating the velocity ratio v/v_f and flow ratio Q/Q_f (where v and Q are part-full velocity and flow-rate, and v_f and Q_f are the full-depth values). These latter lines are slightly dependent on the friction loss equation; the Colebrook-White equation is used here.

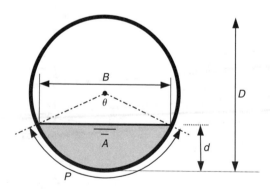

Fig. 8.7 Definition of geometric elements for a circular pipe

Example 8.6

A circular sewer of diameter 300 mm is flowing half full. Determine d, A, P, R, B and d_m.

Solution

Calculations are simply based on the property definitions (Table 8.4) and the geometry of a circle.

$$D = 0.3 \text{ m}$$

So,

$$d = 0.15 \text{ m}$$

$$A = \frac{1}{2} \cdot \frac{\pi D^2}{4} = \frac{1}{2} \cdot \frac{\pi\, 0.3^2}{4} = 0.0353 \text{ m}^2$$

$$P = \frac{\pi D}{2} = \frac{\pi\, 0.3}{2} = 0.471 \text{ m}$$

$$R = \frac{A}{P} = \frac{0.0353}{0.471} = 0.075 \text{ m}$$

$$B = D = 0.3 \text{ m}$$

$$d_m = \frac{A}{B} = \frac{0.0353}{0.3} = 0.118 \text{ m}$$

Table 8.5 Expressions for geometric elements in a part-full circular pipe

Parameter	Expression (θ in radians)
A	$\dfrac{D^2}{8}(\theta - \sin\theta)$
P	$\dfrac{D\theta}{2}$
R	$\dfrac{A}{P} = \dfrac{D}{4}\left[\dfrac{\theta - \sin\theta}{\theta}\right]$
B	$D \sin\dfrac{\theta}{2}$
d_m	$\dfrac{A}{B} = \dfrac{D(\theta - \sin\theta)}{8\sin\theta/2}$

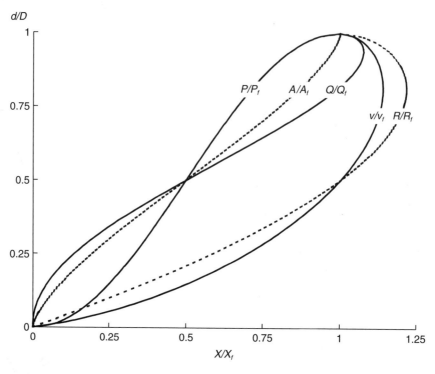

Fig. 8.8 Values of geometric and hydraulic elements for part-full pipe flow

In part-full pipes, maximum flow velocity and flow-rate do not occur when the pipe is running full; they occur when it is slightly less than full. This is because the circular shape affects the relative magnitudes of the flow area and the wetted perimeter (which determines the magnitude of the frictional resistance). At low flows, the wetted perimeter is high compared with flow area, resulting in low velocities. Velocity increases with depth (see Fig. 8.8) until at the highest depths the increase in the wetted perimeter is again high compared with the flow area, and this results in a fall in velocity. It follows that, eventually, the flow-rate will also fall, since flow-rate is the product of cross-sectional area and velocity. Table 8.6 summarises these effects.

The effect of a free surface within a pipe is largely understood, and conventionally the hydraulic radius (R) is substituted for pipe diameter (D) in

Table 8.6 Proportional flow-rate and velocity at various depths

	d/D = 0.5	d/D = 1.0	*Maximum*
Q/Q_f	0.5	1.0	1.08 at d/D = 0.94
v/v_f	1.0	1.0	1.14 at d/D = 0.81

the pipe flow equations to model the varying depth of flow (using equation 8.16). The substitution appears to predict the effects quite accurately, but there is evidence to suggest that it overestimates flow-rate and velocity in rough pipes by a few per cent.

Graphs of the velocity ratio v/v_f and flow-rate ratio Q/Q_f in the form of Fig. 8.8 can be used in conjunction with the Wallingford pipe flow chart to determine v and Q for any part-full case. This approach is appropriate for problems such as:

- given Q, S_f and d/D, find D and v
- given Q, D and S_f, determine d/D and v.

Example 8.7 gives an illustration.

Example 8.7

Given $Q = 70$ l/s, $S_f = 1{:}200$, $d/D = 0.25$ and $k_s = 0.6$ mm, find the necessary diameter D and corresponding velocity.

Solution

From Fig. 8.8, $Q/Q_f = 0.14$ (for $d/D = 0.25$), so $Q_f = 500$ l/s. From Fig 8.5, $D = 600$ mm and $v_f = 1.7$ m/s. From Fig. 8.8 again, $v/v_f = 0.7$ so $v = 1.2$ m/s.

8.4.3 Butler–Pinkerton charts

An alternative, more direct method for part-full pipe problems is provided by the Butler–Pinkerton charts (Butler and Pinkerton, 1987). These are based on a shape correction factor ψ which can be used to modify pipe diameter rather than replace it. Usually D is replaced by $4R$ (from equation 8.16) so incorporating the shape correction factor gives:

$$\psi D \equiv 4R$$

$$\psi = \frac{4R}{D}$$

$$\psi = \frac{(\theta - \sin \theta)}{\theta} \tag{8.17}$$

By substituting the shape correction factor ψ into the Colebrook–White equation, an expression valid for any proportional depth of flow is obtained:

$$v = -2\sqrt{2gS_f\psi D} \, \log_{10}\left(\frac{k_s}{3.7\psi D} + \frac{2.51v}{\psi D\sqrt{2gS_f\psi D}}\right) \tag{8.18}$$

An example of a Butler-Pinkerton chart is given as Fig. 8.9. A separate chart is required for each roughness value, and for each pipe diameter. Once the correct chart has been chosen, definition of any two of the remaining variables allows determination of the other two. This is particularly useful to cope with questions such as:

- given Q, and constraints for minimum v and maximum d/D, find minimum S_f and D (the design case)
- given D and S_f, find Q and v at various values of d/D (the analysis case).

The charts consist of two families of curves: the S-curves representing the modified Colebrook-White equation 8.18 and Q-curves representing the continuity equation:

$$v = \frac{8Q}{\psi\theta D^2} \tag{8.20}$$

The intersection of each S- and Q-curve gives the resulting normal flow depth and steady state velocity (see Example 8.8).

8.4.4 Non-circular sections

Circular pipes are by far the most common in shape. They have the shortest circumference per unit of cross-sectional area and so require least wall material to resist internal and external pressures. They are also easy to

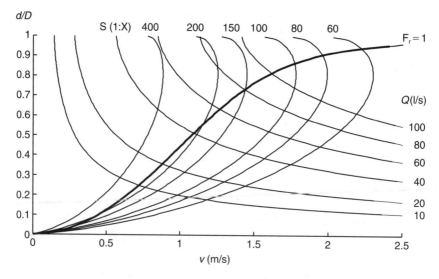

Fig. 8.9 Butler-Pinkerton chart (D 300 mm, k_s 0.6 mm)

Example 8.8

Given $Q = 60$ l/s, $d/D = 0.75$, minimum $v = 0.75$ m/s and $k_s = 0.6$ mm, find the minimum gradient of a 300 mm diameter pipe.

Solution

The Butler-Pinkerton chart (Fig. 8.9) can be used by estimating the point of intersection of the Q-curve (read from the right sloping upwards) with the hydraulic gradient S-curve (read downwards sloping first right then left).
   Thus at $d/D = 0.75$, $S_f = 1:280$ ($v = 1.05$ m/s).

---

manufacture. However, other shapes have been adopted in the past for sewers and drains – most commonly egg-shaped pipes, but also rectangular, trapezoidal, U-shaped, oval, horseshoe (arch) and compound types. Fig. 8.10 shows a range of shapes, their geometry and hydraulic elements.

Provided cross-sectional shape does not differ much from the circular, the basic equations for pipe flow can be utilised by substituting $4R$ for D.

8.4.5 Surcharge

Surcharging refers to pipes designed to run full or part-full, conveying flow under pressure. This can occur, for example, when flood flows exceed the design capacity, and it is therefore likely that all storm sewers will become surcharged at some time during their operational life.

A sewer pipe can surcharge in one of two ways, normally referred to as 'pipe surcharge' and 'manhole surcharge'.

Fig. 8.11(a) shows a longitudinal vertical section along a length of sewer running part-full (without surcharge). As explained in Section 8.4.1, the hydraulic gradient coincides with the water surface (and is parallel to the pipe bed). If there is an increase in flow entering the sewer, the consequence will be that the depth of flow in the pipe will increase.

Now imagine the sewer in Fig. 8.11(b) carrying the maximum flow-rate (just less than full). If there is an increase in flow entering the sewer, the carrying capacity of the pipe can no longer be increased by a simple increase in depth. The capacity of a pipe is a function of diameter, roughness and hydraulic gradient. To increase capacity, the only one of these that can change automatically (in response to 'natural forces') is hydraulic gradient. It follows that the new hydraulic gradient must be greater than the old (equal to the gradient of the pipe), and the result – pipe surcharge – is shown on Fig. 8.11(b). Increased local losses at manholes (Table 8.3) will further increase energy losses.

If inflow continues to increase, the hydraulic gradient will increase. The obvious danger is that the hydraulic gradient will rise above ground level.

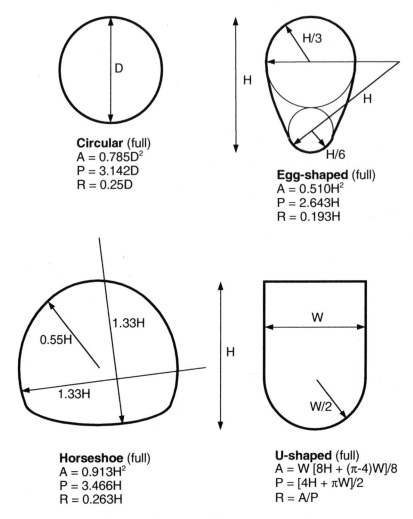

Fig. 8.10 Geometric elements for non-circular sewers

This may cause manhole covers to lift and the flow to flood onto the surface – 'manhole surcharge'.

The transition from conditions in Fig. 8.11(a) to those in Fig. 8.11(b) is sudden. As shown in Table 8.6, maximum flow is carried when the pipe is less than full. If the pipe is running at this maximum level, a further slight increase in flow-rate or small disturbance will result in a sudden increase in pipe flow depth, not only filling the pipe completely, but also establishing a hydraulic gradient in excess of S_o.

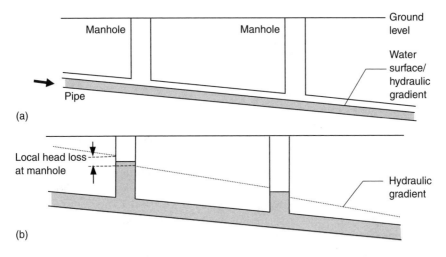

Fig. 8.11 (a) Part-full pipe flow without surcharge;
          (b) Pipe flow with surcharge

8.4.6 Velocity profiles

As discussed briefly in Sections 8.2.2 and 8.3.3, velocity varies over the cross-section of a pipe. Velocity is at a minimum at the boundary and increases towards the centre. Maximum velocity may be at the surface when the flow depth is low, or a little below it when the flow depth is higher (Fig. 8.12). The presence of a sediment bed in the invert of the pipe also affects the profile.

These profiles are significant when considering the transport of types of solids that are found only in specific parts of the cross-section (floating solids close to the surface, or heavy solids close to the bed). This is discussed further in Chapters 16 and 20.

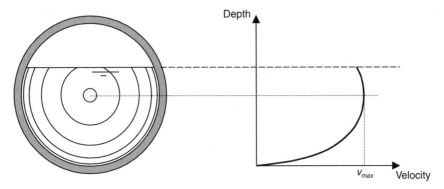

Fig. 8.12 Velocity profile in a part-full pipe

8.4.7 Minimum velocity

It is important for sewers to be able to convey wastewater or stormwater without long-term deposition of solid material. This is normally achieved by specifying a minimum mean velocity, a so-called 'self-cleansing' velocity, at a particular flow condition (e.g. pipe-full capacity) or for a particular frequency of occurrence (e.g. daily).

A common design criterion is to specify a minimum velocity when the pipe runs full. A value of 1.0 m/s is typical. The basis of this is that, although the pipe may never flow precisely full, mean velocities exceed the pipe-full velocity for flow-rates greater than $0.5\,Q_f$ (see Fig. 8.8). This method has the advantage of simplicity in computation, but lacks precision. The other common approach is to specify self-cleansing velocities at some specified depth of flow (e.g. 0.75 m/s at $d/D = 0.75$).

In the past, the velocity criterion has sometimes been relaxed for larger diameter sewers (say >750 mm). More recent research has shown this to be a mistake, and there is even evidence to suggest that higher self-cleansing velocities should be specified for larger diameters (see Chapter 16).

8.4.8 Minimum shear stress

Another potentially important parameter related to solid deposition/erosion is boundary shear stress. As water flows over the rigid boundary of the pipe channel, it exerts an average shear stress or drag τ_o (N/m²) in the direction of flow, given by:

$$\tau_o = \rho g R S_o \tag{8.21}$$

Substituting equation 8.8 and assuming $D = 4R$ gives:

$$\tau_o = \frac{\rho \lambda v^2}{8} \tag{8.22}$$

Example 8.9

The metric equivalent of 'Maguire's Rule' states that an appropriate self-cleansing pipe slope is given by $S_o = 1/D$ (D in mm). Interpret this in terms of a minimum shear stress standard.

Solution

Assuming $\rho = 1000$ kg/m³

$$\tau_o = \rho g R S_o = 1000 g \cdot \frac{D}{4} \cdot \frac{1}{1000 D} \approx 2.5 \text{ N/m}^2$$

---

indicating that the applied shear stress varies linearly with friction factor and as the square of velocity of flow. Shear stress is not uniform around the boundary because of the variations in velocity.

8.4.9 Maximum velocity

Historically, many sewerage systems were designed so that velocity would not exceed a specified maximum. This was no doubt a sensible criterion for early brick sewers with relatively weak lime-mortar joints. However, research has shown that abrasion is not normally a problem with modern pipe materials. Perkins (1977) has suggested that no fixed maximum limit is required, but where velocities are high (>3 m/s) careful attention needs to be given to:

- energy losses at bends and junctions
- formation of hydraulic jumps leading to intermittent pipe choking
- cavitation (see Section 14.3.5) causing structural damage
- air entrainment (significant when $v = \sqrt{5gR}$)
- the possible need for energy dissipation or scour prevention
- safety provisions.

8.5 Open-channel flow

8.5.1 Uniform flow

Part-full pipe flow (covered in the last section) is the most common condition in sewer systems. Design methods, as we have seen, tend to be related to those for pipes flowing full. However, in hydraulic terms, part-full pipe flow is a special case of open-channel flow, the basic principles of which are considered in this section.

The concept of normal depth, and the nature of the energy grade line and hydraulic grade line, explained in Section 8.4.1 for part-full pipe flow, apply to all cases of open-channel flow.

Manning's equation

A number of purely empirical formulae for uniform flow in open-channels have been developed over the years, a common example of which is Manning's equation:

$$v = \frac{1}{n} R^{\frac{2}{3}} S_o^{\frac{1}{2}} \qquad (8.23)$$

n Manning's roughness coefficient; typical values are given in Table 8.7 (units are not usually given, but to balance equation 8.23 the units of 1/n must be $m^{\frac{1}{3}} s^{-1}$)

S_o bed slope (–)

Table 8.7 Typical values of Manning's *n*

Channel material	n range
Glass	0.009–0.013
Cement	0.010–0.015
Concrete	0.010–0.020
Brickwork	0.011–0.018

If Manning's equation is plotted on the Moody diagram, it gives a horizontal line indicating the equation is only applicable to rough turbulent flow.

Ackers (1958) has shown that if k_s/D is in the typical range of 0.001 to 0.01, the values of k_s and n are related (to within 5%) by the relationship:

$$n = 0.012k_s^{\frac{1}{6}} \qquad (8.24)$$

where k_s is in mm, and n is as defined for equation 8.23.

8.5.2 Non-uniform flow

As stated, in uniform free surface flow, when the flow depth is normal, the total energy line, hydraulic grade line and channel bed (or pipe invert) are all parallel. In many situations, however, such as changes in pipe slope, diameter or roughness, non-uniform flow conditions prevail and these lines are not parallel. In sewer systems, it is likely that there will be regions of uniform flow interconnected with zones of non-uniform flow. Methods of predicting conditions in non-uniform flow are presented in the following sections.

8.5.3 Specific energy

If the Bernoulli equation (8.5) is redefined so that the channel bed is used as the datum (in place of a horizontal plane) we have, with reference to Fig. 8.13:

$$\text{total head} = \frac{p}{\rho g} + \frac{v^2}{2g} + z = \frac{\rho g h}{\rho g} + \frac{v^2}{2g} + x = h + x + \frac{v^2}{2g}$$

This gives *specific energy, E*:

$$E = d + \frac{v^2}{2g} \qquad (8.25)$$

or $\quad E = d + \dfrac{Q^2}{2gA^2}$

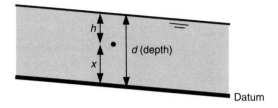

Fig. 8.13 Terms for derivation of specific energy

Thus, for a given flow-rate, E is a function of depth only (as A is a function of d). Depth can be plotted against specific energy (Fig. 8.14) showing that there are two possible depths at which flow may occur with the same specific energy. The depth which actually occurs depends on the channel slope and friction, and on any special physical conditions in the channel. At the critical depth d_c, the specific energy is a minimum for a given Q.

8.5.4 Critical, subcritical and supercritical flow

The non-dimensional Froude number (F_r) is given by:

$$F_r = \frac{v}{\sqrt{gd_m}} \tag{8.26}$$

where d_m is the hydraulic mean depth, as defined in Section 8.4.2.

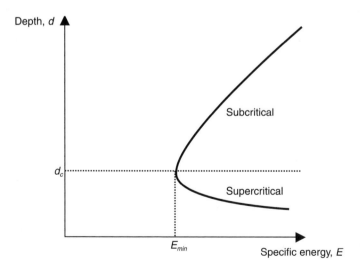

Fig. 8.14 Depth against specific energy, for constant flow

It can be shown that $F_r = 1$ at critical depth. If $F_r < 1$, flow is *subcritical*; the depth is relatively high and the velocity relatively low. This flow is sometimes referred to as 'tranquil' flow. If $F_r > 1$, flow is *super-critical*; velocity is relatively high, and depth low. This flow is also called 'rapid' or 'shooting' flow.

The critical velocity v_c is given by:

$$v_c = \sqrt{gd_m} \qquad\qquad (8.27)$$

Example 8.10

What is the critical depth, velocity and gradient in a 0.3 m circular sewer if the critical flow-rate is 50 l/s? If the pipe is actually discharging 80 l/s, determine the depth of flow (assuming it to be uniform) and comment on the flow conditions. ($k_s = 0.6$ mm).

Solution

The Butler-Pinkerton chart (Fig. 8.9) can be used by estimating the point of intersection of the Q-curve (read from the right sloping upwards) with the $F_r = 1$ curve.

$$Q = 50 \text{ l/s}, D = 300 \text{ m}$$

This gives:

$$d_c/D = 0.57 \qquad v_c = 1.2 \text{ m/s} \qquad S_c = 1{:}200$$

The same charts can be used to find proportional depth of flow which is read at the intersection of the Q-curve and the relevant S-curve (read downwards sloping first right then left).

$$Q = 80 \text{ l/s}, S_f = 1{:}200$$

This gives:

$$d/D = 0.84$$

As the intersection is above the F_r curve, flow must be subcritical.

Critical proportional depth can also be found using Straub's empirical equation:

$$\frac{d_c}{D} = 0.567 \frac{0.05^{0.506}}{0.3^{1.264}} = 0.57$$

In principle, this identity should allow determination of critical depth. However, for circular channels, there is no simple analytical solution. As with the Colebrook-White equation, a solution can be achieved computationally, graphically or by approximation.

Critical conditions ($F_r = 1$) have been plotted on the Butler-Pinkerton chart given (see Fig. 8.9), giving critical depth and critical slope for each flow-rate. Subcritical conditions exist in the region above this line, and supercritical below it.

As a good approximation, the critical depth in a circular pipe (d_c) can be determined from the following empirical equation (Straub, 1978):

$$\frac{d_c}{D} = 0.567 \frac{Q^{0.506}}{D^{1.264}} \tag{8.28}$$

where $0.02 < d_c/D \leq 0.85$ (units for Q, m³/s). See Example 8.10.

Normal depth may be subcritical or supercritical. A *mild* slope is defined as one in which normal depth is greater than critical depth (so uniform flow is subcritical), and a *steep* slope is defined as one in which normal depth is less than critical depth (so uniform flow is supercritical).

Most sewer designs are for subcritical flow. Close to critical depth, flow tends to be unstable and should be avoided if possible. Flow in the supercritical state is stable but has the disadvantage that if downstream conditions dictate the formation of subcritical conditions, a hydraulic jump will form. This effect is described later in the chapter.

8.5.5 Gradually varied flow

When variations of depth with distance must be taken into account, detailed analysis is required. This is done by splitting the channel length into smaller segments and assuming that the friction losses can still be accurately calculated using one of the standard equations such as Colebrook-White.

The general equation of gradually varied flow can be derived as:

$$\frac{d(d)}{dx} = \frac{S_o - S_f}{1 - F_r^2} \tag{8.29}$$

d depth of flow (m)
x longitudinal distance (m)
S_o bed slope (–)
S_f friction slope, h_f/L as defined in Section 8.3.2 (–)
F_r Froude number (–)

Examples of gradually varied flow in sewer systems are shown in Fig. 8.15. Fig. 8.15(a) shows flow ending at a 'free overall' – a sudden drop at the

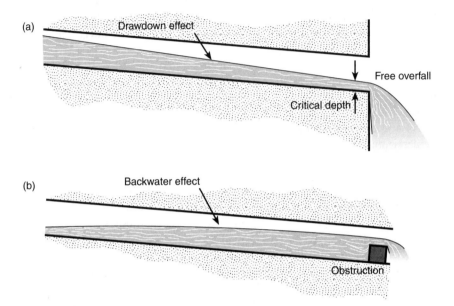

Fig. 8.15 Drawdown and backwater effects (in a pipe)

end of the pipe or channel such as the inflow to a pumping station. Close to the end of the pipe, conditions are critical, and for a long distance upstream the depth will be subject to a 'drawdown' effect (provided flow is subcritical). The effect is most pronounced for flatter pipes.

Fig. 8.15(b) shows flow backing up behind an obstruction. As flow approaches the obstruction, the depth increases: a 'backwater' effect.

8.5.6 Rapidly varied flow

When supercritical flow meets subcritical flow, a discontinuity called a *hydraulic jump* is formed (Fig. 8.16) at which there may be considerable energy loss.

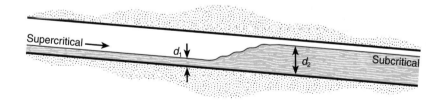

Fig. 8.16 Hydraulic jump (in a pipe)

There is no convenient analytical expression for the relationship between d_1 and d_2 on Fig. 8.16 in a part-full pipe. Straub (1978) however has developed an empirical approach using an approximate value for Froude number:

$$F_{r1} = \left(\frac{d_c}{d_1}\right)^{1.93}$$

(8.30)

where F_{r1} is the upstream Froude number. For cases where $F_{r1} < 1.7$ the depth d_2 is given by:

$$d_2 = \frac{d_c^2}{d_1}$$

(8.31)

for $F_{r1} > 1.7$:
$$d_2 = \frac{d_c^{1.8}}{d_1^{0.73}}$$

(8.32)

Hydraulic jumps are generally avoided in drainage systems because they have the potential to cause erosion of sewer materials. If they are unavoidable,

---

Example 8.11

A 600 mm pipe flowing part-full has a slope of 1.8 in 100 (k_s = 0.6 mm). Flow depth (in uniform conditions) is 0.12 m. Confirm that flow is supercritical using equation 8.28. An obstruction causes the flow downstream to become subcritical and, therefore, a hydraulic jump forms. Determine the depth immediately downstream of the jump.

Solution

$$\frac{d}{D} = \frac{0.12}{0.6} = 0.2 \quad \text{From Fig. 8.8,} \quad \frac{Q}{Q_f} = 0.1$$

From Fig. 8.5, Q_f = 900 l/s, so Q = 90 l/s = 0.09 m³/s

$$\frac{d_c}{D} = 0.567 \frac{Q^{0.506}}{D^{1.264}} = 0.567 \frac{0.09^{0.506}}{0.6^{1.264}} = 0.32 \quad \text{so } d_c = 0.19 \text{ m}$$

$d < d_c$, so flow is supercritical. From (8.30):

$$F_{r1} = \left(\frac{d_c}{d_1}\right)^{1.93} = \left(\frac{0.19}{0.12}\right)^{1.93} = 2.43$$

so from (8.32): $d_2 = \dfrac{d_c^{1.8}}{d_1^{0.73}} = \dfrac{0.19^{1.8}}{0.12^{0.73}} = 0.24 \text{ m}$

their position should be determined so that suitable scour protection can
be provided.

Problems

8.1 A pipe flowing full, under pressure, has a diameter of 300 mm and
roughness k_s of 0.6 mm. The flow-rate is 100 l/s. Use the Moody
diagram to determine the friction factor λ and the nature of the turbulence (smooth, transitional or rough). Determine the friction losses in
a 100 m length. Check this by determining the hydraulic gradient
using the appropriate Wallingford chart. (Assume kinematic viscosity
of 1.14×10^{-6} m²/s.)   [0.024, transitional, 0.8 m]

8.2 A pipe is being designed to flow by gravity. When it is full, the flow-rate
should be at least 200 l/s and the velocity no less than 1.0 m/s. Use a
Wallingford chart to determine the minimum gradient for a 600 mm
diameter pipe (k_s 0.6 mm). What will the pipe-full flow-rate actually be?
At what part-full depth would velocity go below 0.8 m/s?
                              [0.18 in 100, 300 l/s, 180 mm]

8.3 A surcharged manhole with a 30° bend has a local loss constant
$k_L = 0.5$. Determine the pipe length, L_E, equivalent to this local loss
(assuming that it is independent of velocity) for a pipe with a diameter
of 450 mm and k_s of 1.5 mm. If velocity is 1.3 m/s, is the assumption
above valid? (Assume kinematic viscosity $= 1.14 \times 10^{-6}$ m²/s.)
                                          [8.7 m, yes]

8.4 A gravity pipe has a diameter of 600 mm, slope of 1 in 200, and when
flowing full has a flow-rate of 610 l/s and velocity of 2.2 m/s. Flowing
part-full at a depth of 150 mm, what is the velocity, flow-rate, area of
flow, wetted perimeter, hydraulic radius and applied shear stress?
          [1.5 m/s, 80 l/s, 0.055 m², 0.63 m, 0.09 m, 4.4 N/m²]

8.5 A 300 mm diameter pipe is being designed for the following:
maximum flow-rate 80 l/s, minimum allowable velocity 1.0 m/s,
roughness k_s 0.6 mm. Determine, using the Butler–Pinkerton chart:
a) the gradient required based on the pipe running full
b) the depth at which it will actually flow at that gradient
c) the minimum velocity that will be achieved if the working flow
    rate is 10 l/s
d) the gradient at which the sewer would need to be constructed to
    just ensure that the minimum velocity is achieved at that flow-rate.
                          [1:190, 250 mm, 0.78 m/s, 1:95]

8.6 A pipe, diameter 450 mm, k_s 0.6 mm, slope 1.5 in 100, is flowing part-
full with a water depth of 100 mm. Are conditions subcritical or
supercritical?   [super]

8.7 If a hydraulic jump takes place in the pipe in Problem 8.6, such that
conditions upstream of the jump are as in 8.6, what would be the
depth downstream of the jump?   [0.18 m]

Key sources

Chadwick, A. and Morfett, J. (1998) *Hydraulics in Civil and Environmental Engineering*, 3rd edn, E & FN Spon.

Hydraulics Research (1990) *Charts for the hydraulic design of channels and pipes*, 6th edn, Hydraulics Research, Wallingford.

References

Ackers, P. (1958) *Resistance of fluids in channels and pipes*, Hydraulics research paper No. 2, HMSO.

Barr, D.I.H. (1975) Two additional methods of direct solution of the Colebrook-White function, TN128. *Proceedings of the Institution of Civil Engineers*, Part 2, 59, December, 827–835.

Butler, D. and Pinkerton, B.R.C. (1987) *Gravity Flow Pipe Design Charts*, Thomas Telford.

Flaxman, E.W. and Dawes, N.J. (1983) Developments in materials and design techniques for sewerage systems. *Water Pollution Control*, Part 2, 164–178.

Hamill, L. (2001) *Understanding Hydraulics*, 2nd edn, Palgrave.

Kay, M. (1999) *Practical hydraulics*, E & FN Spon.

Perkins, J.A. (1977) *High velocities in sewers*, Report No. IT165, Hydraulics Research Station, Wallingford.

Straub, W.O. (1978) A quick and easy way to calculate critical and conjugate depths in circular open channels. *Civil Engineering* (ASCE), December, 70–71.

HR Wallingford, and Barr, D.I.H. (1998) *Tables for the hydraulic design of pipes, sewers and channels*, 7th edn, Volume 1, Thomas Telford.

9 Hydraulic features

9.1 Flow controls

Flow controls can be used to limit the inflow to, or outflow from, elements in an urban drainage system. Typical uses include restricting the continuation flow at a CSO to the intended setting (Chapter 12), and controlling water level in tanks to ensure that the storage volume is fully exploited (Chapter 13). Flow controls can also be used to limit the rate at which stormwater actually enters the sewer system in the first place, deliberately backing up water in planned areas like car parks to prevent more damaging floods downstream in a city centre (Chapter 21).

Flow controls can be fixed, always imposing the same relationship between flow-rate and water level, or adjustable, where the relationship can be changed by adjustment of the device.

9.1.1 Orifice plate

The simplest way of controlling inflow to a pipe is by an orifice plate. This forces the flow to pass through an area less than that of the pipe (Fig. 9.1).

An orifice plate is fixed to the wall of the chamber where the inlet to the pipe is formed, and it usually either creates a smaller circular area (Fig. 9.2(a)) or covers the upper part of the pipe area (Fig. 9.2(b)). The area of the opening can only be changed by physically detaching and replacing or repositioning the plate.

Hydraulic analysis of an orifice is a simple application of the Bernoulli equation (Chapter 8). Comparing the total head at points 1 and 2 on Fig. 9.1(a), and assuming there is no loss of energy, we can write

$$\frac{p_1}{\rho g} + \frac{v_1^2}{2g} + z_1 = \frac{p_2}{\rho g} + \frac{v_2^2}{2g} + z_2$$

Now $p_1 = p_2 = 0$ (gauge pressure) and $z_1 - z_2 = H$, so, assuming that the velocity at 1 is negligible, we have:

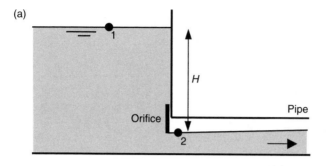

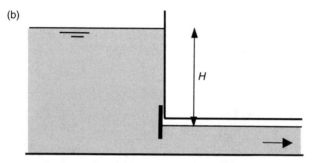

Fig. 9.1 Orifice plate (vertical section)
(a) normal;
(b) drowned

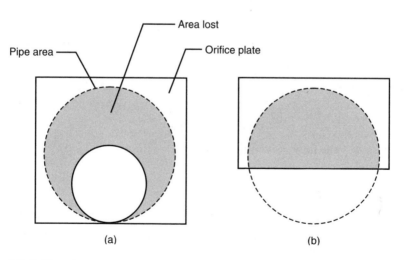

Fig. 9.2 Orifice plate arrangements

$$H = \frac{v_2^2}{2g}$$

or:

$$v_2 = \sqrt{2gH}$$

So flow-rate, $Q = A_o\sqrt{2gH}$

where A_o is the area of the orifice (m²).

The assumptions made above affect the accuracy of the answer, and this is compensated for by an 'orifice coefficient', C_d giving:

$$Q_{actual} = C_d A_o \sqrt{2gH} \tag{9.1}$$

This is sometimes written as $Q = CA_o\sqrt{gH}$, where C includes the $\sqrt{2}$.

Conditions downstream may cause the orifice to be 'drowned' – the downstream water level to be above the top of the orifice opening. H in equation 9.1 should now be taken as the difference in the water levels, as on Fig. 9.1(b). The minimum value of H for which the orifice will be not drowned (H_{min}) can be determined from Fig. 9.3 (in which D_o is the diameter of the orifice, and D is the diameter of the pipe). Use of Fig. 9.3 requires the value of the water level in the pipe downstream (d), which can be calculated using the properties of part-full pipe flow described in Section 8.4. Example 9.1 demonstrates the calculation. For $H < H_{min}$ the orifice will be drowned.

For an orifice that is not drowned, C_d in equation 9.1 generally has a value between 0.57 and 0.6. For a drowned orifice, C_d can be estimated from:

$$C_d = \frac{1}{1.7 - \left(\dfrac{A_o}{A}\right)} \tag{9.2}$$

where A is the flow area in the pipe (m²).

9.1.2 Penstock

A penstock is an adjustable gate that creates a reduction in area at the inlet to a pipe in the manner of Fig. 9.2(b). The position of the penstock can be raised or lowered either manually or mechanically, by means of a wheel or a motorised actuator turning a spindle.

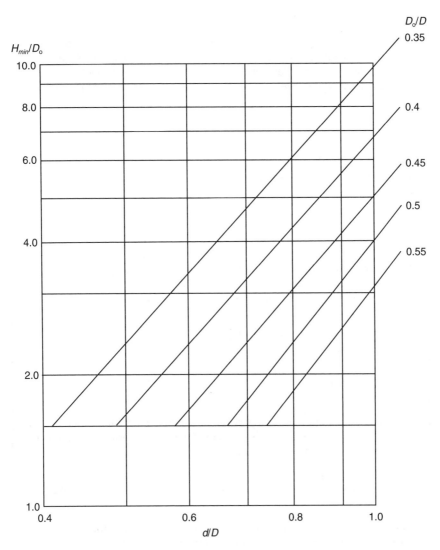

Fig. 9.3 Chart to determine H_{min} for non-drowned orifice (based on Balmforth *et al.* [1994] with permission of Foundation for Water Research, Marlow)

A penstock is more elaborate than an orifice plate. The advantage is that it can be adjusted to suit conditions – either, in the case of manual adjustment, to an optimum position to suit operational requirements, or, in the case of mechanical adjustment with remote control, to respond to changing requirements, perhaps as part of a real-time control system (described in Chapter 22).

Example 9.1

The following arrangement is proposed. A tank will have an outlet pipe with diameter 750 mm, slope 0.002, and k_s 1.5 mm. Flow to the outlet pipe will be controlled by a circular orifice plate, diameter 300 mm. Determine the flow-rate when water level in the tank is 2 m above the invert of the outlet.

Solution

First assume that the orifice is not drowned.
So $H = 2.0 - 0.3 = 1.7$ m (see Fig. 9.1(a))
Equation 9.1 gives $Q_{actual} = C_d A_o \sqrt{2gH}$
assuming $C_d = 0.6$,

$$Q_{actual} = 0.6\pi \frac{0.3^2}{4} \sqrt{2g1.7} = 0.244 \text{ m}^3/\text{s or 244 l/s}$$

Now check assumption that orifice is not drowned, using Fig. 9.4.
What is the flow-rate in outlet pipe flowing full?
Use chart for $k_s = 1.5$ mm, or Table 8.1 .... $Q_f = 492$ l/s

this gives $\frac{Q}{Q_f} = 0.5$ so (from Fig. 8.8) $\frac{d}{D} = 0.5$

for the orifice, $\frac{D_o}{D} = \frac{0.3}{0.75} = 0.4$

Note that the depth of uniform part-full flow in the outlet pipe would be above the top of the orifice. This does not mean that the orifice is necessarily drowned since conditions are non-uniform. For

$\frac{d}{D} = 0.5$ and $\frac{D_o}{D} = 0.4$, Fig. 9.3 gives $\frac{H_{min}}{D_o} = 1.7$, so H_{min} is 0.51 m,

which is less than the actual H of 1.7 m, and so the orifice is not drowned. Therefore the flow-rate calculated above (244 l/s) applies.

Blockage is a potential problem with both an orifice plate and a penstock, and both should be designed to allow a 200 mm diameter sphere to pass.

9.1.3 Vortex regulator

In a similar way to an orifice plate or penstock, a vortex regulator constricts flow, usually with the purpose of exploiting a storage volume; the magnitude of the flow-rate passing through the device depends on the upstream water depth. The regulator consists of a unit (see Fig. 9.4) into which flow is guided tangentially, creating (at sufficiently high flow-rates) a rotation of liquid inside the chamber. This creates a vortex with high peripheral velocities and large centrifugal forces near the outlet. These forces increase with upstream head until an air core occupies most of the outlet orifice creating a back-pressure opposing the flow.

This type of device has a distinctive head-discharge curve as shown in Fig. 9.5. The 'kickback' occurs during the formation of a stable vortex in rising flow. The shape of the curve depends on the detailed geometry of the regulator and the downstream conditions, but cases have been reported of a linear or near-linear relationship after vortex formation (Green, 1988; Parsian and Butler, 1993). The reason for this is, as the head increases, the air core diameter decreases proportionately, effectively producing a variable diameter orifice. During falling flow, the vortex collapses but does not reproduce the kickback phase.

The main advantage of the vortex regulator is that it provides a degree of throttling not possible with an orifice no smaller than 200 mm. Hence, regulators can avoid problems of blockage or ragging that would occur on small diameter orifices. In addition, it has been demonstrated that the discharge through a vortex regulator is not directly related either to its inlet

Fig. 9.4 Vortex regulator

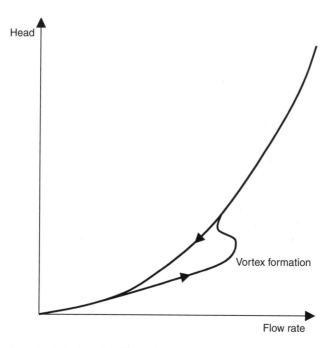

Fig. 9.5 Head-discharge relationship for vortex regulator

or outlet cross-sectional area (Butler and Parsian, 1993). Therefore, the impact of any ragging of the openings is less pronounced than might be expected in comparison with an orifice. The same study also showed that, in all cases, the retention of single solids within the device led to an increase in the discharge until the solid was eventually ejected. An additional advantage is that, since the head-discharge curve is initially flatter than an equivalent orifice, some savings can be made in the volume of storage required for flow balancing.

9.1.4 Throttle pipe

With a throttle pipe, it is the pipe itself that provides the flow control. Flow-rate through the pipe depends on its inlet design, length, diameter and hydraulic gradient. If the pipe is short, or has a steep slope or large diameter, it may be 'inlet controlled'; the flow is controlled by an orifice equivalent to the diameter of the pipe. However, if the pipe is long, the friction loss along its length will be the governing factor. This condition is known as outlet control.

A common throttle pipe application is as the continuation pipe of a stilling pond CSO (to be described in detail in Chapter 12). Fig. 9.6 shows that, when the weir is operating, the throttle pipe will be surcharged and thus flow-rate will be related to the hydraulic gradient (not the pipe

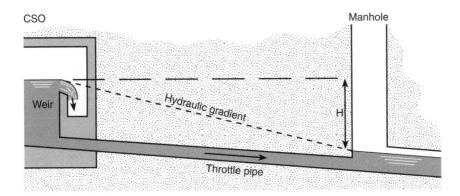

Fig. 9.6 Throttle pipe (vertical section)

gradient). There will also be local losses (not shown on Fig. 9.6) which may be significant. So, with reference to Fig. 9.6,

$$H = S_f L + k_L \frac{v^2}{2g}$$ (9.3)

S_f friction slope, given by pipe design chart/table (–)

$k_L \dfrac{v^2}{2g}$ local losses (as defined in Section 8.3.7 and Table 8.3)

In throttle pipe calculations, it is sometimes convenient to represent local losses by an equivalent pipe length, as explained in Section 8.3.7 (and demonstrated in Example 9.2).

To prevent blockage, the diameter of the throttle pipe should not be less than 200 mm. Clearly the length of the throttle pipe plays an important part in creating the flow control, and in design cases where it is inappropriate to reduce the diameter, the desired hydraulic control may be achieved by increasing the length (subject to restrictions in site layout). The diameter of an outlet-controlled throttle pipe will certainly be larger than that of the orifice plate giving equivalent flow control.

9.1.5 Flap valve

A flap valve is a hinged plate at a pipe outlet that restricts flow to one direction only. A typical application is at an outfall to receiving water with tidal variation in level. When the level of the receiving water is below the outlet, the outflow discharges by lifting the flap (Fig. 9.7(a)). When the outlet is flooded, the flap valve prevents tidal water entering the sewer (Fig. 9.7(b)). In these circumstances, any flow in the sewer will back up in

Example 9.2

A throttle pipe carrying the continuation flow from a stilling pond CSO will have a length of 28 m. When the weir comes into operation, the water level in the CSO will be 1.8 m above the water level at the downstream end of the throttle pipe, 1.4 m above the soffit at the pipe inlet. Under these conditions the continuation flow (in the throttle pipe) should be as close as possible to 72 l/s. The roughness, k_s, of the pipe material is assumed to be 1.5 mm, and local losses are taken as $1.4 \dfrac{v^2}{2g}$. Determine an appropriate diameter for the throttle pipe. Confirm that this throttle pipe is not 'inlet controlled'.

   If, as an alternative, there were no throttle pipe and flow control was achieved by an orifice, what would be its diameter (assuming that it would not be drowned)?

Solution

Solve by trial and error. . . . 200 mm pipe gives the following.
Represent local losses by equivalent length:

$$\frac{k_s}{D} = \frac{1.5}{200} = 0.0075$$

assume rough turbulent, Moody diagram (Fig. 8.4) gives $\lambda = 0.034$

so from equation 8.15, $\dfrac{L_E}{D} = \dfrac{k_L}{\lambda} = \dfrac{1.4}{0.034} = 41$ therefore $L_E = 8$ m

total length $= 28 + 8 = 36$ m
hydraulic gradient $= 1.8/36 = 0.05$
Chart for $k_s = 1.5$ mm, or Table 8.1, gives flow-rate (for 200 mm dia) of 75 l/s. So 200 mm diameter is suitable.
If the throttle pipe is inlet controlled, control comes solely from the inlet acting as an orifice. Apply orifice formula (assuming $C_d = 0.59$):

$$Q_{actual} = C_d A_o \sqrt{2gH} = 0.59 \times \pi \frac{0.2^2}{4} \sqrt{2g1.4} = 97 \text{ l/s}$$

so control does not solely come from the inlet: the pipe is not inlet controlled. Consider use of an orifice plate

$$Q_{actual} = C_d A_o \sqrt{2gH}$$

What orifice diameter would give the same control as the throttle pipe?

$$0.075 = 0.59 \times \pi \frac{D_o^2}{4} \sqrt{2g1.4} \quad \text{giving } D_o = 0.175 \text{ m (unacceptably small)}$$

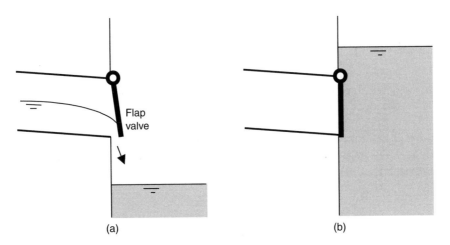

Fig. 9.7 Flap valve operation

the pipe, and if the energy grade line rises above the tidal water level, there will be outflow. The flap (which may have considerable weight) will then create a local head loss. Methods of estimating this loss are proposed by Burrows and Emmonds (1988).

9.1.6 Summary of characteristics of flow control devices

Table 9.1 gives a summary of the characteristics of the flow control devices considered above (excluding the flap valve, which has a different function from the other devices).

9.2 Weirs

9.2.1 Transverse weirs

Standard analysis, using the Bernoulli equation, of flow over a rectangular weir gives the theoretical equation for the relationship between flow-rate

Table 9.1 Summary of characteristics of flow control devices

Orifice plate	Simple, cheap.
	Flow control can only be adjusted by physically detaching and replacing or repositioning the plate.
Penstock	Easily adjusted. When automated, can be used for real-time control.
Vortex regulator	Controls flow with larger opening than equivalent orifice.
Throttle pipe	Larger opening than equivalent orifice.
	Significant construction costs.

and depth as:

$$Q_{theor} = \frac{2}{3}b\sqrt{2g}H^{\frac{3}{2}}$$

Q_{theor} flow-rate (m³/s)
b width of weir (Fig. 9.8) (m)
H height of water above weir crest (Fig. 9.8) (m)

Several assumptions are made in the analysis and it is necessary to introduce a discharge coefficient to relate the theoretical result to the actual flow-rate:

$$Q = C_d\frac{2}{3}b\sqrt{2g}H^{\frac{3}{2}} \tag{9.4}$$

where C_d = discharge coefficient. With this form of equation, C_d has a value between 0.6 and 0.7; C_d is sometimes written so that it incorporates some of the other constants in the equation.

The value of C_d and the accuracy of the equation depend partly on whether the weir crest fills the whole width of a channel or chamber, or is a rectangular notch which forces the flow to converge horizontally. For the former, an empirical relationship by Rehbock can be used:

$$Q = C_d\frac{2}{3}b\sqrt{2g}[H + 0.0012]^{\frac{3}{2}} \tag{9.5}$$

in which $C_d = 0.602 + 0.0832\dfrac{H}{P}$

and where P is the height of weir crest above channel bed (Fig. 9.8) (m).

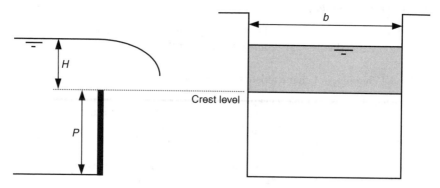

Fig. 9.8 Rectangular weir

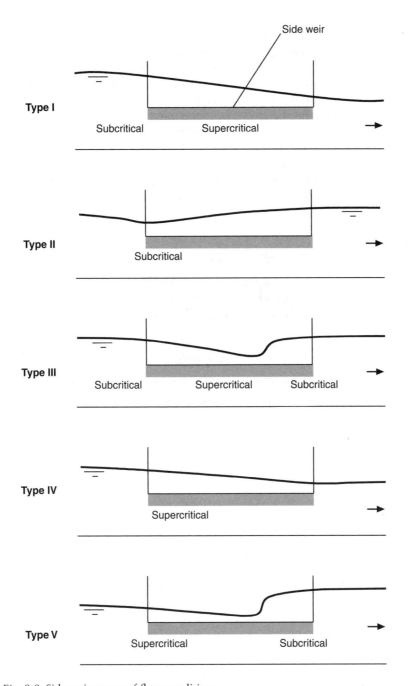

Fig. 9.9 Side weir: types of flow condition

9.2.2 *Side weirs*

The flow arrangements for side weirs are more complex than for transverse weirs because flow-rate in the main channel is decreasing with length (as some flow is passing over the weir) and conditions are non-uniform.

The possible flow conditions at a side weir are normally classified into 5 types as illustrated on Fig. 9.9. These conditions can be analysed by assuming that specific energy is constant along the main channel. The standard curve of depth against specific energy (for constant flow-rate), introduced as Fig. 8.14, is reproduced as Fig. 9.10 with the curve for a slightly decreased flow-rate added.

The classification of flow types is based partly on the slope of the channel. *Mild* and *steep* slopes have been defined in Section 8.5.4.

Type I

> *Channel slope:* mild   *Weir crest* below *critical depth*

Depth along the weir is supercritical as a result of the fact that the weir crest is below critical depth. As the flow-rate decreases, we move from point 1 to 2 on Fig. 9.10 and the depth (*d*) decreases.

Type II

> *Channel slope:* mild   *Weir crest* above *critical depth*

Depth along the weir is subcritical as a result of the fact that the weir crest

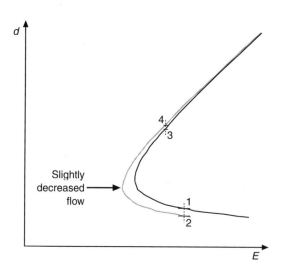

Fig. 9.10 Flow parallel to side weir: depth against specific energy

is above critical depth. As the flow-rate decreases, we move from point 3 to 4 on Fig. 9.10 and the depth (*d*) increases.

Type III

Channel slope: mild   Weir crest below *critical depth*

At the start of the weir, conditions are as Type I. However, conditions downstream are such that a hydraulic jump forms before the end of the weir.

Type IV

Channel slope: steep   Weir crest below *critical depth*

Conditions are similar to those for Type I, except that supercritical conditions would prevail in the main channel in any case because it is steep.

Type V

Channel slope: steep   Weir crest below *critical depth*

Conditions are similar to those for Type III, except that supercritical conditions prevail before the start of the weir because the channel is steep.

For all types, the variation of water depth with distance, derived from standard expressions for spatially-varied flow, is given by:

$$\frac{d(d)}{dx} = \frac{Q_c d\left[-\dfrac{dQ_c}{dx}\right]}{gB^2 d^2 - Q_c^2} \tag{9.6}$$

d depth of flow (m)
x longitudinal distance (m)
Q_c flow-rate in the main channel (m³/s)
B width of the main channel (m)

$\left[-\dfrac{dQ_c}{dx}\right]$ is the rate at which flow-rate in the main channel is decreasing –

that is, the rate at which flow passes over the weir per unit length of weir. Therefore, equation 9.4 for flow over a weir can be adapted to give:

$$-\frac{dQ_c}{dx} = C_d \frac{2}{3}\sqrt{2g}H^{\frac{3}{2}} \tag{9.7}$$

Note that H is water depth relative to the weir crest, whereas d is water depth relative to the channel bed. In a double side weir arrangement (one weir on either side of the main channel), the right-hand side of equation 9.7 is doubled. It has been found that the Rehbock expression, equation 9.5, gives appropriate values of C_d for side weirs, even though it was originally proposed for transverse weirs.

For methods of solution of these equations see Chow (1959) and Balmforth and Sarginson (1978). More recently May *et al.* (2003) have presented a simple formula for total flow discharged over a side weir, backed up by charts for determining coefficients. They also provide general guidance on design and construction. For high side weir overflows (Type II flow conditions), design calculations can be based on charts presented by Delo and Saul (1989). One of these is given as Fig. 9.11. Its use is demonstrated in Example 12.3 in Chapter 12. Symbols on Fig. 9.11 are:

Q_u	inflow (m³/s)
Q_d	continuation flow-rate (m³/s)
B_u	upstream chamber width (m)
B_d	downstream chamber width (m)
P_u	upstream weir height (m)
P_d	downstream weir height (m)
Y_u	upstream water depth (m)
Y_d	downstream water depth (m)
L	Length of weir (m)

9.3 Inverted siphons

Inverted siphons carry flows under rivers, canals, roads, etc (for example, Fig. 9.12). They are necessary when this crossing cannot be made by means of a pipe-bridge, or by having the whole sewer length at a lower level. Unlike normal siphons, inverted siphons do not require special arrangements for filling; they simply fill by gravity. However, they do present some problems and are avoided where possible.

Inverted siphons are an interesting case from a hydraulic point of view, and are dissimilar from most other flow conditions in sewers. As we have seen in Chapter 8, the majority of sewers flow part-full, and when the flow-rate is low, the depth is low. When sewer flows are pumped, the pipe flows full and the pumps tend to deliver the flow at a fairly constant rate, but not continuously (as will be described in Chapter 14). In contrast, inverted siphons flow full and they flow continuously. At low flow, the velocity can be very low which, unfortunately, creates the ideal conditions for sediment deposition.

The most important aim in design is to minimise silting. Some silting is virtually inevitable at low flows, but at higher flows the system should be self-cleansing. It is normally assumed that this will be achieved if the velocity is greater than 1 m/s (this subject will be considered further in Chapter 16). The higher the velocity, the lower the danger of silting.

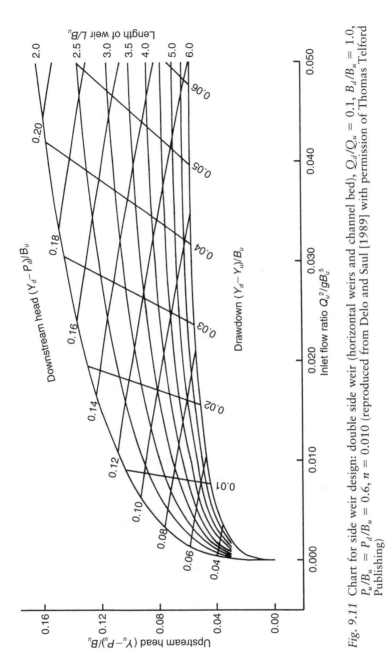

Fig. 9.11 Chart for side weir design: double side weir (horizontal weirs and channel bed), $Q_d/Q_u = 0.1$, $B_d/B_u = 1.0$, $P_u/B_u = P_d/B_u = 0.6$, $n = 0.010$ (reproduced from Delo and Saul [1989] with permission of Thomas Telford Publishing)

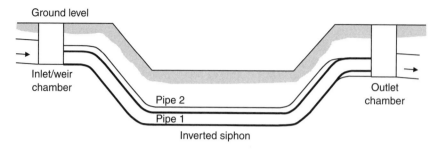

Ground level

Inlet/weir
chamber

Outlet
chamber

Pipe 2

Pipe 1

Inverted siphon

Fig. 9.12 Inverted siphon for wastewater, vertical section (schematic)

Many siphons consist of multiple pipes as a means of minimising siltation. The low flows will be carried by one pipe, smaller than the sewers on either side of the siphon. At higher flows, this pipe will be self-cleansing and an arrangement of weirs will allow overflow into other pipes. In separate systems, two pipes for wastewater are usually enough (Fig. 9.12); in combined sewers, a third much larger pipe is usually needed.

Other devices for avoiding siltation are sometimes needed. On small systems, a penstock upstream can be used to back up the flow and create an artificial flushing wave. Silt can be removed directly by providing penstocks or stop boards for isolating sections of pipe, and access for removing silt. An independent washout chamber can be provided, and used in conjunction with a system for pumping out silt.

Example 9.3

An existing single-pipe inverted siphon, carrying wastewater only, is to be replaced with a twin-barrelled siphon because of operational problems caused by sedimentation. The required length is 70 m; available fall (invert to invert) is 0.85 m. Determine the pipe sizes required for an average dry weather flow (DWF) of 90 l/s and a peak flow of 3 DWF. Assume the inlet head loss is 150 mm, the self-cleansing velocity is 1.0 m/s, and $k_s = 1.5$ mm.

Solution

One approach: use one pipe to carry DWF, second pipe to carry excess.
Available hydraulic gradient = (0.85 − 0.15)/70 = 0.01.
Use chart for $k_s = 1.5$ mm, or Table 8.1 for pipe calculations.
300 mm pipe carries 98 l/s at velocity 1.38 m/s. Velocity is sufficient.
Excess flow = (3 × 90) − 98 = 172 l/s.
375 mm pipe carries 177 l/s at velocity 1.6 m/s. Velocity is sufficient.
So, use pipe diameters 300 mm and 375 mm.

9.4 Gully spacing

Several approaches to establishing the required spacing of road gullies have been proposed. The simplest have been mentioned in Chapter 7, but in this section more sophisticated methods are outlined.

9.4.1 Road channel flow

The typical geometry of flow in a road channel is as given in Fig. 9.13.

For channels of shallow triangular section, Manning's equation (8.23) can be simplified by assuming the top width of the channel flow (B) equals the wetted perimeter (P), to give:

$$Q = 0.31Cy^{\frac{8}{3}} \tag{9.8}$$

where Q is the channel flow-rate with 'channel criterion' C (fixed for the road):

$$C = \frac{zS_o^{\frac{1}{2}}}{n} \tag{9.9}$$

y flow depth (m)
z side slope (1:z)
S_o longitudinal slope ($-$)
n Manning's roughness coefficient ($m^{-1/3}$ s)

Manning's n for roads ranges from 0.011 for smooth concrete to 0.018 for asphalt with grit. Example 9.4 demonstrates use of these equations.

9.4.2 Gully hydraulic efficiency

The hydraulic efficiency of a gully depends on the depth of water in the channel immediately upstream, the width of flow arriving and the geometry of the grating. A typical efficiency curve is given in Fig. 9.14. This shows that at low flows, gullies are approximately 100% efficient and all flow is

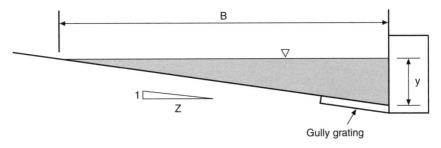

Fig. 9.13 Geometry of road channel flow (exaggerated vertical scale)

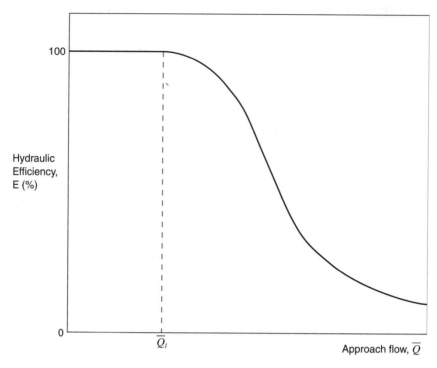

Fig. 9.14 Typical gully efficiency curve (after Davis *et al.*, 1996)

Example 9.4

Determine the flooded width of a concrete road ($n = 0.012$) when the flow rate is 20 l/s. The road has a longitudinal gradient of 1% and a crossfall of 1:40.

Solution

From equation 9.9, calculate the channel criterion:

$$C = \frac{40 \times 0.01^{\frac{1}{2}}}{0.012} = 333.3$$

Rearranging 9.8 gives:

$$y = \left(\frac{Q}{0.31C}\right)^{\frac{3}{8}} = \left(\frac{0.02}{0.31 \times 333.3}\right)^{\frac{3}{8}} = 0.041 \text{ m}$$

Thus the depth of flow is 41 mm leading to a width of flow $B = yz = 0.041 \times 40 = 1.62$ m

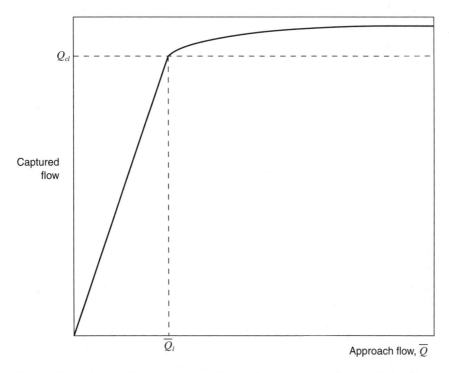

Fig. 9.15 Relationship between approaching and captured flow for a typical gully
(after Davis *et al.*, 1996)

captured. Once the approach flow $\overline{Q}$ exceeds $\overline{Q}_l$, efficiency drops off
rapidly. When approach flow is plotted against captured flow (as in Fig.
9.15), it is clear that the captured flow $\overline{Q}_l$ corresponding to 100% efficiency is not the maximum flow that the gully can capture. Higher
approach flows result in an increase in captured flow due to the greater
flow depths over the grating. Thus, the capacity of a gully Q_c can be
increased by allowing a small bypass flow. May (1994) suggests an
optimum value is 20%.

Thus, the hydraulic capture efficiency E for an individual gully grating
is:

$$E = \frac{Q_c}{\overline{Q}} \tag{9.10}$$

where E is a function of grating type, water flow width, road gradient and
crossfall.

Data on the efficiency of a number of grating types can be found in

Table 9.2 Example gully efficiencies (*E*) at standard 1:20 crossfall (adapted from Hydraulics Research Station, 1994)

Flow width	Longitudinal gradient (1:X)				
B (m)	20	30	50	100	300
0.5	100	100	100	100	100
0.75	87	94	97	99	100
1.0	63	75	82	93	96
1.5	33	43	47	60	76

TRRL Contractor Report CR2 (Hydraulics Research Station, 1984). An example is given in Table 9.2.

9.4.3 Spacing

The basic approach to gully hydraulic design is to make sure that they are sufficiently closely spaced to ensure that the flow-spread in the road channel is lower than the allowable width (*B*). Fig. 9.16 shows a schematic of the flow conditions along a road of constant longitudinal gradient and crossfall subject to constant inflow. Gullies are spaced at a distance L apart, except the first gully, that is at a distance of L_1. The inflow per unit length q is generated by constant intensity rainfall. The flow bypassing each gully must be included in the flow arriving at the next inlet.

Intermediate gullies

The maximum flooded width B and flow rate $\overline{Q}$ occurs just upstream of a road gully and consists of the sum of the runoff $Q_r = qL$ and the bypass flow Q_b from the previous gully:

$$\overline{Q} = Q_b + Q_r$$

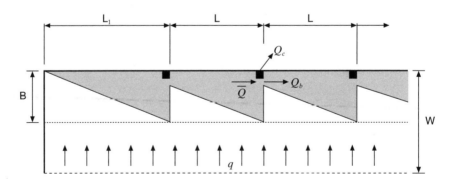

Fig. 9.16 Spacing of initial and intermediate gullies

And from the Rational Method equation (described more fully in Chapter 11):

$$Q_r = iWL$$

where i is the rainfall intensity for a storm duration equal to the time of entry and assuming complete imperviousness (runoff coefficient, $C = 1$), and W is the road width contributing flow to the gully. The flow arriving at the gully is either captured or bypasses it, so:

$$\overline{Q} = Q_b + Q_c$$

$$Q_c = Q_r$$

Thus the captured flow is equal to the runoff generated between gullies. Hence substituting equation 9.10 gives:

$$E\overline{Q} = iWL$$

$$L = \frac{E\overline{Q}}{iW} \tag{9.11}$$

Initial gullies

The most upstream gully in the system is a special case as it does not have to handle carry over from the previous gully, thus $Q_b = 0$ and $\overline{Q} = Q_r$, so:

$$L_1 = \frac{\overline{Q}}{iW} \tag{9.12}$$

Example 9.5 shows how gully spacing can be calculated.

Example 9.5

Determine the spacing of the initial and subsequent gullies on a road in the London area. The road is 5 m wide with a crossfall of 1:20 and a longitudinal gradient of 1%. The road surface texture suggests a Manning's n of 0.010 should be used. A design rainfall intensity of 55 mm/h is to be used at which the flood width should be limited to 0.75 m.

Solution

Allowable flow depth = 0.75/20 = 0.0375 m

Channel criterion (9.9),

$$C = \frac{20 \times 0.01^{\frac{1}{2}}}{0.010} = 200$$

Maximum flow rate (9.8),

$$\overline{Q} = 0.31 \times 200 \times 0.0375^{\frac{5}{3}} = 0.010 \text{ m}^3/\text{s}$$

Thus the spacing of the initial gully should be (9.12):

$$L_1 = \frac{0.010 \times 3600 \times 10^3}{55 \times 5} = 131 \text{ m}$$

Read from Table 9.2, E = 0.99

$$L = EL_1 = 130 \text{ m}$$

Allow a 20% reduction of capacity for potential blockage.
Maximum gully spacing is approximately 100 m.

A second special case is the terminal gully that can have no carryover. These act as weirs under normal conditions and as orifices under large water depths. Methods to design such gullies are given in Contractor Report CR2 (Hydraulics Research Station, 1984).

Problems

9.1 An orifice plate is being designed for flow control at the outlet of a detention tank. The outlet pipe has a diameter of 450 mm, slope of 0.0015, and roughness k_s of 0.03 mm. Water level in the tank at the design condition varies between 1.5 and 1.7 m above the outlet invert, and the desired outflow is 100 l/s. Select an appropriate orifice diameter. (Assume orifice $C_d = 0.6$; check that the orifice will not be drowned.)   [200 mm, not drowned]

9.2 A throttle pipe to control outflow (to treatment) from a CSO is being designed. The pipe will have a length of 25 m, and diameter 200 mm. A check is being carried out to see how well the pipe will control the flow when it is new (i.e. when it has the roughness of the clean pipe material, $k_s = 0.03$ mm). When the difference between the water level at the upstream and downstream end is 1.5 m, what will be the flow-rate in the pipe? (Neglect local losses.) In this condition, is the pipe inlet-controlled (assume $C_d = 0.6$)? [125 l/s, yes]

9.3 A rectangular transverse weir in a CSO has a width equal to the width of the chamber itself: 2.2 m. The weir crest is 1.05 m above the floor of the chamber. When the water level is 0.15 m above the crest, determine the flow-rate over the weir. [0.235 m³/s]

9.4 Estimate the flow-rate in the channel of a road with a longitudinal gradient of 0.5% and a crossfall of 1:40 if the width of flow is 2.5 m. Assume $n = 0.013$. [42 l/s]

References

Balmforth, D.J. and Sarginson, E.J. (1978) A comparison of methods of analysis of side weir flow. *Chartered Municipal Engineer*, 105, October, 273–279.

Burrows, R. and Emmonds, J. (1988) Energy head implications of the installation of circular flap gates on drainage outfalls. *Journal of Hydraulic Research*, 26(2), 131–142.

Butler, D. and Parsian, H. (1993) The performance of a vortex flow regulator under blockage conditions. *Proceedings of the 6th International Conference on Urban Storm Drainage, Niagara Falls, Canada*, 1793–1798.

Chow, V.T. (1959) *Open-channel hydraulics*, McGraw-Hill.

Davis, A., Jacob, R.P. and Ellett, B. (1996) A review of road-gully spacing methods. *Journal of Chartered Institution of Water and Environmental Management*, 10, April, 118–122.

Delo, E.A. and Saul, A.J. (1989) Charts for the hydraulic design of high side-weirs in storm sewage overflows. *Proceedings of the Institution of Civil Engineers*, Part 2, 87, June, 175–193.

Green, M.J. (1988) Flow control evaluation, in *Proceedings of Conflo 88: Attenuation Storage and Flow Control for Urban Catchments*, Oxford.

Hydraulics Research Station (1984) *The Drainage Capacity of BS Road Gullies and a Procedure for Estimating their Spacing*. TRRL Contractor Report CR2, Transport and Road Research Laboratory, Crowthorne.

May, R.W.P. (1994) Alternative hydraulic design methods for surface drainage. Road Drainage Seminar, H.R. Wallingford, Wallingford, November.

May, R.W.P., Bromwich, B.C., Gasowski, Y. and Rickard, C.E. (2003) *Hydraulic Design of Side Weirs*, Thomas Telford.

Parsian, H. and Butler, D. (1993) Laboratory investigation into the performance of an in-sewer vortex flow regulator. *Journal of the Institution of Water and Environmental Management*, 7, April, 182–187.

10 Foul sewers

10.1 Introduction

Separate foul sewers form an important component of many urban drainage systems. The emphasis in this chapter is on the design of such systems. In particular the distinction is made between large and small foul sewers and their different design procedures. Analysis of existing systems using computer-based methods is covered in Chapters 19 and 20. Design of non-pipe-based systems is discussed in Chapter 23.

Flow regime

All foul sewer networks physically connect wastewater sources with treatment and disposal facilities by a series of continuous, unbroken pipes. Flow into the sewer results from random usage of a range of different appliances, each with its own characteristics. Generally, these are intermittent, of relatively short duration (seconds to minutes) and are hydraulically unsteady. At the outfall, however, the observed flow in the sewer will normally be continuous and will vary only slowly (and with a reasonably repeatable pattern) throughout the day. Fig. 10.1 gives an idealised picture of these conditions.

The sewer network will have zones with continuously flowing wastewater, as well as areas that are mostly empty but are subject to flushes of flow from time-to-time. It is unlikely, even under maximum continuous flow conditions, that the full capacity of the pipe will be utilised. Intermittent pulses feed the continuous flow further downstream, and this implies that somewhere in the system there is an interface between the two types of flow. As the usage of appliances varies throughout the day, the interface will not remain at a single fixed location.

10.2 Design

This chapter shows how foul sewers can be designed to cope with conditions described above. A general approach to foul (and storm) sewer

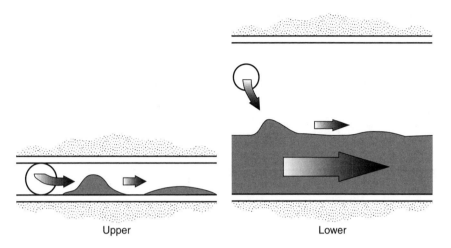

Upper Lower

Fig. 10.1 Hydraulic conditions in foul sewers in dry weather (schematic)

design is illustrated in Fig. 10.2. This should be read in conjunction with Fig. 7.8.

Design is accomplished by first choosing a suitable *design period* and *criterion of satisfactory service*, appropriate to the foul *contributing area* under consideration. The type and number of buildings and their population (the maximum within the design period) are then estimated, together with estimates of the unit water consumption. This information is used to calculate *dry weather flows* in the main part of the system. Flows in building drains and small sewers are assessed in a probability-orientated discharge unit method, based on usage of domestic appliances. *Hydraulic design* of the pipework is based on safe transportation of the flows generated using the principles presented in Chapter 8. Broader issues of sewer layout including horizontal and vertical alignment are covered in Chapter 7.

10.2.1 Choice of design period

Urban drainage systems have an extended life-span and are typically designed for conditions 25–50 years into the future. They may well be in use for very much longer. The choice of design period will be based on factors such as:

- useful life of civil, mechanical and electrical components
- feasibility of future extensions of the system
- anticipated changes in residential, commercial or industrial development
- financial considerations.

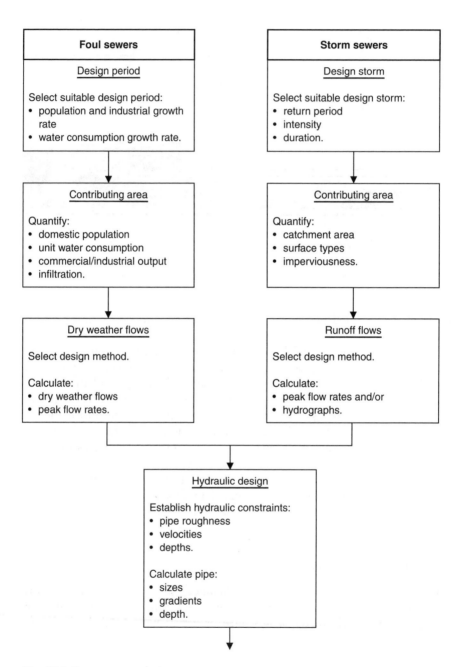

Fig. 10.2 Sewer system design

It is necessary to make estimates of conditions throughout the design period that are as accurate as possible.

10.2.2 Criterion of satisfactory service and risk

The degree of protection against wastewater 'backing-up' or flooding is determined by consideration of the specified criterion of satisfactory service. This protection should be consistent with the cost of any damage or disruption that might be caused by flooding. In practice, cost–benefit studies are rarely conducted for ordinary urban drainage projects; a decision on a suitable criterion is made simply on the basis of judgement and precedent. Indeed, this decision may not even be made explicitly, but nevertheless it is built into the design method chosen.

The design choice of the peak-to-average flow ratio implicitly fixes the level of satisfactory service in large foul sewers. For small sewers, the criterion can be used explicitly to determine flows, though standard (and therefore fixed) values are routinely used.

10.3 Large sewers

In this book, a distinction is drawn between large and small foul sewers. This is only for convenience as there is no precise definition to demarcate between the two types. Indeed, the same pipe may act as both large and small at different times of the day (measured in hours) or at different times in its design period (measured in years).

Flow in large foul sewers is mostly open channel (although in exceptional circumstances this may not be the case), continuous and quasi-steady. Changes in flow that do occur will be at a relatively slow rate and in a reasonably consistent diurnal pattern. In large sewers, we can say that the inflows from single appliances are not a significant fraction of the capacity of the pipe and that there is substantial base-flow (see Fig. 10.1).

10.3.1 Flow patterns

The pattern of flow follows a basic diurnal pattern, although each catchment will have its own detailed characteristics. Generally, low flows occur at night with peak flows during the morning and evening. This is related to the pattern of water use of the community, but also has to do with the location at which the observation is made. Fig. 10.3 illustrates the impact of three important effects. The inflow hydrograph (A) represents the variation in wastewater generation that will, in effect, be similar all around the catchment (see Chapter 4). If the wastewater were collected at one point and then transported from one end of a long pipe to the other, flow *attenuation* due to in-pipe storage would cause a reduction in peak flow, a lag in time to peak and a distortion of the basic flow pattern (B). Normal sewer

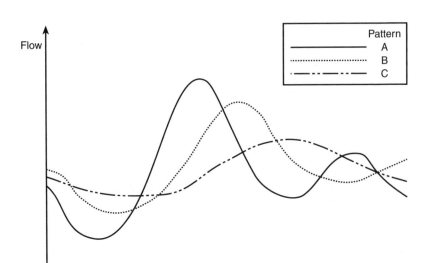

Fig. 10.3 Diurnal wastewater flow pattern modified by attenuation and diversification effects

catchments are not like this, and consist of many-branched networks with inputs both at the most distant point on the catchment and adjacent to the outfall. Thus the time for wastewater to travel from the point of input to the point under consideration is variable and this *diversification* effect causes a further reduction in peak and distortion in flow pattern (C). Additional factors that can influence the flow pattern are the degree of infiltration and the number and operation of pumping stations. These effects can be predicted in existing sewer systems using computational hydraulic models (as described in Chapter 19), but also need to be predicted in the design of new systems.

The flow is usually defined in terms of an average flow (Q_{av}) – or *dry weather flow* (DWF) – and peak flow. The magnitude of the peak flow can then be related to the average flow (see Fig. 10.4). A minimum value can also be defined.

Large sewer design therefore entails estimating the average dry weather flow in the sewer by assuming a daily amount of wastewater generated per person (or per dwelling, or per hectare of development) contributing to the flow, multiplied by the population to be served at the design horizon. Commercial and industrial flows must also be estimated at the design horizon. Allowance should be made for infiltration. The peak flow can be found by using a suitable multiple or peak factor.

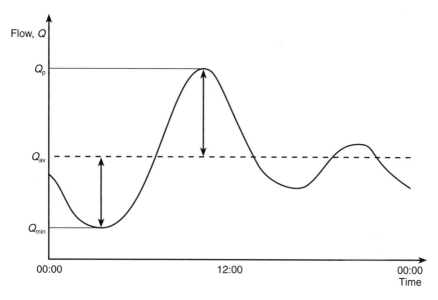

Fig. 10.4 Definition of diurnal wastewater flow pattern

10.3.2 Dry weather flow

When the wastewater is mainly domestic in character, DWF is defined as:

> The average daily flow ... during seven consecutive days without rain (excluding a period which includes public or local holidays) following seven days during which the rainfall did not exceed 0.25 mm on any one day.
>
> (IWEM, 1993)

If the flow contains significant industrial flows, DWF should be measured during the main production days. Ideally, flows during summer and winter periods should be averaged to obtain a representative DWF.

DWF is therefore the average rate of flow of wastewater not immediately influenced by rainfall; it includes domestic, commercial and industrial wastes, and infiltration, but excludes direct stormwater inflow. The quantity is relevant both to foul and combined sewers. It can be expressed simply in the following manner (Ministry of Housing, 1970):

$$DWF = PG + I + E \tag{10.1}$$

DWF dry weather flow (litres/day)
P population served

G   average *per capita* domestic water consumption (l/hd.d)
I   infiltration (l/d)
E   average industrial effluent discharged in 24 hours (l/d)

10.3.3 Domestic flow (PG)

The domestic component of dry weather flow is the product of the population and the average *per capita* water consumption.

Population (P)

A useful first step in predicting the contributing population that will occur at the end of the design period is to obtain as much local, current and historical information as possible. Official census information is often available and can be of much value. Additional data can almost certainly be obtained at the local planning authority, and officers should be able to advise on future population trends, and also on the location and type of new industries. Housing density is a useful indicator of current or proposed population levels.

Per capita water consumption (G)

In Chapter 4 we have already discussed in detail the relationship between water use and wastewater production. We have also considered typical (UK) *per capita* values and discussed that there will be changes in *per capita* water consumption that are independent of population growth.

Where typical discharge figures for developments similar to those under consideration are available, these should be used. In the absence of such data, the European Standard on *Drain and Sewer Systems Outside Buildings* (BS EN 752–4: 1998) states a daily *per capita* figure of between 150 and 300 l should be used. A figure of 220 l (200 l + 10% infiltration) has been widely used in the UK.

Specific design allowance can be made for buildings such as schools and hospitals as given in Table 10.1. See also Example 10.1.

Table 10.1 Daily volume and pollutant load of wastewater produced from various sources

Category	Volume (l/day)	BOD_5 load (g/day)	Per
Day schools	50–100	20–30	Pupil
Boarding schools	150–200	30–60	Pupil
Hospitals	500–750	110–150	Bed
Nursing homes	300–400	60–80	Bed
Sports centre	10–30	10–20	Visitor

10.3.4 Infiltration (I)

The importance of groundwater infiltration and the problems it can cause have been discussed in Section 4.4. As mentioned above, the conventional approach in design is to specify infiltration as a fraction of DWF – namely 10%. Thus, for a design figure of 200 l/hd.d, 20 l/hd.d would be specified. More recent evidence (Ainger *et al.*, 1997) suggests this may be too low. The suggestion is made that for new systems in high groundwater areas, infiltration figures as high as 120 l/hd.d should be used.

There is a difficulty, however, in making such a large design allowance for infiltration. If an allowance is used, this will increase the design flow rate and may in turn increase the required pipe diameter. A bigger sewer will have a larger circumference and joints, potentially allowing more infiltration to enter the system. Thus, the allowance may well have actually caused more infiltration!

Is there a solution to this dilemma? It is suggested that rather than building-in large design allowances that may cause larger pipes to be chosen, it would be a better investment to ensure high standards of pipe manufacture, installation and testing.

10.3.5 Non-domestic flows (E)

Background information on non-domestic wastewater flows can be found in Section 4.3. In design, probably the most reliable approach is to make allowance for flows on the basis of experience of similar commerce or industry elsewhere. If this data is not available, or for checking what is known, the following information can be used. Table 10.2 shows examples of daily wastewater volume produced by a variety of commercial sources. Table 10.3 provides areal allowance for broad industrial categories. Henze *et al.* (1997) present data for a wide range of industries.

Most commercial and industrial premises will have a domestic waste component of their waste and, ideally, the estimation of this should be based on a detailed survey of facilities and their use. Mann (1979) suggests that a figure of 40–80 l/hd. (8 hour shift) is appropriate.

Table 10.2 Daily volume and pollutant load of wastewater produced from various commercial sources

Category	Volume (l/day)	BOD_5 load (g/day)	Per
Hotels, boarding houses	150–300	50–80	Bed
Restaurants	30–40	20–30	Customer
Pubs, clubs	10–20	10–20	Customer
Cinema, theatre	10	10	Seat
Offices	750	250	100 m²
Shopping centre	400	150	100 m²
Commercial premises	300	100	100 m²

Table 10.3 Design allowances for industrial wastewater generation

Category	Volume (l/s.ha)	
	Conventional	*Water saving**
Light	2	0.5
Medium	4	1.5
Heavy	8	2

* Recycling and reusing water where possible

10.3.6 Peak flow

Two approaches to estimating peak flows are used. In the first, typically used in British practice, a fixed DWF multiple is used. In the second, a variable peak factor is specified. Both methods aim to take account of diurnal peaks and the daily and seasonal fluctuations in water consumption together with an allowance for extraneous flows such as infiltration.

BS EN 752–4 recommends a multiple up to 6 is used. This figure is most appropriate for use in sub-catchments subject to relatively little attenuation and diversification effects. For larger sewers, a value of 4 is more realistic. A still lower figure (2.5) is relevant for predicting dry weather flows in combined sewers, because this flow will determine velocity not capacity.

Sewers for Adoption (WRc, 1995) suggests that a design flow of 4000 l/unit dwelling.day (0.046 l/s per dwelling) should be used for foul sewers serving residential developments. This approximates to 3

Example 10.1

Estimate the average daily wastewater flow (l/s) and BOD_5 concentration (mg/l) for an urban area consisting of: residential housing (100 000 population), a secondary school (1000 students), a hospital (1000 beds) and a central shopping centre (50 000 m²).

Solution

Area	Magnitude	Unit flow	Flow rate	Unit BOD_5 load	BOD_5 load
		(m³/unit.d)	*(m³/d)*	*(kg/unit.d)*	*(kg/d)*
Residential	100 000 pop.	0.20	20 000	0.06	6000
School	1000 students	0.10	100	0.03	30
Hospital	1000 beds	0.75	750	0.15	150
Shopping	50 000 m²	0.004	200	0.0015	75
Total			21 050		6255

Average daily wastewater flow = (21 050 × 1000)/(24 × 3600) = 244 l/s
Average BOD_5 concentration = (6255 × 1000)/21 050 = 297 mg/l

persons/property discharging 200 l/hd.day with a peak flow multiple of 6.0 and 10% infiltration.

Opinions and practice differ on whether the DWF to be multiplied should include or exclude infiltration. If DWF is determined from equation 10.1, the most satisfactory form of applying a multiple of 4 (for example) is: 4(DWF − *I*) + *I*.

Peak flows may also be determined by the application of variable peak factors. Fig. 10.3 shows that attenuation and diversification effects tend to reduce peak flows, and so the ratio of peak to average flow generally decreases from the 'top' to the 'bottom' of the network. Thus, peak factor varies depending on position in the network (see Fig. 10.5). Location is usually described in terms of population served or the average flow-rate at a particular point.

The relationship between peak factor (P_F) and population can be described algebraically with equations of the form:

$$P_F = \frac{a}{P^b} \tag{10.2}$$

P population drained in 1000s
a,b constants

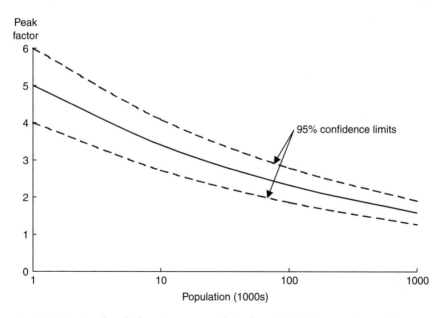

Fig. 10.5 Ratio of peak flows to average daily flow (with 95 percentile confidence limits)

However, there are a number of other such equations and some of the most well known are listed in Table 10.4.

Example 10.2 illustrates that, the numerical values produced by different equations can vary significantly. Thus, any of the formulae available should be used with caution.

One of the reasons for the disparity in the peak factor predictions is the general variability in diurnal flow patterns. The degree of uncertainty is also illustrated by the confidence limits (dashed lines) in Fig. 10.5.

10.3.7 Design criteria

Capacity

Foul sewers should be designed (in terms of size and gradient) to convey the predicted peak flows. It is common practice to restrict depth of flow (typically to $d/D = 0.75$) to ensure proper ventilation.

Self-cleansing

Once the pipe size has been chosen based on capacity, the pipe gradient is selected to ensure a minimum 'self-cleansing' velocity is achieved. The self-cleansing velocity is that which avoids long-term deposition of solids, and should be reached at least once per day. BS EN 752–4 recommends a minimum of 0.7 m/s for sewers up to DN300. Higher velocities may be needed in larger pipes (see Chapter 16). *Sewers for Adoption* requires a

Table 10.4 Peak factors

Reference	Method	Notes	Equation
Harman (1918)	$1 + \dfrac{14}{4 + \sqrt{P}}$	1	10.3a
Gifft (1945)	$\dfrac{5}{P^{1/6}}$	1	10.3b
Babbitt (1952)	$\dfrac{5}{P^{1/5}}$	1	10.3c
Fair & Geyer (1954)	$1 + \dfrac{18 + \sqrt{P}}{4 + \sqrt{P}}$	1	10.3d
–	$4Q^{-0.154}$	2	10.3e
Gaines (1989)a	$2.18Q^{-0.064}$	3	10.3f
Gaines (1989)b	$5.16Q^{-0.060}$	3	10.3g
BS EN 752–4	6	–	

[1]Population P in 1000s, [2]Flow Q in 1000 m³/d, [3]Flow Q in l/s

---

Example 10.2

A separate foul sewer network drains a domestic population of 250 000. Estimate the peak flow rate of wastewater at the outfall (excluding infiltration) using both Babbitt's and Gaines's formula. The daily *per capita* flow is 145 l.

Solution

Average daily flow = (250 000 × 145)/(3600 × 24) = 420 l/s
Babbitt (equation 10.3c):

$$P_F = \frac{5}{p^{1/5}} = \frac{5}{250^{1/5}} = 1.66$$

Peak flow = 1.66 × 420 = 697 l/s
Gaines (equation 10.3f):

$$P_F = 2.18Q^{-0.064} = 2.18 \times 420^{-0.064} = 1.48$$

Peak flow = 1.48 × 420 = 622 l/s

---

velocity of 0.75 m/s to be achieved at the typical diurnal peak of one-third the design flow (i.e. 2 DWF). Some engineers prefer to specify a higher self-cleansing velocity to be achieved at full-bore flow. Fig. 8.8 shows how this allows for the reduction in velocity that occurs in pipes that are flowing less than half full. In practice, the pipe size and gradient are manipulated together to obtain the best design.

Roughness

For design purposes, it is conservatively assumed that the pipe roughness is independent of pipe material. This is because in foul and combined sewers, all materials will become slimed during use (see Chapter 8). BS EN 752–4: 1998 recommends a k_s value of 0.6 mm (for use in the Colebrook-White equation) where the peak DWF exceeds 1.0 m/s, and 1.5 mm where it is between 0.76 and 1.0 m/s.

Minimum pipe sizes

The minimum pipe size is generally set at DN75 or DN100 for house drains and DN100 to DN150 for the upper reaches of public networks, and is based on experience.

10.3.8 Design method

The following procedure should be followed for foul sewer design:

1   Assume pipe roughness (k_s)
2   Prepare a preliminary layout of sewers, including tentative inflow locations
3   Mark pipe numbers on the plan according to the convention described in Chapter 7
4   Define contributing area DWF to each pipe
5   Find cumulative contributing area DWF
6   Estimate peak flow (Q_p) based on average DWF and peak factor/ multiple
7   Make a first attempt at setting gradients and diameters of each pipe
8   Check $d/D < 0.75$ and $v_{max} > v > v_{min}$
9   Adjust pipe diameter and gradient as necessary (given hydraulic and physical constraints) and return to step 5.

Example 10.3 illustrates the design of a simple foul sewer network.

10.4 Small sewers

As we have seen earlier, small sewers are subject to random inflow from appliances as intermittent pulses of flow, such that peak flow in the pipe is a significant fraction of the pipe capacity and there is little or no baseflow.

As an appliance empties to waste, a relatively short, highly turbulent pulse of wastewater is discharged into the small sewer. As the pulse travels down the pipe, it is subject to attenuation resulting in a reduction in its flow-rate and depth, and an increase in duration and length (see Fig. 10.1).

10.4.1 Discharge Unit Method

Building drainage and small sewerage schemes are often designed using the Discharge Unit Method as an alternative to the methods previously described. Using the principles of probability theory, discharge units are assigned to individual appliances to reflect their relative load-producing effect. Peak flow-rates from groups of mixed appliances are estimated by addition of the relevant discharge units. The small sewer can then be designed to convey the peak flow. This approach is now explained in more detail.

Probabilistic framework

Consider a single type of appliance discharging identical outputs that have an initial duration of t' and a mean interval between use of T'. Hence, the

Example 10.3

A preliminary foul sewer network is shown in Fig. 10.6. Design the network using fixed DWF multiples (6 for domestic flows, and 3 for industrial) based on the availability of an average grade of 1:100. The inflow, Q_a is 30 l/s at peak. For the sake of simplicity, infiltration can be neglected.

Data from the network is contained in the shaded portion of the Table. Maximum proportional depth is 0.75 and minimum velocity is 0.75 m/s. Pipes roughness is $k_s = 1.5$ mm.

Solution

Using the raw data on land use, peak inflow rates are calculated. It is assumed that the commercial and industrial rates specified are peak rates.

(1) Pipe number	(2) No. of houses	(3) Peak flow rate $(Q_3)^*$ (l/s)	(4) Commercial area (ha)	(5) Peak flow rate $(Q_5)^{**}$ (l/s)	(6) Industrial area & type (ha)	(7) Peak flow rate $(Q_7)^+$ (l/s)	(8) Total peak flow rate $(Q_3+Q_5+Q_7)$ (l/s)
1.0	200	8.4	–	0	1.65 M	19.8	28.2
2.0	250	10.5	–	0	1.70 L	10.2	20.7
1.1	140	5.9	1.10	1.1	0.60 L	3.6	10.6
1.2	500	21.0	2.80	2.8	–	0	23.8

* Based on 3 persons per house, 200 l/hd.d and DWF multiple of 6 ($Q_3 = 0.042$ l/s.house)
** Based on 300 l/d.100 m² and DWF multiple of 3 ($Q_5 = 1$ l/s.ha)
+ Based on 2 and 4 l/s.ha for Light and Medium industry receptively and DWF multiples of 3 ($Q_7 = 6$ or 12 l/s.ha)

Pipe velocities and depths are calculated using the Colebrook-White equation or can be read from Butler-Pinkerton charts (e.g. Fig. 8.9). The pipe/gradient combination chosen is shown in bold.

(1) Pipe number	(2) Peak flow [l/s]	(3) Cumulative peak flow [l/s]	(6) Assumed pipe size (mm)	(7) Minimum gradient (1:x)	(8) Proportional depth of flow	(9) Velocity (m/s)	Comments
1.0	28.2	58.2	250	90	0.75	1.45	Depth-limited
			300	240	0.75	1.04	Depth-limited
			375	600	0.67	0.75	Velocity-limited
2.0	20.7	20.7	150	47	0.75	1.45	
			225	270	0.64	0.75	
1.1	10.6	89.5	300	95	0.75	1.60	
			375	320	0.75	1.02	
1.2	23.8	113.3	375	200	0.75	1.27	
			450	500	0.75	0.90	

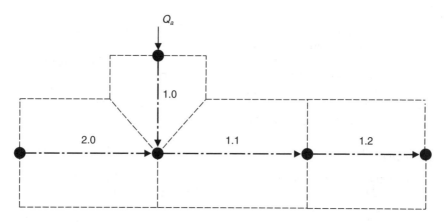

Fig. 10.6 System layout (Example 10.3)

probability p that the appliance will be discharging at any instant is given by:

$$p = \frac{duration\ of\ discharge}{mean\ time\ between\ discharges} = \frac{t'}{T'} \tag{10.4}$$

Example 10.4

Calculate the probability of discharge of a single WC that discharges for 10 seconds every 20 minutes at peak times. What percentage of time will the WC be loading the system?

Solution

From equation 10.4:

$p_{WC} = 10/1200 = 0.0083$

The WC will be loading the system 0.8% of the time (at peak) and hence will *not* be discharging for 99.2% of the time.

In most systems, however, there will be more than one appliance. How can we answer a question such as 'what is the probability that r from a total of N appliances will discharge *simultaneously*?' Application of the binomial distribution states that if p is the probability that an event will happen in any single trial (i.e. the probability of success) and $(1-p)$ is the probability that it will fail to happen (i.e. the probability of failure) then the probability that the event will occur exactly r times in N trials $(P(r,N))$ is:

$$P(r,N) = {}^NC_r p^r (1 - p)^{N-r} \qquad (10.5a)$$

or

$$P(r,N) = \frac{N!}{r!(N - r)!} p^r (1 - p)^{N-r} \qquad (10.5b)$$

Thus, to use the binomial probability distribution in this application, we must assume:

- each trial has only two possible outcomes – success or failure; that is, an appliance is either discharging or it is not
- the probability of success (p) must be the same on each trial (i.e. independent events), implying that t' and T' are always the same.

Neither of these assumptions is fully correct for discharging appliances, but they are close enough for design purposes. Example 10.5 illustrates the basic use of equation 10.5.

Example 10.5

What is the probability that 20 from a total of 100 WCs ($p = 0.01$) will discharge simultaneously?

Solution

N = number of trials = total number of connected appliances = 100
p = probability of success = probability of discharge = 0.01

Using the binomial expression with the above data gives (equation 10.5b):

$$P(20,100) = \frac{100!}{20!80!} 0.01^{20} 0.99^{80} = 2.4 \times 10^{-20}$$

In other words, this eventuality is extremely unlikely.

Design criterion

Whilst this type of basic information is of interest, it is not of direct use. In design, we are concerned with establishing the probable number of appliances discharging simultaneously against some agreed standard. Practical design is carried out using a confidence level approach or 'criterion of satisfactory service' (J) as introduced in Section 10.2.2. For small sewers, this is defined as the percentage of time that up to c appliances out of N will be discharging. So:

$$\sum_{r=0}^{c} P(r,N) \geq J \qquad (10.6)$$

In design terms, we are trying to establish the value of c for a given J. A typical value for J would be 99%, implying actual loadings will only exceed the design load for less than 1% of the time (see Example 10.6).

Example 10.6

For a criterion of satisfactory service of 99%, determine the number of water widgets discharging simultaneously from a group of 5, if their probability of discharge is 20% (unusually high, but used for illustrative purposes). If each widget discharges $q = 0.5$ l/s, find the design flow.

Solution

Now, $N = 5$, $p = 0.2$ and $J = 0.99$. Using equation 10.5 for increasing values of r we get:

r	$P(r,N)$	$\Sigma P(r,N)$	Σq (l/s)
0	0.327	0.327	0
1	0.410	0.737	0.5
2	0.204	0.941	1.0
3	0.051	0.992	1.5

So, since at $r = 3$, $\Sigma P(r,N) > 0.99$, up to 3 water widgets will be found discharging 99% of the time and more than 3 will discharge just 1% of the time (i.e. during one peak period every hundred days). Design for $c = 3$ simultaneous discharges, $q = 1.5$ l/s.

At a given criterion of satisfactory service, each individual appliance will therefore have a unique relationship between:

- the number of connected appliances and the number discharging simultaneously
- the number of connected appliances and flow-rate (because the discharge capacity of each appliance is known, and assumed constant).

Fig. 10.7 illustrates the relationship between number of connected appliances and simultaneous discharge for three common devices, prepared

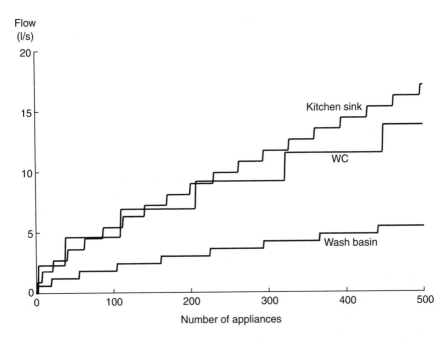

Fig. 10.7 Simultaneous discharge of WC, sink and basin at 99% criterion of satisfactory service

using the binomial distribution and data from Table 10.5. The stepped appearance of the plots does not reflect the resolution of the calculations used to produce them, but is inherent in the calculations.

Mixed appliances

In a practical design situation, there will be a mix of appliance types rather than the single types previously discussed. The basic binomial distribution does not take into account the interactions in a mixed system between appliances of different frequency of use, discharge duration and flow-rate.

Table 10.5 Typical UK appliance flow and domestic usage data (adapted from Wise and Swaffield, 2002)

Appliance	Flow-rate q *(l/s)*	Duration t' *(s)*	Recurrence use interval T' *(s)*	Probability of discharge p
WC (9 l)	2.3	5	1200	0.004
Wash basin	0.6	10	1200	0.008
Kitchen sink	0.9	25	1200	0.021
Bath	1.1	75	4500	0.017
Washing machine	0.7	300	15 000	0.020

To overcome this problem, the *Discharge Unit (DU) method* has been developed, itself an extension of the earlier *fixture unit* method (Hunter, 1940) used to calculate water supply loads. This is based on the premise that the same flow-rate may be generated by a different number of appliances depending on their type. DUs are, therefore, attributed uniquely to each appliance type, and the value will depend on:

- the rate and duration of discharge
- the criterion of satisfactory service.

Recommended values are given in Table 10.6.

Therefore, it is possible to express all appliances in terms of DUs using a family of design curves, based only on intensity of use. BS EN 752–4: 1998 recommends a power law is used to approximate the relationship between design flow-rate Q and the cumulative number of discharge units DU, so:

$$Q = k_{DU} \sqrt{\Sigma n_{DU}}$$ (10.7)

Q peak flow (l/s)
k_{DU} dimensionless frequency factor
n_{DU} number of discharge units

The value of k_{DU} depends on the intensity of usage of the appliance(s) and is given in Table 10.7. Design curves are given in Fig. 10.8, and are used in Example 10.7.

Table 10.6 Discharge unit ratings for domestic appliances

Appliance	Discharge units, DU	
	BS EN 12056–2	BS EN 752–4
WC (9 l)	1.6–2.1	1.2–2.5
Wash basin	0.3	0.3–0.6
Kitchen sink	1.3	0.8–1.3
Bath	1.3	0.8–1.3
Washing machine (up to 6 kg)	0.6	0.5–0.8

Table 10.7 Frequency of use factors (BS EN 752–4: 1998)

Frequency of use	k_{DU}
Intermittent: dwellings, guest houses, offices	0.5
Frequent: hospitals, schools, restaurants	0.7
Congested: public facilities	1.0

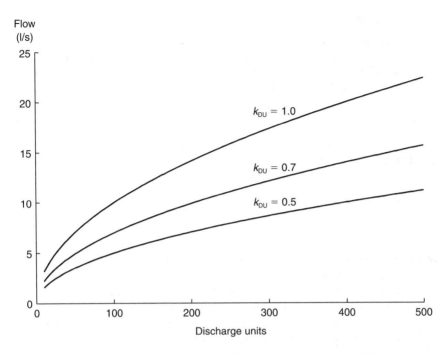

Fig. 10.8 Relationship between appliance discharge units and design flow-rate

10.4.2 Design criteria

In small sewers and drains, design criteria relate principally to the capacity of the pipe and the requirements of self-cleansing. Sewers are normally designed (BS EN 752–4) so that the design flow (at the relevant confidence level) can be conveyed with a proportional depth $d/D < 0.7$. This is done assuming steady, uniform flow conditions as described in Chapter 8.

In small sewers, where solids are transported by being pushed along the pipe invert, self-cleansing is difficult to assess on a theoretical basis (as considered further in Section 10.5). Even if flow is assumed to be steady and uniform (which it is not), such low flows may require quite steep gradients to achieve self-cleansing velocities. At the heads of runs, the pipe gradient is usually based on 'accepted practice' and can be 'relaxed' somewhat (as shown in Table 10.8) to a minimum gradient and number of connected

Table 10.8 BS EN 752–4 deemed to satisfy self-cleansing rules for small sewers

Design flow (l/s)	DN (mm)	Gradient	Connected WCs
<1	≤100	≥1:40	–
>1	100	≥1:80	1
	150	≥1:150	5

Example 10.7

A residential block is made up of 20 flats, each fitted with a WC, wash basin, sink, bath and washing machine. It is estimated that in any one flat, between 08:00 and 09:00, all of the appliances are likely to be in use on a Monday. Calculate the design flow-rate using the Discharge Unit method.

Solution

The discharge units for all appliances $= 1.9 + 0.3 + 1.3 + 1.3 + 0.6 = 5.4$. Hence, for 20 flats the total discharge units is 108. Assuming $k_{DU} = 0.5$, from equation 10.7 or Figure 10.8:

$$Q = 0.5\sqrt{108} = 5.2\,l/s$$

WCs, depending on the required pipe size. This is in recognition of the flush wave produced by the WC in transporting solids. However, there is evidence to suggest that such steep slopes are not really necessary and that very flat sewers can work perfectly well (Lillywhite and Webster, 1979).

The implication of Table 10.8 is that, for a public sewer with diameter 150 mm or greater, the maximum gradient that need be used is 1:150, provided there are at least 5 connected dwellings. *Sewers for Adoption* (WRc, 2001) recommends 10 connected dwellings. The Protocol on Design, Construction and Adoption of Sewers in England and Wales (DEFRA, 2002) allows a minimum diameter of 100 mm to be used for pipes serving up to 10 dwellings.

The major factors influencing minimum pipe diameter are its ability to carry gross solids and its ease of maintenance. Large solids frequently find their way into sewers, either accidentally or deliberately, particularly via the WC and property access points. The minimum pipe size is as set out in Section 10.3.7.

An application of the small sewer design method is given in Example 10.8.

10.4.3 Choice of methods

As mentioned earlier in the chapter, the two different design methods (for large and small sewers) represent the different flow regimes in foul sewers. If a large network is to be designed in detail, there comes a point where a change must be made from one method to another. The point at which the change takes place depends on local circumstances, but its location is important as it has considerable impact on pipe sizes and gradients, and hence cost.

BS EN 752–4: 1998 suggests the population method should be used if the probability method gives a pipe size larger than DN 150.

Example 10.8

Design the foul sewer diameter and gradients for the small housing estate shown in Fig. 10.9. Data on the network is shown in the shaded portion of the table. Use the following design data:

Minimum diameter (mm):	150
Minimum velocity (m/s):	0.75
Minimum gradient:	1:150 (provided number of WCs ≥5)
Maximum proportional depth of flow:	0.75
Pipe roughness (mm):	0.6

Solution

For each sewer length, use the minimum pipe diameter and calculate the minimum gradient required to achieve: the necessary capacity + self-cleansing. Use Tables 10.6–10.8, equation 10.7 and the Butler-Pinkerton charts.

Assume each dwelling has (WC + basin + sink) DUs = 1.9 + 0.3 + 1.3 = 3.5
For individual pipe lengths draining at least 5 dwellings, reduce the gradient to 1:150. Take $k_{DU} = 0.5$.

(1) Pipe number	(2) No. of houses	(3) No. of discharge units	(4) Cumulative no. of discharge units	(5) Design flow rate (l/s)	(6) Assumed pipe size (mm)	(7) Minimum gradient (1:x)	(8) Proportional depth of flow	(9) Velocity (m/s)	Comments
1.1	4	14	14	1.9	150	55	0.21	0.75	
1.2	9	31.5	45.5	3.4	150	85	0.32	0.75	*
						150	0.37	0.62	
2.1	10	35	35	3.0	150	75	0.28	0.75	*
						150	0.34	0.59	
1.3	1	3.5	84	4.6	150	100	0.37	0.75	*
						150	0.43	0.65	
3.1	6	21	21	2.3	150	70	0.23	0.75	*
						150	0.29	0.54	
1.4	5	17.5	122.5	5.5	150	120	0.43	0.75	*
						150	0.46	0.69	
1.5	2	7	129.5	5.7	150	125	0.45	0.75	*
						150	0.47	0.70	

* Gradient relaxed to 1:150 as $Q > 1$ l/s, WCs ≥ 5

10.5 Solids transport

It is surprising that the transport of gross solids is not routinely and explicitly considered in the design of large or small sewers. In recent years, research has begun to fill the gaps in our understanding of the movement of solids in the different hydraulic regimes encountered, and is giving some important feedback to practical design and operation.

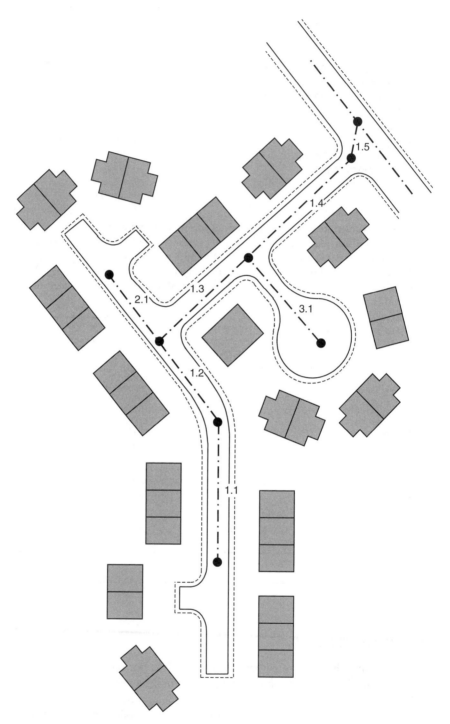

Fig. 10.9 System layout and catchment plan (Example 10.8)

The main characteristics of gross solids transport in sewers are as follows.

- There is a wide variety of solids, and the physical condition of some types varies widely, influencing the way they are transported.
- Some solids change their condition as they move through the system, as a result of physical degradation and contact with other substances in the sewer.
- In some hydraulic conditions solids are carried with the flow, yet at lower flow-rates they may be deposited.
- During movement, solids do not necessarily move at the mean water velocity.
- Some solids affect the flow conditions within the sewer.

10.5.1 Large sewers

When solids are advected (moved whilst suspended in the flow) in large sewers, forces acting on the solids position them at different flow depths depending on their specific gravity and on the hydraulic conditions. Fig. 10.10 indicates how some solids can be carried along at levels where the local velocity is greater than the mean velocity (v). This means that solids may 'overtake' the flow and arrive at CSOs and WTPs before the peak water flow.

Fig. 10.11 shows laboratory results for a solid plastic cylinder (artificial faecal solid) plotted as longitudinal solid velocity against mean water velocity, for two contrasting gradients. A linear relationship fits all this

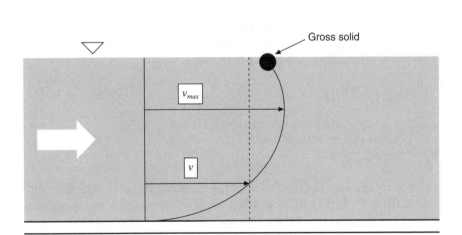

Fig. 10.10 Movement of gross solids in large sewers

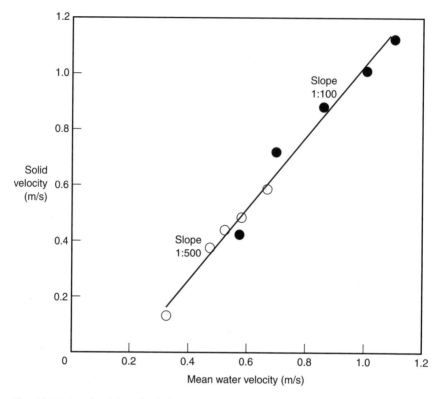

Fig. 10.11 Artificial faecal solid velocity versus mean velocity, with linear fit (after
 Butler *et al.*, 2003)

data well ($R^2 = 0.98$), and this was found to be the case for all the artificial solids studied and for various 'real' gross solids (Butler *et al.*, 2003).
This linear relationship can be expressed as:

$$v_{GS} = \alpha v + \beta \tag{10.8}$$

v_{GS} velocity of a particular gross solid (m/s)
v mean water velocity (m/s)
α, β coefficients

Laboratory results indicate β to typically be small enough to neglect, but α
varies from 0.98 to 1.27 depending on solid type, with lower specific-
gravity solids generally having the higher values. It has also been recommended (Davies *et al.*, 1996) that, for the modelling of solids movement in
unsteady flow, the relationship between the mean water velocity and the

average velocity of any solid type can be assumed to be the same in unsteady (gradually varied) flow as it is in steady (uniform) flow.

Generally, solid size has not been found to be an important variable, except at low flow depths. In this case, larger solids tend to be retarded more than smaller ones by contact with the pipe wall.

Under certain hydraulic conditions (typically low flows, such as overnight), solids may be deposited. Davies *et al.* (1996) found that a solid's propensity to deposit is based on critical hydraulic parameters of flow depth and mean velocity. They argued that (at least for modelling purposes) deposition of solids takes place when the value of *either* depth or mean velocity goes below the critical value, and re-suspension takes place when that level is subsequently exceeded. Fig. 10.12 shows a graph of mean velocity against depth, with points representing the conditions for deposition of a sanitary towel observed in a laboratory study. The dotted lines indicate suitable values for the critical depth (vertical) and velocity (horizontal). Above and to the right of the dotted lines are conditions in which these types of solid are carried by the flow (both depth and velocity exceeding the critical value). Below or to the left of the dotted lines are conditions in which they would be deposited. Table 10.9 gives depth/velocity values results for various gross solid types.

10.5.2 Small sewers

The movement of solids in small sewers is somewhat different to that in large sewers. Laboratory experiments demonstrate that there are two main

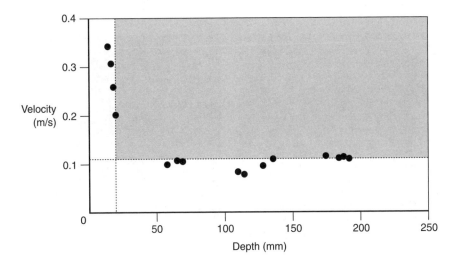

Fig. 10.12 Hydraulic conditions for deposition of solids (sanitary towels) (after Butler *et al.*, 2003)

Table 10.9 Critical depth/velocity for various solid types

Solid type	Critical depth (mm)	Critical velocity (m/s)
Solid plastic cylinders:		
Length 80 mm, dia. 37 mm	30	0.20
Length 44 mm, dia. 20 mm	22	0.13
Length 22 mm, dia. 10 mm	20	0.10
Cotton wool wipe	10	0.08
Sanitary towel	20	0.11

modes of solid movement: 'floating' and 'sliding dam'. The 'floating' mechanism occurs when the solid is small relative to the pipe diameter and flush wave input. The solid moves with a proportion of the wave velocity and has little effect on the wave itself. Solids which are large compared with the flush wave and pipe diameter move with a sliding dam mechanism (Littlewood and Butler, 2003). In this case, the flush wave builds up behind the solid, which acts as a dam in the base of the pipe. When the flow's hydrostatic head and momentum overcome the friction between solid and pipe wall, the solid begins to move along the pipe invert. The amount of movement that occurs depends on how 'efficient' the solid is as a dam: the higher the efficiency, the further the solid will move for the same flush wave. The two modes of movement are illustrated in Fig. 10.13. Photograph (a) shows toilet tissue alone in the flow and photograph (b) shows toilet tissue and an artificial faecal solid in combination. Note the pool of water forming behind the solid and propelling it along. The role of the toilet tissue in forming the 'dam' is also noteworthy. Solids tend to move furthest in the sliding dam mode.

 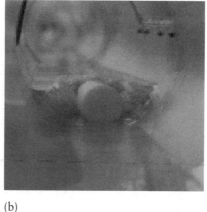

(a)   (b)

Fig. 10.13 Floating (a) and sliding dam (b) mechanisms of solid movement (courtesy of Dr Richard Barnes)

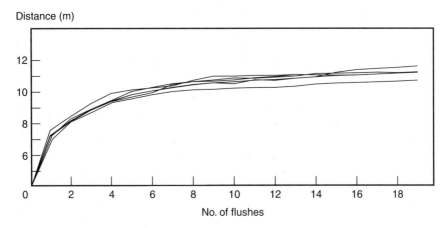

Fig. 10.14 Limited solid transport distance for a gross solid in a 150 mm diameter pipe

Eventually, whichever mode of movement prevails, the solid will deposit on the pipe invert, some distance away from its entry point. It will remain there until another wave enters the pipe, travels along to meet the stranded solid, and resuspends it. The solid will move further downstream, but for a distance less than the initial movement. The distance moved under the influence of each subsequent flush decreases, until the solid is no longer moved at all by the attenuated flush wave (Swaffield and Galowin, 1992). Thus each solid, flush wave and pipe diameter has a 'limiting solid transport distance'. Fig. 10.14 shows that, under repeated tests, the solid is not moved more than 13 m even after 20 flush waves have been passed down the pipe. In fact, very little further movement is noted beyond 10 flushes.

Problems

10.1 Explain how you would go about the preliminary investigation and design of a foul sewer network for a large housing development.

10.2 What are the main differences in the hydraulic regime between large and small foul sewers? What implications do these have on the design procedures adopted?

10.3 Explain the main factors affecting the shape of the dry weather flow diurnal profile.

10.4 An urban catchment is drained by a separate foul sewer network and has an area of 500 ha and a population density of 75 hd/ha. At the outfall of this catchment, calculate:

a)   the average dry weather flow (in l/s) assuming water consumption is 160 l/hd.d, trade effluent is 10 m³/ha.d over 10% of the catchment and infiltration is 20 l/hd.d   [84 l/s]

b)   the peak dry weather flow using Babbitt's formula.   [203 l/s]

10.5   If the outfall sewer in Problem 10.4 is 500 mm in diameter with a gradient of 1:200, calculate:

a)   the depth of peak flow, assuming k_s = 1.5 mm   [325 mm]

b)   the additional population that could be served, assuming that proportional depth does not exceed 0.75.   [2922]

10.6   Redesign the foul sewer network specified in Example 10.3 on a steep site with an inflow of Q_a = 45 l/s.

10.7   Explain how the binomial probability distribution forms the basis of the Discharge Unit small sewer design method.

10.8   It has been estimated that, in an office block, each WC is used at peak times every 5 minutes and discharges for 10 seconds. In a group of 5 WCs, calculate the maximum number discharging simultaneously at the 99.9% confidence level.   [2]

10.9   Redesign the foul sewer network specified in Example 10.3 to serve the residential housing only, using the Discharge Unit method.

10.10   Calculate the total number of dwellings that can be drained by a 150 mm diameter pipe (k_s = 1.5 mm) running with a proportional depth of 0.75 at a gradient of 1:300, using both large and small sewer design methods. Assume 3.5 DUs or 0.046 l/s per dwelling.

[174, 73]

10.11   Explain the main differences in the way gross solids are transported in large and small sewers.

Key sources

ASCE/WPCF (1982) *Gravity Sanitary Sewer Design and Construction*, ASCE Report No. 60, WPCF Manual No. FD-5.

Bartlett, R.E. (1979) *Public Health Engineering – Sewerage*, 2nd edn, Applied Science Publishers.

Butler, D. and Graham, N.J.D. (1995) Modeling dry weather wastewater flow in sewer networks. *American Society of Civil Engineers, Journal of Environmental Engineering Division*, **121**(2), Feb, 161–173.

Stanley, G.D. (1975) *Design Flows in Foul Sewerage System*s, DoE Project Report No. 2.

Swaffield, J.A. and Galowin L.S. (1992) The *Engineered Design of Building Drainage Systems*, Ashgate.

References

Ainger, C.M., Armstrong, R.A. and Butler, D. (1997) *Dry Weather Flow in Sewers*, Report R177, CIRIA, London.

Babbitt, H.E. (1953) *Sewerage and Sewage Treatment*, 7th edn, John Wiley & Sons.

BS EN 752–4: 1998 *Drain and Sewer Systems Outside Buildings. Part 4: Hydraulic Design and Environmental Considerations.*

BS EN 12056–2: 2000 *Gravity Drainage Systems Inside Buildings. Part 2: Sanitary Pipework, Layout and Calculation.*

Butler, D., Davies, J.W., Jefferies, C. and Schütze, M. (2003) Gross solids transport in sewers. *Proceedings of Institution of Civil Engineers, Water, Maritime & Energy,* **156** (WM2), 165–174.

Davies, J.W., Butler, D. and Xu, Y.L. (1996) Gross solids movement in sewers: laboratory studies as a basis for a model. *Journal of the Institution of Water and Environmental Management,* **10**(1), 52–58.

DEFRA (2002) Protocol on Design, Construction and Adoption of Sewers in England and Wales (2002), DEFRA Publications, London. www.defra.gov.uk/environment/water/industry/sewers.

Fair, J.C. and Geyer, J.C. (1954) *Water Supply and Waste-water Disposal,* John Wiley & Sons.

Gaines, J.B. (1989) Peak sewage flow rate: prediction and probability. *Journal of Pollution Control Federation,* **61**, 1241.

Gifft, H.M. (1945) Estimating variations in domestic sewage flows. *Waterworks and Sewerage,* **92**, 175.

Harman, W.G. (1918) Forecasting sewage in Toledo under dry-weather conditions. *Engineering News-Record,* **80**, 1233.

Henze, M., Harremoës, P., Jansen, J.C. and Arvin, E. (1997) *Wastewater Treatment – Biological and Chemical Processes,* 2nd edn, Springer-Verlag.

Hunter, R.B. (1940) *Methods of Estimating Loads in Plumbing Systems,* BMS 65 and BMS 79, National Bureau of Standards, Washington, DC.

IWEM (1993) *Glossary. Handbooks of UK Wastewater Practice,* The Institution of Water and Environmental Management.

Lillywhite, M.S.T. and Webster, C.J.D. (1979) Investigations of drain blockages and their implications for design. *The Public Health Engineer,* **7**(2), 53–60.

Littlewood, K. and Butler, D. (2003) Movement mechanisms of gross solids in intermittent flow. *Water Science & Technology,* **47**(4), 45–50.

Mann, H.T. (1979) *Septic Tanks and Small Sewage Treatment Works,* WRc Report No. TR107.

Metcalf and Eddy, Inc. (1991) *Wastewater Engineering. Treatment, Disposal and Re-use,* 3rd edn, McGraw-Hill.

Ministry of Housing and Local Government (1970) *Technical Committee on Storm Overflows and the Disposal of Storm Sewage,* Final Report, HMSO.

Wise, A.F.E. and Swaffield, J.A. (2002) *Water, Sanitary and Waste services for Buildings,* 5th edn, Butterworth-Heinemann.

WRc (2001) *Sewers for Adoption – a Design and Construction Guide for Developers,* 5th edn, Water UK.

11 Storm sewers

11.1 Introduction

This chapter deals with the properties and the design of pipe-based systems for carrying stormwater. Computer-based analysis of existing systems is covered in Chapters 19 and 20. Design of non-pipe-based systems is covered in Chapters 21 and 23.

Flow regime

All storm sewer networks physically connect stormwater inlet points (such as road gullies and roof downpipes) to a discharge point, or outfall, by a series of continuous and unbroken pipes. Flow into the sewer results from the random input over time and space of rainfall-runoff. Generally, these flows are intermittent, of relatively long duration (minutes to hours) and are hydraulically unsteady.

Separate storm sewers (more than foul sewers) will stand empty for long periods of time. The extent to which the capacity is taken up during rainfall depends on the magnitude of the event and conditions in the catchment. During low rainfall, flows will be well below the available capacity, but during very high rainfall the flow may exceed the pipe capacity inducing pressure flow and even surface flooding. Unlike in foul sewer design (see Chapter 10), no distinction is made between large and small sewers in the design of storm systems.

11.2 Design

The magnitude and frequency of rainfall is unpredictable and cannot be known in advance, so how are drainage systems designed? The general method has been illustrated in Fig. 10.2 (Chapter 10) as a flow chart, and should be read in conjunction with Fig. 7.8.

Design is accomplished by first choosing a suitable *design storm*. The physical properties of the storm *contributing area* must then be quantified. A number of methods of varying degrees of sophistication have been developed to estimate the *runoff flows* resulting from rainfall. *Hydraulic design* of the pipework, using the principles presented in Chapter 8, ensures sufficient, sustained capacity. Broader issues of sewer layout including horizontal and vertical alignment have been covered in Chapter 7.

11.2.1 Design storm

The concepts of statistically analysed rainfall and the design storm were introduced in Chapter 5. These give statistically representative rainfall that can be applied to the contributing area and converted into runoff flows. Once flows are known, suitable pipes can be designed.

The choice of design storm return period therefore determines the degree of protection from stormwater flooding provided by the system. This protection should be related to the cost of any damage or disruption that might be caused by flooding. In practice, cost–benefit studies are rarely conducted for ordinary urban drainage projects, a decision on design storm return period is made simply on the basis of judgement and precedent.

Standard practice in the UK (WRc, 2001) is to use storm return periods of 1 year or 2 years for most schemes (for steeper and flatter sites respectively) with 5 years being adopted where property in vulnerable areas would be subject to significant flood damage. Higher periods up to 25 years may be adopted for city centre sewers. Flooding from combined sewers into housing areas is likely to be more hazardous than storm runoff flooding of open land, so the type of flooding likely to occur will influence selection of a suitable return period.

Although we can assess and specify design rainfall return period, our greatest interest is really in the return period of flooding. It is normally assumed that the frequency of *rainfall* is equivalent to the frequency of *runoff*. However, this is not completely accurate. For example, antecedent soil moisture conditions, areal distribution of the rainfall over the catchment and movement of rain all influence the generation of stormwater runoff (see Chapters 5 and 6). These conditions are not the same for all rainfall events, so rainfall frequency cannot be identical to runoff frequency. However, comprehensive storm runoff data is less common than rainfall records, and so the assumption is usually the best reasonable approach available.

It is certainly *not* the case, however, that frequency of *rainfall* is equivalent to the frequency of *flooding*. Sewers are almost invariably laid at least 1 m below the ground surface and can, therefore, accommodate a considerable surcharge before surface flooding occurs (see Chapter 8). Hence, the capacity of the system under these conditions is increased above the design capacity, perhaps even doubled. Inspection of Fig. 5.2 in Chapter 5 illustrates that a 10 year storm will give a rainfall intensity approximately twice that of a 1 year storm for most durations. It follows, therefore, that where sewers have been designed to a 1 year standard, surcharge may increase that capacity up to an equivalent of a 10 year storm without surface flooding.

Table 11.1 shows the recommendations made by the relevant European Standard (BS EN 752–4: 1998) for design storm frequency or return period related to the location of the area to be drained. It suggests that a design check should be carried out to ensure that adequate protection against flooding is provided at specific sensitive locations. Design flooding frequencies are also given in the table.

11.2.2 Flooding

As part of the design process it is usual to assess the broad implications of surface flooding (manhole surcharge) from the piped system. The simplest approach is to identify points in the system prone to manhole surcharge using design storms of return period equal to the design flooding return period (see Table 11.1). Different storm *durations* are assessed to determine the worst case. This duration will normally be greater than or equal to the time of concentration of the point where the flow exits from the system (Orman, 1996). More sophisticated analysis can be undertaken using long-term historical or synthetic time–series rainfall data to calculate the predicted frequency of flooding (see Chapter 19).

Care should be taken by the designer to consider and define the potential route of sewer flooding (WRc, 2001). Ideally this requires a digital ground model of the catchment levels, identification of all points of entry/exit to the drainage system plus location of all effective flow barriers such as kerbs, walls and other relevant urban features. However, the cost of data collection at the required level of detail is unlikely to justify such an approach (Orman, 1996) and models of above-ground flow in urban areas are not yet routinely available. In many cases, the effects of flooding can be minimised by the careful positioning of buildings in relation to the topography and by the sympathetic design of landscaping features.

Table 11.1 Recommended design frequencies (adapted from BS EN 752–4: 1998)

Location	Design storm return period (yr)	Design flooding return period (yr)
Rural areas	1	10
Residential areas	2	20
City centres/industrial/ commercial areas:		
• with flooding check	2	30
• without flooding check	5	–
Underground railways/underpasses	10	50

Major–minor systems (dual drainage)

The philosophy of designing surface features for overland flood flows has been formalised in some countries into a *major–minor* system approach (Wisner and Kassem, 1982). The minor system consists of the traditional drainage hardware such as kerbs, gutters and sewers, to control more frequent storm flows. The major system will generally mimic the natural drainage pattern prior to urbanisation but consist of an arrangement of pavements, road central reservations, swales, flood-ways, retention basins and flood-relief channels acting as a continuous overland flow path or floodway system to safely accommodate more severe flood events.

11.2.3 *Return period and risk*

As mentioned in Chapter 5, the *T* year return period of an annual maximum rainfall event is defined as the long-term average of the intervals between its occurrence or exceedance. Of course, the actual interval between specific occurrences will vary considerably around the average value *T*, some intervals being much less than *T*, others greater.

The risk that an annual event will be exceeded during the lifetime of the drainage system is derived as follows. The probability that, in any one year, the annual maximum storm event of magnitude X is greater than or equal to the *T* year design storm of magnitude x is:

$$P(X \geq x) = \frac{1}{T} \tag{11.1}$$

So, the probability that the event will not occur in any one year is:

$$P(X < x) = 1 - P(X \geq x) = 1 - \frac{1}{T}$$

and the probability it will not exceed the design storm in N years must be:

$$P^N(X < x) = \left(1 - \frac{1}{T}\right)^N$$

The probability or risk r that the event will equal or exceed the design storm at least once in N years is therefore:

$$r = 1 - \left(1 - \frac{1}{T}\right)^N \tag{11.2}$$

If the design life of a system is N years, there is a risk r that the design storm event will be exceeded at some time in this period. The magnitude of the risk is given by equation 11.2. Example 11.1 explains how they may be used.

11.3 Contributing area

The following characteristics of a contributing area are significant for storm sewers: physical area, shape, slope, soil type and cover, land-use, roughness, wetness and storage. Of these, the catchment area and land-use are the most important for good prediction of stormwater runoff.

11.3.1 Catchment area measurement

The boundaries of the complete catchment to be drained can be defined with reasonable precision either by field survey or use of contour maps.

Example 11.1

What is the probability that at least one 10 year storm will occur during the first 10 year operating period of a drainage system? What is the risk over the 40 year lifetime of the system?

Solution

First ten years: $T = 10$, $N = 10$.
The answer is not:

$r = 1/T = 0.1,$

nor

$r = 10 \times 1/T = 1.0$

It is (equation 11.2):

$r = 1-(1-0.1)^{10} = 0.651$

Thus, there is a 65% probability that at least one 10 year design storm will occur within 10 years. In fact, it can be shown that, for large T, there is 63% risk that a T-year event will occur within a T-year period.
Lifetime: $T = 10$, $n = 40$.

$$r = 1 - \left(1 - \frac{1}{10}\right)^{40} = 0.985$$

In general, a very high return period is required if risk is to be minimised over the lifetime of the system.

They should be positioned such that any rain that falls within them will be directed (normally under gravity) to a point of discharge or outfall.

After the preliminary sewer layout has been produced, the catchment can be divided up into sub-catchment areas draining towards each pipe or group of pipes in the system. The sub-areas can then be measured by planimeter if using paper maps, or automatically if using a GIS-based package. Aerial photographs may also be used. For simplicity, it is assumed that all flow to a sewer length is introduced at its head (that is, at the upstream manhole).

11.3.2 Land-use

Once the total catchment area has been defined, estimates must be made of the extent and type of surfaces that will drain into the system. The *percentage imperviousness* (PIMP) of each area is measured by defining impervious surfaces as roads, roofs and other paved surfaces (equation 6.5). Measurement can be done manually from maps or automatically from aerial photographs (Finch *et al.*, 1989; Scott, 1994). Table 11.2 and Fig. 11.1 illustrate a land-use classification in London.

Alternatively, the percentage impermeable area (PIMP) can be related approximately to the density of housing development using the following relationship:

$$PIMP = 6.4\sqrt{J} \qquad 10 < J < 40 \qquad (11.3)$$

where J is the housing density (dwellings/ha).

11.3.3 Runoff coefficient

The dimensionless *runoff coefficient* C has already been defined in Chapter 6 as the proportion of rainfall that contributes to runoff from the surface. Early workers such as Lloyd-Davies (1906) assumed that 100% runoff came from impervious surfaces and 0% from pervious surfaces, so $C = PIMP/100$ and this assumption is still commonly adopted in the UK.

Table 11.2 Approximate percentage imperviousness of land-use types in London

Land-use category	PIMP
Dense commercial	100
Open commercial	65
Dense housing	55
Flats	50
Medium housing	45
Open housing	35
Grassland	<10
Woodland	<10

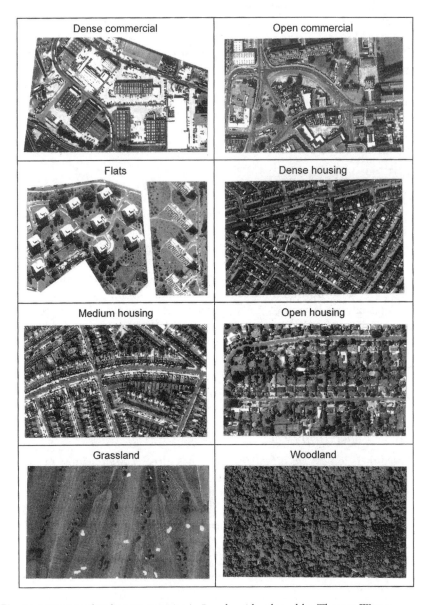

Fig. 11.1 Various land use categories in London (developed by Thames Water Utilities Ltd on their Beckton and Crossness sewerage modelling projects in association with consulting engineers BGP Reid Crowther Ltd and Montgomery Watson Ltd and reproduced with permission)

However, the coefficient actually accounts for the initial runoff losses (e.g. depression storage), continuing losses (e.g., surface infiltration) and implicitly accounts for the hydrodynamic effects encountered as the water flows over the catchment surface. Therefore, C must be related to *PIMP*, but not necessarily equal to it – some runoff will come from pervious surfaces, for example. Equation 6.4 in Chapter 6 shows clearly that $C = PR/100$ is related to *PIMP* plus soil type and antecedent conditions. So considerable knowledge of the catchment is required for accurate determination.

For design purposes, standard values of C such as those in Table 11.3 are often used. Weighted average coefficients are needed for areas of mixed land use.

11.3.4 Time of concentration

An important term used in storm sewer design is *time of concentration* t_c. It is defined as the time required for surface runoff to flow from the remotest part of the catchment area to the point under consideration. Each point in the catchment has its own time of concentration. It has two components, namely the overland flow time, known as the *time of entry* t_e, and the channel or sewer flow time, the *time of flow* t_f. Thus:

$$t_c = t_e + t_f \tag{11.4}$$

Time of entry

The time of entry will vary with catchment characteristics such as surface roughness, slope and length of flow path together with rainfall characteristics. Table 11.4 shows ranges of values dependent on storm return period; rarer, heavier storms produce more water on the catchment surface and, hence, faster overland flow.

Table 11.3 Typical values of runoff coefficient in urban areas (adapted from Urban Water Resources Council, 1992)

Area description	Runoff coefficient	Surface type	Runoff coefficient
City centre	0.70–0.95	Asphalt and concrete paving	0.70–0.95
Suburban business	0.50–0.70	Roofs	0.75–0.95
Industrial	0.50–0.90	Lawns	0.05–0.35
Residential	0.30–0.70		
Parks and gardens	0.05–0.30		

Table 11.4 Time of entry (after DoE/NWC, 1981)

Return period (yr)	Time of entry (min)
1	4–8
2	4–7
5	3–6

Time of flow

Velocity of flow in the sewers can be calculated from the hydraulic properties of the pipe, using one of the methods described in Chapter 8. Pipe-full velocity is normally used as a good approximation over a range of proportional depths. If sewer length is known or assumed, time of flow can be calculated.

11.4 Rational Method

The Rational Method has a long history dating back to the middle of the 19th century. The Irish engineer Mulvaney (1850) was probably the first to publish the principles on which the method is based, although Americans tend to credit Kuichling (1889) and the British credit Lloyd-Davies (1906) for the method itself. The method and its further development are described below.

11.4.1 Steady state runoff

Consider a simple, flat, fully impervious rectangular catchment with area A. A depth of rain, I, falls in a time, t. If there were also an impervious wall along the edges of the catchment, and no sewers, this rain would simply build up over the area to a depth, I. The volume of water would be $I \times A$.

Now imagine that the runoff is flowing into a sewer inlet at point X with steady state conditions: water is landing on the area, and flowing away, at the same rate. The sewer will carry the volume of rain ($I \times A$) at a steady, constant rate over the time (t) of the rainfall. So for flow-rate (Q):

$$Q = \frac{IA}{t}$$

and since the intensity of rain, $i = I/t$:

$$Q = iA$$

Now, since catchments are not totally impervious, and there will be initial and continuing losses, the runoff coefficient C can be introduced, to give:

$$Q = CiA \tag{11.5a}$$

Adjusting for commonly used units gives:

$$Q = 2.78CiA \qquad\qquad (11.5b)$$

Q maximum flow rate (l/s)
i rainfall intensity (mm/h)
A catchment area (ha)

11.4.2 Critical rainfall intensity

For this method to be used for design purposes, the rainfall intensity that causes the catchment to operate at steady state needs to be known. This should give the maximum flow from the catchment. The Rational Method states that a catchment just reaches steady state when the duration of the storm (and hence intensity i) is equal to the time of concentration of the area.

But why is this? Fig. 11.2(a) shows a hydrograph resulting from uniform rainfall with duration less than the time of concentration. Fig. 11.2(b) gives the hydrograph for the same catchment, resulting from the same uniform rainfall intensity, but this time with infinite duration. The peak flow on Fig. 11.2(a) is the lower because the entire catchment is not contributing together (at steady state): contributions from remote parts of the catchment are still arriving after contributions from near parts have ceased. The maximum flow is reached when all the catchment contributes together, i.e. when time is equal to or greater than the time of concentration t_c, as in Fig. 11.2(b).

The basis of the Rational Method is, therefore, an engineering 'worst case'. The duration of the storm must be at least the time of concentration; otherwise, the maximum flow would not be reached. However, it should not be longer, because storms with longer durations have statistically

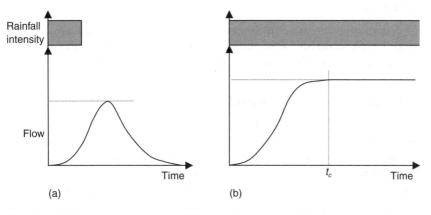

Fig. 11.2 Hydrograph response to different duration rainfall of same intensity

lower intensities (see Fig. 5.2). Therefore the worst case is when the duration is equal to the time of concentration (see Example 11.2).

Example 11.2

A new housing estate is to be drained by a separate storm sewer network. The estate is rectangular in plan, 1200 × 900 m and will consist of approximately 30% paved and roofed surfaces. Determine the maximum capacity required of the sewer carrying stormwater from the whole area. The longest branch leading to this point is 1350 m.

Assume that the average velocity is 1.5 m/s and the time of entry is 4 minutes. Use the 'Ministry of Health' formula (equation 5.2) for rainfall calculation.

Solution

$$t_c = 4 + \frac{1350}{1.5 \times 60} = 19 \text{ min} \qquad i = \frac{750}{19 + 10} = 26 \text{ mm/h}$$

PIMP = 30%

Assume 100% runoff from impervious areas, 0% from pervious areas hence
C = PIMP/100 = 0.3

$$Q = CiA$$

$$Q = 0.3 \frac{26 \times 10^{-3}}{60 \times 60} 1200 \times 900 = 2.34 \text{ m}^3/\text{s}$$

Small areas

A fixed rainfall intensity of 50 mm/h is often used for small areas (main sewer length <200 m). This avoids using inappropriately high intensities calculated using very low concentration times.

11.4.3 Modified Rational Method

Increased understanding of the rainfall-runoff process has led to further development of the Rational Method. The Modified Rational Method is recommended in the *Wallingford Procedure* (DoE/NWC, 1981), and shown to be accurate for UK catchment sizes up to 150 ha.

In this approach, the runoff of rainfall is disaggregated from other routing effects. Thus, the runoff coefficient C is considered to consist of two components:

$$C = C_v C_R \tag{11.6}$$

C_v volumetric runoff coefficient (–)
C_R dimensionless routing coefficient (–)

Volumetric runoff coefficient (C_v)

This is the proportion of rainfall falling on the catchment that appears as surface runoff in the drainage system. The value of C_v depends on whether the whole catchment is being considered, or just the impervious areas alone. Assuming the latter:

$$C_v = \frac{PR}{PIMP} \tag{11.7}$$

where *PR* is given by equation 6.4 in Chapter 6 and *PIMP* is given by equation 6.5. Under summer rainfall conditions, C_v ranges from 0.6–0.9, with the lower values pertaining to rapidly-draining soils and higher values to heavy clay soils. Note that in this method C_v is calculated not assumed.

Dimensionless routing coefficient (C_R)

The dimensionless routing coefficient C_R varies between 1 and 2, and accounts for the effect of rainfall characteristics (e.g. peakedness) and catchment shape on the magnitude of peak runoff. A fixed value of 1.30 is recommended for design. So, for peak flow Q_p:

$$Q_p = 2.78 \times 1.30 C_v i A_i$$

where A_i is the impervious area (ha).

$$Q_p = 3.61 C_v i A_i \tag{11.8}$$

11.4.4 Design criteria

It is good practice to follow a number of basic criteria during the design process.

Capacity

Storm sewers should be designed (size and gradient) to convey the predicted peak flows. It is conventional to design the pipes to just run full (for example, to $d/D = 1.0$) but not surcharged. The small extra capacity that can be achieved at just below full flow (see Chapter 8) is neglected.

Self-cleansing

In addition to capacity, the pipe should also be designed to achieve self-cleansing. This is achieved by ensuring a specific velocity is reached at the design flow (1.0 m/s is typically used). In practice, the pipe size and gradient are manipulated together to obtain the best design.

For sewers designed for capacity with long return period storms (e.g. 10+ years), a more conservative approach is to design for self-cleansing with more frequent events (e.g. 1 year).

Roughness

As with foul sewers, it is conservatively assumed for design purposes that the pipe roughness will be independent of pipe material, although sliming is not a major issue. BS EN 752–4 recommends a k_s value of 0.6 mm for storm sewers.

Minimum pipe sizes

The minimum pipe size is generally set at similar levels to those for foul sewers (see Chapter 10).

11.4.5 Design method

The following procedure should be followed for the Modified Rational Method:

1 Assume design rainfall return period (T), pipe roughness (k_s), time of entry (t_e) and volumetric runoff coefficient (C_v).
2 Prepare a preliminary layout of sewers, including tentative inlet locations.
3 Mark pipe numbers on the plan according to the convention described in Chapter 7.
4 Estimate impervious areas for each pipe.
5 Make a first attempt at setting gradients and diameters of each pipe.
6 Calculate pipe-full velocity (v_f) and flow-rate $(Q_f = \pi D^2 v_f/4)$.
7 Calculate time of concentration from equation 11.4. For downstream pipes, compare alternative feeder branches and select the branch resulting in the maximum t_c.
8 Read rainfall intensity from IDF curves (see Chapter 5) for $t = t_c$ (for design T).
9 Estimate the cumulative contributing impervious area.
10 Calculate Q_p from equation 11.8.
11 Check $Q_p < Q_f$ and $v_{max} > v_f > v_{min}$.
12 Adjust pipe diameter and gradient as necessary (given hydraulic and physical constraints) and return to step 5.

This is essentially a manual calculation procedure, although software packages are available to automate the repetitive calculations (e.g. WinDes by Micro Drainage). Example 11.3 illustrates how the method may be used to design a simple storm sewer network.

Example 11.3

A simple storm sewer network is shown in Fig. 11.3. Appropriate rainfall data is given in Table 11.5 and network data in the shaded portion of Table 11.6. Assume pipe gradients are fixed.

Design the network using the Modified Rational Method for a 2 year return period storm using a volumetric runoff coefficient of 0.9 and a time of entry 5 min. Pipe roughness is 0.6 mm.

Table 11.5 Rainfall intensities (mm/h) at network site

Duration (min)	Return period (yr)		
	1	2	5
6.0	50.0	61.7	81.6
6.2	49.3	60.7	80.4
6.4	48.5	59.8	79.2
6.6	47.8	58.9	78.1
6.8	47.1	58.0	77.0
7.0	46.4	57.2	75.9
7.2	45.8	56.4	74.9
7.4	45.2	55.6	73.9
7.6	44.5	54.8	72.9
7.8	44.0	54.1	71.9
8.0	43.4	53.4	71.0
8.2	42.8	52.7	70.1
8.4	42.3	52.0	69.3
8.6	41.8	51.4	68.4
8.8	41.2	50.7	67.6
9.0	40.8	50.1	66.8

Solution

Pipe-full velocities and capacities are calculated using the Colebrook-White equation. The design is completed in Table 11.6.

Two points about the methodology are stressed. Firstly, it is important that calculations are carried out for each pipe in turn, and that area and time of concentration refer to the whole upstream contributing area not just to the local sub-catchment area. The second point is that each pipe will be designed for a *different* (critical) design storm, with shorter duration, higher intensity storms used for upstream pipes (because they have a shorter time of concentration) and longer duration, lower intensity storms used for downstream sections.

Table 11.6 Design table for Modified Rational Method

Return period 2 yr
Pipe roughness 0.6 mm
Time of entry 5.0 min
Coefficient, C_v 0.9

Pipe number	Pipe length (m)	Pipe gradient (1:x)	Impervious area (ha)	Sum impervious area (ha)	Assumed diameter (mm)	Pipe-full velocity (m/s)	Pipe capacity (l/s)	Time of flow (min)	Time of concentration (min)	Rainfall intensity (mm/h)	Calculated discharge (l/s)	Comments
1.0	120	200	0.4	0.4	225	0.9	35	2.2	7.2	56.4	73.3	$Q_p > Q_f$
					300	1.1	80	1.8	6.8	58.0	75.4	OK
2.0	100	200	0.6	0.6	300	1.1	80	1.5	6.5	59.4	115.8	$Q_p > Q_f$
					375	1.3	140	1.3	6.3	60.3	117.5	OK*
1.1	150	400	0.8	1.8	600	1.2	350	2.1	8.9	50.4	295	OK*

* More cost-effective solutions could be arrived at by varying pipe gradient

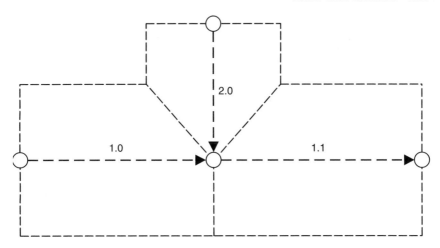

Fig. 11.3 System layout (Example 11.3)

11.4.6 Limitations

The Rational Method is based on the following assumptions.

1 The rate of rainfall is constant throughout the storm and uniform over the whole catchment.
2 Catchment imperviousness is constant throughout the storm.
3 Contributing impervious area is uniform over the whole catchment.
4 Sewers flow at constant (pipe-full) velocity throughout the time of concentration.

Assumption 1 can underestimate, as can assumption 3 (this will be explored further in the next section). On the other hand, assumption 2 tends to overestimate, as does asumption 4 – sewers do not always run full, and storage effects reduce peak flow. Fortunately, in many cases, these inaccuracies cancel each other out, producing a reasonably accurate result. Thus the Rational Method, and its modified version, are simple, widely used approaches suitable for first approximations in most situations and appropriate for full design in small catchments (<150 ha).

11.5 Time–area Method

11.5.1 The need

Area is treated as a constant in the Rational Method. In reality, the contributing area is not constant. For example, during the beginning of rainfall the area builds up with time, closest surfaces contributing first, more distant ones later. In many cases, the Rational Method is appropriate even though it does not take this type of effect into account; in others it is not. This can be demonstrated by considering the three simple cases below.

Case (i)

The main storm sewers for a proposed industrial estate are shown in plan on Fig. 11.4(i). The capacity required at X is calculated using the Rational Method. Assume a catchment of area $A = 100\,000$ m², $C = 0.6$, $t_e = 4$ minutes, $v_f = 1.5$ m/s and length of the longest sewer is 450 m. Now, by using the 'Ministry of Health' formula (equation 5.2 in Chapter 5) for rainfall intensity (i) related to duration (D):

$$i = \frac{750}{D + 10}$$

We have:

$$t_c = 4 + [450/(1.5 \times 60)] = 9 \text{ mins}$$

$$i = \frac{750}{9 + 10} = 39 \text{ mm/h.}$$

$$Q = \frac{39}{1000 \cdot 60 \cdot 60} \times 100\,000 \times 0.6 = 0.65 \text{ m}^3\text{/s}$$

Case (ii)

An alternative layout (Fig. 11.4(ii)) is now considered in which the area is increased and one of the sewers is extended. How much additional flow will there be at X?

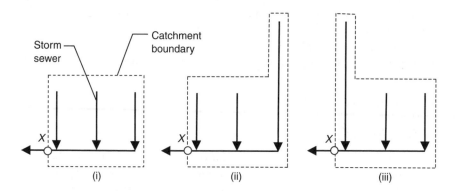

Fig. 11.4 Layout of storm sewers

The same rainfall formula is used, C is still 0.6, the assumed velocity and time of entry are unchanged. However, now A has been increased to 112 000 m^2 and the length of longest sewer increased to 720 m. This gives:

$$t_c = 4 + 720/(1.5 \times 60) = 12 \text{ mins}$$

$$i = \frac{750}{12 + 10} = 34 \text{ mm/h.}$$

$$Q = \frac{34}{1000 \cdot 60 \cdot 60} \times 112\,000 \times 0.6 = 0.63 \text{ m}^3\text{/s}$$

It appears that the flow at X is now less, even though the area has increased! So, it is the increase in the time of concentration, and consequent reduction in i, rather than the increase in area, which has had the greatest effect on Q.

In this case, the Rational Method is inappropriate because it does not consider the 'worst case'. If rainfall lasting 9 minutes fell on this catchment, the whole of the original 100 000 m^2 would contribute together, and a flow of 0.65 m^3/s, as determined for Case (i) would be produced. The value for flow of 0.63 m^3/s would underestimate the capacity required.

Case (iii)

Now consider another alternative. In Fig. 11.4(iii), A has been increased to 112 000 m^2 (again), but because of the shape of the extended catchment, the length of longest sewer is no greater than it was for Case (i). Therefore i remains as 39 mm/hr.

$$\text{So } Q = \frac{39}{1000 \cdot 60 \cdot 60} \times 112\,000 \times 0.6 = 0.73 \text{ m}^3\text{/s}$$

This is an increase on 0.65 m^3/s, which makes more sense! The difference is that in Case (iii) the extra area contributes rapidly (because it is close to X) and does not increase the time of concentration. So the key to whether the Rational Method gives appropriate answers or not is *the way the contributing area builds up with time*.

11.5.2 Basic diagram

A time–area diagram attempts to overcome one of the main limitations of the basic Rational Method by representing the rate of contribution of area (CA). Now for Case (i) above, if the contributing area discharging towards point X builds up at a constant rate, the time–area diagram will be as shown in Fig. 11.5(i).

It is possible to determine whether or not the Rational Method is appropriate for any particular shape of time–area diagram by re-arranging the basic rational expression:

$$Q = CAi = CA\left[\frac{750}{D + 10}\right] = 750\left[\frac{CA}{D + 10}\right]$$

Thus, the higher the value of $\left[\dfrac{C.A}{D + 10}\right]$, the greater the value of Q.

How can we visualise $\left[\dfrac{C.A}{D + 10}\right]$? The gradient of the straight line representing the time–area relationship on Figure 11.5(i) is $\left[\dfrac{C.A}{D}\right]$, but if the point $(D = -10, A = 0)$ is connected to any point on the time–area relationship, a line is produced with a gradient of $\left[\dfrac{C.A}{D + 10}\right]$. The highest value of this gradient, and therefore the highest value of Q, is given by the dashed line on Fig. 11.5(i).

The time–area diagram for Case (ii) is given on Fig. 11.5(ii). If the dashed line is drawn through point 1 rather than point 2, it will have the steepest gradient, and therefore the value of Q will be greatest.

The line through 1 looks like a tangent, and is the basis of a method called the *Tangent Method* (Reid, 1927), which has now been largely superseded. However, it still provides a means for determining whether the Rational Method gives the worst case for a particular time–area diagram. When a tangent of this type can be drawn, it means that the Rational Method has *not* given the 'worst case'.

The time–area diagram for Case (iii) is given in Fig. 11.5(iii). No tangent can be drawn in this case, therefore the Rational Method *does* give the worst case. (Note that the tangents in Figs 11.5(i)–(iii) were drawn from time $= -10$ minutes only because of the particular formula used for rainfall.)

11.5.3 Diagram construction

The diagram is used for storm sewer design by assuming that the time–area plot for each individual pipe sub-catchment is linear. However, the design of each pipe is not concerned just with the local sub-catchment (in a similar way to the Rational Method) but also with the 'concentrating' flows from upstream pipes. The combined time–area diagram for each pipe

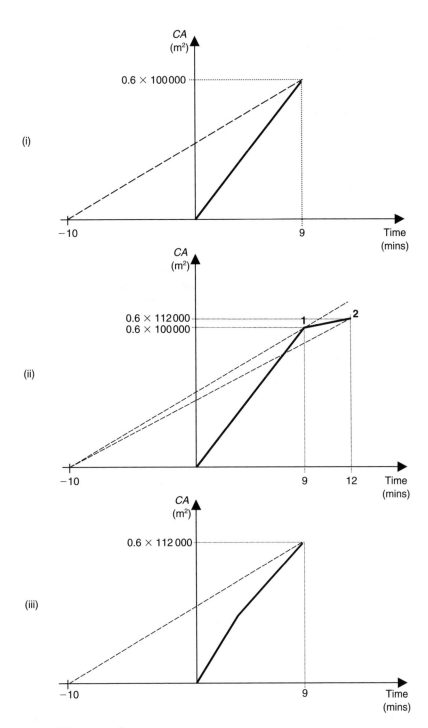

Fig. 11.5 Time–area diagrams

can be produced using the principle of linear superposition. This is illustrated by the case of a simple two-pipe system (and assuming $C = 1$) in Fig. 11.6a.

In this network, the time of concentration of pipe 1 is $t_{c(1)}$, and this is plotted directly onto the time–area plot (Fig. 11.6b). Pipe 2 has a sub-catchment time of concentration of $t_{c(2)}$, relative to its own outfall. However, this diagram is not directly overlaid on the previous one, but is lagged by the time of flow in pipe 1, $t_{f(1)}$. The ordinates of the two separate diagrams are added to produce the complete diagram.

The resulting time–area diagram is made up of linear segments but, as more individual pipes are added, the shape tends to become non-linear. Example 11.4 illustrates a more complex network.

11.6 Hydrograph methods

One of the major weaknesses of the Rational Method is that it only produces worst-case design flow and not a hydrograph of flow against time. Hydrograph methods have been developed to overcome this limitation.

11.6.1 Time–area Method

The Time–area Method uses the time–area diagram to produce not only a peak design flow, but also a flow hydrograph. The method also allows straightforward use of time-varying rainfall – the design storm (see Chapter 5).

Equation 6.8 from Chapter 6 is repeated below, and this gives the basic equation for finding flow $Q(t)$ when a continuous time–area diagram is combined with rainfall depth increments, $I_1, I_2, \ldots I_N$.

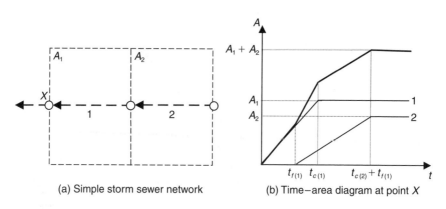

(a) Simple storm sewer network   (b) Time–area diagram at point X

Fig. 11.6 Example construction of time–area diagram: (a) Simple storm sewer network; (b) Time–area diagram for point X

Example 11.4

A storm sewer network has been designed initially with the layout shown in Fig. 11.7, and data given (in shaded section) in Table 11.7. Construct the time–area diagram at the network outfall assuming a time of entry of 2 mins and a Manning's n of 0.010.

Solution

For simplicity, assume the pipes run full to calculate flow velocities and, hence, time of flow.

Table 11.7

Pipe number	Diameter (mm)	Pipe length (m)	Pipe gradient (1:x)	Impervious area (ha)	Pipe-full velocity (m/s)	Time of flow (min)
1.0	375	1000	300	2.5	1.2	13.9
1.1	450	500	400	0	1.2	7.1
2.0	300	620	250	2.5	1.1	9.2
1.2	600	300	500	1.0	1.3	4.0

Time of flow data and time of entry information is used to derive the time–area diagram (Fig. 11.8):

1.2 $t_{c(1.2)}$ = 6.0 mins,   flow starts to contribute @ t = 0 mins
2.0 $t_{c(2.0)}$ = 11.2 mins, $t_{f(1.2)}$ = 4.0 mins
1.0 $t_{c(1.0)}$ = 15.9 mins, $t_{f(1.1)} + t_{f(1.2)}$ = 11.1 mins

$$Q(t) = \sum_{\omega=1}^{N} \frac{dA(j)}{dt} I_\omega \qquad (11.9)$$

$Q(t)$ runoff hydrograph ordinate at time t (m³/s)
$dA(j)/dt$ slope of the time–area diagram at time j (m²/s)
I_ω is the rainfall depth in the ωth of N blocks of duration Δt (m)
j $t-(\omega-1)\,\Delta t(s)$

Assuming linear incremental change in the time–area diagram, ΔA_1, $\Delta A_2 \ldots \Delta A_j \ldots$ over rainfall time blocks, $\Delta t_1, \Delta t_2 \ldots \Delta t_j \ldots$, runoff is given by:

$$Q(t) = \sum_{\omega=1}^{N} \Delta A_j i_{e\omega} \qquad (11.10)$$

where i_e is $I/\Delta t$.

This can be expanded to give:

$$Q(1) = A_1 i_1$$
$$Q(2) = A_2 i_1 + A_1 i_2$$
$$Q(3) = A_3 i_1 + A_2 i_1 + A_1 i_3$$
$$\ldots$$

The method is summarised as follows:

* Select a suitable integer time interval Δt, typically $t_c/10$.
* Prepare a suitable rainfall hyetograph using Δt as the time interval. For each design point under consideration:
* Produce a time–area diagram.

Calculate outflow by reading off the relevant rainfall intensity from the hyetograph (i_l) and contributing area (A_j) from the time–area diagram for each time increment (Δt). These need to be accumulated for each time step according to equation 11.10.

Example 11.5 illustrates how the method may be used to design a simple storm sewer network.

Limitations

This method has the advantage over the Rational Method in that it takes some account of the shape of the catchment, allowing an output hydrograph to be produced and to include the effects caused by time-varying rainfall. However, the method still only allows linear translation of the flood wave through the catchment and makes no allowance for storage effects.

11.6.2 TRRL Method

Dissatisfaction with the Rational and Time–area Methods led to the development of the Transport and Road Research Laboratory (TRRL) Method (Watkins, 1962). During development, data from some 286 storms on 12 varied catchments was analysed. The TRRL method is based on the time–area approach to runoff estimation and includes the Lloyd-Davies assumptions of considering 100% runoff from impermeable areas and using pipe-full velocity as the routing velocity. However, it has improvements in the calculation of the pipe hydraulics. Specifically, storage in the network is taken into account in a relatively simple way.

Example 11.5

Using the time–area diagram developed in Fig. 11.8, evaluate the outfall flow hydrograph for the sewer network in Fig. 11.7 under the following design storm.

Time (min)	Effective rainfall depth (mm)
0–5	3
5–10	6
10–15	3

Solution

Using a 5 minute time increment ($\Delta t = 5$ min), read off the cumulative contributing area at each time step and then the incremental area. Convert the rainfall depths to intensities.

Δt	Time (mins)	ΣA (ha)	ΔA (ha)	i_e (mm/h)
1	5	1.4	1.4	36
2	10	2.6	1.2	72
3	15	4.2	1.6	36
4	20	4.9	0.7	
5	25	5.7	0.8	
6	30	6.0	0.3	
			Σ 6.0	

From equation 11.10:

$Q(1) = 1.4 \times 36$ $= 50.4$

$Q(2) = 1.2 \times 36 + 1.4 \times 72$ $= 144.0$

$Q(3) = 1.6 \times 36 + 1.2 \times 72 + 1.4 \times 36$ $= 194.4$

$Q(4) = 0.7 \times 36 + 1.6 \times 72 + 1.2 \times 36$ $= 183.6$

$Q(5) = 0.8 \times 36 + 0.7 \times 72 + 1.6 \times 36$ $= 136.8$

$Q(6) = 0.3 \times 36 + 0.8 \times 72 + 0.7 \times 36$ $= 93.6$

$Q(7) = \qquad\quad 0.3 \times 72 + 0.8 \times 36$ $= 50.4$

$Q(8) = \qquad\qquad\qquad\quad 0.3 \times 36$ $= 10.8$

The ordinates should be multiplied by 2.78 to obtain the hydrograph flow in l/s, from which the peak flow is:

$Q_p \approx 550$ l/s @ 17 minutes

Note: this exceeds the capacity of the pipe.

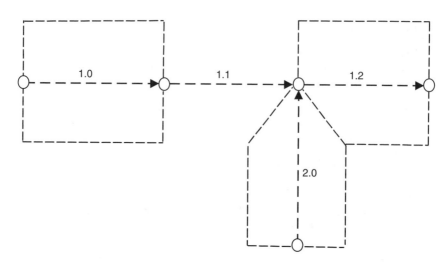

Fig. 11.7 System layout (Example 11.4)

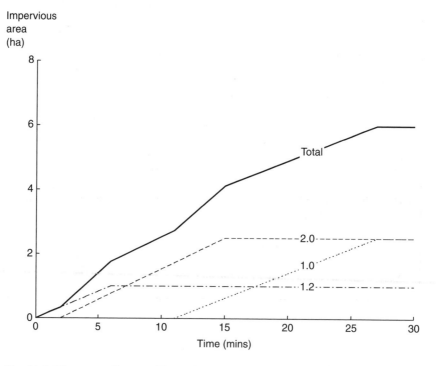

Fig. 11.8 Time–area diagram (Example 11.4)

Flow routing assumption

The flow routing undertaken in the TRRL method is a 'level pool' or 'reservoir' routing technique (see Chapter 13). The method assumes that the retained water acts as a level reservoir of liquid with outflow being uniquely related to water level or storage. This is accomplished in a pipe network by assuming that proportional depth of flow is identical at all points in the system. This assumption is valid if:

• the system is designed with a reasonable degree of taper
• all the pipes in the system are geometrically similar.

Both these requirements should be satisfied in new systems.

Consider an individual circular pipe of length L carrying flow with proportional depth d/D. Thus, $A = f_1(d/D)$ and $V = f_2(d/D)$ where A is the cross-sectional area of flow in the pipe and V is the volume of water stored. Now, if d/D is constant everywhere, $S = f_3(d/D)$ where S is the whole system retention (storage volume). For a given slope and pipe roughness, outflow at the design point, $O = f_4(d/D)$. Therefore:

$$O = f_5(S) \text{ or } S = f_6(O) \tag{11.11}$$

Thus, there is a unique relationship between total volume of water stored (system retention) and the outflow rate at the design point.

So, knowing the inflow I, flow routing can be performed using the basic storage equation to estimate O. Example 11.6 illustrates application of the method.

Example 11.6

Calculate the outflow hydrograph and peak for the sewer network in Fig. 11.7 accounting for in-pipe storage.

Solution

Initially, derive a relationship between S and O for circular pipes. Table 8.5 gives:

$$A = \frac{D^2}{8}(\theta - \sin \theta) = \frac{\pi D^2}{4}\left(\frac{\theta - \sin \theta}{2\pi}\right)$$

where θ is defined in equation 8.17.

Now $\pi D^2/4$ is the cross-sectional area of each pipe. So if L is the length of each pipe of diameter D:

$$S = \left(\Sigma L \frac{\pi D^2}{4}\right)\left(\frac{\theta - \sin \theta}{2\pi}\right)$$

Hence, storage can be derived from the system data:

Pipe number	Diameter (mm)	Pipe length (m)	LD^2
1.0	375	1000	140.6
1.1	450	500	101.3
2.0	300	620	55.8
1.2	600	300	108.0
			Σ 405.7

$$S = 50.7(\theta - \sin \theta)$$

Manning's equation 8.23 gives:

$$O = \frac{1}{n} AR^{2/3}S_o^{1/2}$$

Therefore:

$$O = \frac{1}{n} \frac{D^2}{8}(\theta - \sin \theta)\left(\frac{D}{4}\left[\frac{\theta - \sin \theta}{\theta}\right]\right)^{2/3} S_o^{1/2}$$

$$O = \frac{1}{n} \frac{D^{8/3}}{20.16} \frac{(\theta - \sin \theta)^{5/3}}{\theta^{2/3}} S_o^{1/2} \tag{11.12}$$

For the outfall pipe (1.2), $D = 0.6$ m, $S_o = 0.002$ and $n = 0.010$ can be substituted into equation 11.12. So, by varying θ from $0 \rightarrow 2\pi$ (i.e. d/D: $0 \rightarrow 1$), S and O can be plotted together as in Fig. 11.9(a):

(a)

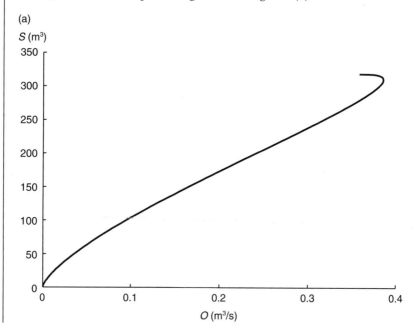

(b)

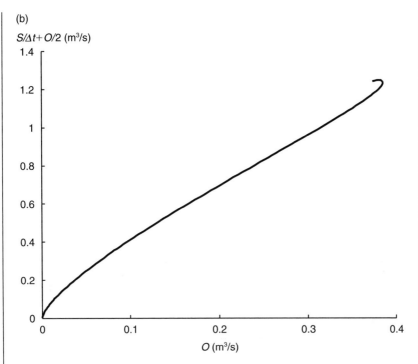

Fig. 11.9 (a) Graph of S against O (Example 11.6);
(b) Graph of $S/\Delta t + O/2$ against O (Example 11.6)

Using a time step of 5 minutes, construct the relationship between $\dfrac{S}{\Delta t} + \dfrac{O}{2}$ and O, also shown (Fig. 11.9(b)).

The calculation now follows the procedure in Example 13.2, with the result shown in Figure 11.10.

The 'routed' peak flow is seen to be:

$$Q_p \approx 370 \text{ l/s @ 25 minutes}$$

which is a considerable reduction from the previous estimate, and now within the capacity of the pipe.

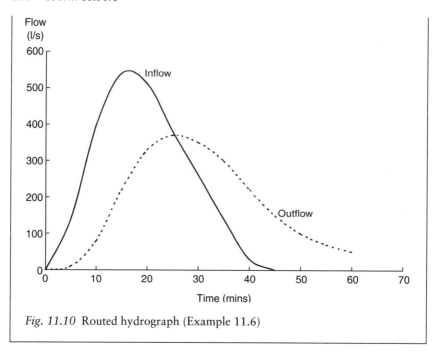

Fig. 11.10 Routed hydrograph (Example 11.6)

Limitations

The TRRL is an advance over the basic time–area method in that it takes pipe storage effects directly into account. However, the routing assumption is relatively crude, and indeed no account is taken of surface storage effects, and surcharge effects cannot be handled.

Coverage of these aspects effectively requires flow simulation models that have greater power than any of the design-orientated tools described in this chapter, including TRRL. These are described in Chapter 19.

Problems

11.1   Describe and justify the main stages in storm sewer design.

11.2   'The frequency of rainfall is neither equivalent to the frequency of runoff nor of flooding.' Discuss this statement with reference to recommended design storm return periods.

11.3   If a storm sewer surcharges during heavy rainfall, was it under-designed? Explain your answer.

11.4   A storm sewer network has been designed based on a 2-year return period storm for a 50 year design life. What is the probability that the network will:
   a)   surcharge at least once in 2 years (the risk of failure)?
   b)   surcharge at least once during its design life?

c) flood in any one year, assuming flooding is caused by the 10-year return period storm?

d) flood at least once in 10 years? [0.75, 1.00, 0.10, 0.65]

11.5 What is percentage imperviousness and how is it related to the runoff coefficient?

11.6 Explain what you understand by time of entry, time of flow and time of concentration. Why is the duration of the design storm in the Rational Method taken as the time of concentration?

11.7 Explain the concept of the Rational Method. What are its main limitations?

11.8 A small separate storm sewer network has the following character-istics (Fig. 11.11):

Sewer	Length (m)	Contributing impervious area (m²)
1.0	180	2000
2.0	90	6000
3.0	90	9000
1.1	90	4000

Use the Rational Method to determine the capacity required for each pipe in the network. Assume a time of entry of 4 minutes, that the pipe-full velocity in each pipe is 1.5 m/s and that design rainfall intensities can be determined from 'Ministry of Health' formulae. Further, assume 100% runoff from impervious areas.

[26, 83, 125, 257 l/s]

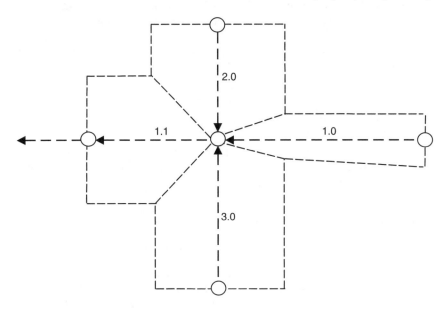

Fig. 11.11 System layout (Problem 11.8)

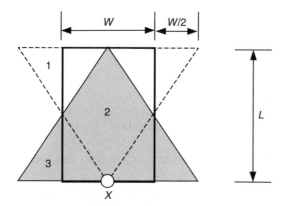

Catchment 1: dashed triangle
Catchment 2: rectangle
Catchment 3: shaded triangle

Fig. 11.12 Three catchments (Problem 11.10)

11.9   What is a time–area diagram? Explain how it is used in the Time–Area Method.

11.10 Construct the time–area diagram (at point X) for each of the equally sized and graded catchments (1–3) shown.

11.11 For the network in Problem 11.8, draw a time–area diagram and use it to check the capacity of pipe 1.1 using the Tangent Method.

[267 l/s]

11.12 Estimate the hydrograph at the outlet to pipe 1.1 (Problem 11.8) for a short storm with the following profile.

Time (min)	0–1	1–2	2–3
Intensity (mm/h)	20	28	64

Use the Time–Area Hydrograph Method.   [Peak 127 l/s]

11.13 Describe the basis of the TRRL method and the reason for its development. What are its limitations?

Key sources

Bartlett, R.E. (1981) *Surface Water Sewerage*, 2nd edn, Applied Science.

Colyer, P.J. and Pethick, R.W. (1976) *Storm Drainage Design Methods. A Literature Review*, Report INT 154, Hydraulics Research Station.

DoE/NWC (1981) *Design and Analysis of Urban Storm Drainage. The Wallingford Procedure*, Four volumes, Standing Technical Committee Report No. 28.

Hall, M.J. (1984) *Urban Hydrology*, Elsevier.

Rickard, C. (2002) Urban drainage, in *Flood Risk Management* (ed., G. Fleming), Thomas Telford, 91–114.

Stephenson, D. (1981) *Stormwater Hydrology and Drainage*, Elsevier.

References

BS EN 752–4: 1998 *Drain and Sewer Systems Outside Buildings. Part 4: Hydraulic Design and Environmental Considerations.*

Finch, J., Reid, A. and Roberts, G. (1989) The application of remote sensing to estimate land cover for urban drainage catchment modelling. *Journal of the Institution of Water and Environmental Management*, 3, Dec, 558–563.

Kuichling, E. (1889) The relation between the rainfall and the discharge of sewers in populous district. *Transactions of American Society of Civil Engineers*, 20, 1–56.

Lloyd-Davies, D.E. (1906) The elimination of storm water from sewerage systems. *Proceedings of Institution of Civil Engineers*, 164(2), 41–67.

Mulvaney, T.J. (1850) On the use of self-registering rain and flood gauges in making observations on the relation of rainfall and flood discharges in a given catchment. *Transactions of Institution of Civil Engineers Ireland*, 4(2), 18.

Orman, N. (1996) *Predicting flooding using hydraulic models*, WaPUG User Note No. 29. www.wapug.org.uk.

Reid, J. (1927) The estimation of storm-water discharge, *Journal of Institution of Municipal Engineers*, 53, 997–1021.

Scott, A. (1994) Low-cost remote-sensing techniques applied to drainage area studies. *Journal of Institution of Water and Environmental Management*, 8, Oct, 498–501.

Transport and Road Research Laboratory (1976) *A Guide for Engineers to the Design of Storm Sewer Systems*, 2nd edn, Road Note 35, Department of the Environment.

Urban Water Resources Council (1992) *Design and Construction of Urban Stormwater Management Systems*, ASCE Manual No. 77, WEF Manual FD-20.

Watkins, L.H. (1962) *The Design of Urban Storm Sewers*. Road Research Technical Paper No. 55, HMSO.

Wisner, P.E. and Kassem, A.M. (1982) Analysis of dual drainage systems by OTTSWMM, in *Urban Drainage Systems* (eds, Featherstone, R.E. and James, A.), Pitman Books, 2/93–2/108.

WRc (2001) *Sewers for Adoption – a Design and Construction Guide for Developers*, 5th edn, Water UK.

12 Combined sewers and combined sewer overflows

12.1 Background

Combined sewer systems have already been discussed extensively in Chapter 2. A significant percentage of sewer systems in many countries are combined, which makes them an important topic in urban drainage. Indeed, there are few parts of this book that do not relate in some way to combined sewers. The essential features of combined sewers are that they carry both wastewater and stormwater in the same pipe, that it is not usually feasible for this pipe to be designed to carry the full combined flow at all times to treatment, and that, therefore, at high flow-rates it is necessary for some of the flow to be discharged to a watercourse at a combined sewer overflow (CSO), as illustrated in Fig. 2.2. This chapter deals with the special characteristics of combined sewers, and in particular with combined sewer overflows.

12.2 System flows

The inflow to a combined sewer system consists of both wastewater (see Chapter 4) and stormwater (see Chapter 6). At the point of inflow, the flow-rates can be calculated from the methods given in Chapters 10 and 11.

A typical layout of a small combined sewer system including a CSO is given in Fig. 12.1. All connections of stormwater and wastewater are made to the single combined sewer. Upstream of the CSO, the pipe carries the full combination of stormwater and wastewater from the upstream catchment. If the combined flow does not exceed the setting of the CSO, all continues to the wastewater treatment plant. If the combined flow does exceed the CSO setting, there will be overflow to the stream and the flow retained in the system downstream will be determined by the CSO setting. Therefore, at different points throughout a combined sewer system in storm conditions, there can be dramatic differences in both the rate and composition of flow.

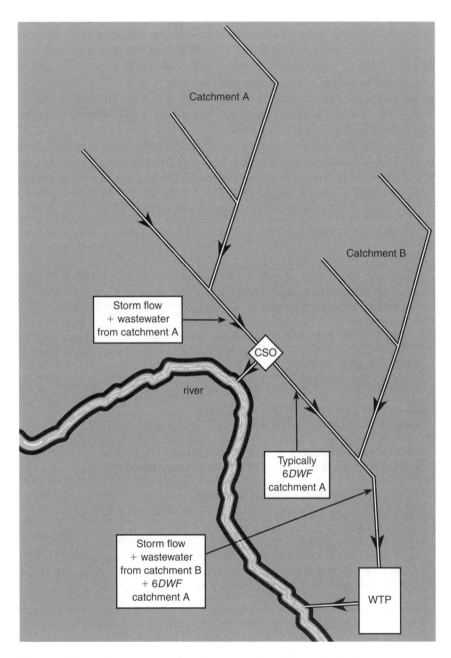

Fig. 12.1 Typical layout of combined sewer system (schematic plan)

Low flow-rates

A combined sewer pipe has a significantly larger diameter than the foul sewer in a separate system draining a catchment of the same size (since the combined sewer must have capacity to carry stormwater too). This means that in dry weather, when wastewater flow-rate is relatively low, the combined sewer (compared with the foul sewer) will have lower flow depths and greater contact between the liquid and the pipe wall – both leading to a greater risk of sediment deposition.

It is partly for this reason that egg-shaped sewers were popular in the past – with a smaller diameter in the lower part of the cross-section for the low flows, and greater area above for storm flows. Egg-shaped sewers are quite common in older systems.

The effect is illustrated and investigated in Example 12.1.

Example 12.1

Determine depth and velocity for a part-full flow-rate of 15 l/s in (i) a 225 mm diameter foul sewer, and (ii) a 600 mm diameter combined sewer. For both take the gradient as 1 in 100, and roughness k_s as 1.5 mm.

Solution

(i) Chart/table gives flow-rate full $Q_f = 45$ l/s
$Q/Q_f = 0.33$, and from Fig. 8.11 $d/D = 0.375$, so depth = 85 mm

Chart/table gives velocity (full) $V_f = 1.1$ m/s
from Fig. 8.11 for $d/D = 0.375$, $v/v_f = 0.9$, so velocity = 0.99 m/s

(ii) Chart/table gives flow-rate full $Q_f = 610$ l/s
$Q/Q_f = 0.025$, and from Fig. 8.11 $d/D = 0.125$, so depth = 75 mm

Chart/table gives velocity (full) $V_f = 2.15$ m/s
from Fig. 8.11 for $d/D = 0.125$, $v/v_f = 0.45$, so velocity = 0.97 m/s.

Note that the larger pipe size does lead to slightly lower depth (and greater area of contact with pipe wall), but the effect on velocity is negligible. This is a general result, and demonstrates that egg-shaped sewers do not have such a significant effect on low-depth velocities as might be expected.

12.3 The role of CSOs

12.3.1 Flow and pollutants

The main function of a CSO is hydraulic: to take an inflow and divide it into two outflows, one to the wastewater treatment plant (the continuation flow, or flow retained) and one to the watercourse (the spill flow) – see Fig. 2.2. The normal means of achieving this is a weir. If the surface of the flow passing through the CSO is below the crest of the weir, flow continues to the WTP only. As the flow-rate increases, so does the level of the water surface, to provide an increase in the hydraulic gradient along the continuation pipe. When the water surface is above the weir crest, some flow passes over the weir while the rest continues to the WTP. Flow-rate over a weir is related to the depth of water above the crest, so if the water surface continues to rise, so does the spill flow. The continuation flow is also likely to rise slightly as a result of the increase in head.

Hydraulic design of a CSO requires care (as will be considered in more detail later). There could be a number of effects of poor hydraulic design. If spill took place prematurely, the capacity of the continuation pipe would be under-used, and an unnecessarily large volume of polluted flow would be discharged to the watercourse. But if the weir was set too high, excessive surcharge of the upstream system might be caused, and too much flow might be forced down the continuation pipe causing flooding elsewhere in the sewer system. In a good hydraulic design, spill will take place at the optimum level, and the continuation flow will not increase greatly while the spill flow is increasing with rising water level.

The other main function of a CSO is related to pollution. The ideal would be that all pollutants continued to the WTP (i.e., were retained within the sewer system), but this is not achieved. Various designs of CSO demonstrate some success in retaining larger solids, but fine suspended and dissolved material tends to be split between continuation flow and spill flow in the same proportion as the split in the flows.

The impact of CSO discharges on receiving waters has been considered in Section 3.4. These impacts are likely to be most serious when CSOs are poorly designed or operating ineffectively. Also, sewers that back up as a result of sediment deposition problems may cause CSOs to operate prematurely (before the inflow has reached the CSO setting), or, in extreme cases, to spill even in dry weather conditions. This may cause serious pollution of receiving waters.

12.3.2 First foul flush

In some systems, a significant feature is the first flush in early storm flows, which may contain particularly high pollutant loads. These are likely to have been derived from the following.

1   *Catchment surface washoff and gully pots.* A first flush from this
    source would be expected as a result of the early rainfall washing off
    pollutants accumulated on the catchment surface and in gully pots
    since the last rainfall.
2   *Wastewater flow.* Since the storm wave moves faster than the waste-
    water 'baseflow', the front of the wave can consist of an ever-increas-
    ing volume of overtaken undiluted baseflow. However, this is
    normally diluted to some extent by the inflow from intermediate
    branches.
3   *Near-bed solids.* In many sewer systems, high concentrations of
    organic solids have been observed in a layer moving just above the
    bed. The added turbulence as the storm flow increases causes these
    solids to become mixed with the stormwater.
4   *Pipe sediments.* Increasing storm flows provide suitable conditions for
    re-erosion of the deposited material.

A first foul flush can be identified on hydrographs and pollutographs
recorded in the system. An obvious sign would be a sharp increase in pollutant concentration near the start of a storm. In fact, even if concentration remained constant as flow-rate increased, this would signify an
increase in pollutant load-rate. A first flush can also be identified by plotting cumulative load against cumulative flow-volume (Fig. 12.2). The 45°

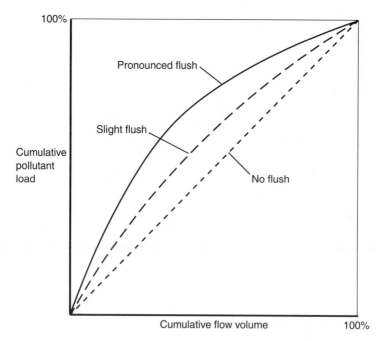

Fig. 12.2 Representation of first foul flush

line indicates that pollutants are uniformly distributed throughout the storm. If the line for a particular storm is above the 45° line, a first flush is suggested. Flushes from different conditions or catchments can be compared in this way. In some catchments, the effect is pronounced, and in others it is not observed at all.

12.3.3 CSO design: problems and solutions

As an introduction to the operation and design of CSOs, this section will briefly consider the problems that might be caused by a poorly designed CSO, and how these can be solved in good designs. There is more detail on the development and design of CSOs in Section 12.5.

Problems

To introduce the general approaches to design, let us first consider a fictitious example of a bad design. Some poor overflows simply consist of an outlet pipe placed at a high level in a manhole to relieve local flooding. This may have been a 'short-term' measure, which has still not been replaced by a long-term solution.

Another example of poor practice is illustrated in Fig. 12.3: a vertical section looking in the direction of flow. The overflow has a single side weir. This arrangement may carry out its hydraulic function satisfactorily, but it is clear that floating solids will tend to flow over the weir and out to

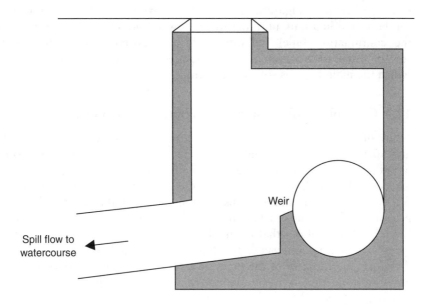

Fig. 12.3 Low side weir CSO

the watercourse. It is almost as if this overflow has been designed as a hydraulic device to skim off the floating solids in this way. In fact, of course, this is the opposite of what would be desirable. Other solids may be retained in the sewer; for example, sinking solids may remain close to the bottom of the pipe as they pass through the structure. But any turbulence or disruption of the flow, caused perhaps by the operation of the overflow, will tend to lift sinking solids into the higher parts of the flow area and potentially over the weir. In fact, many wastewater solids have about the same density as water, and are therefore likely to be overflowed. This design of overflow could not, therefore, be expected to have much success in retaining solids.

Solutions

A number of approaches to the design of CSOs have been developed, and the main ones are described in Section 12.5. Many of the ideas have been simple ones; for example, the problem of floating solids simply flowing straight over the weir described above can be reduced by placing a vertical plate – a scumboard – parallel to the weir, just in front of it, extending vertically above and below the weir crest, so that floating solids are prevented from flowing directly over the weir. The problem of turbulence lifting up the heavier solids can be reduced by creating a chamber in which the flow is slowed down a little, or stilled, to encourage heavy solids to move into the continuation pipe rather than over the weir. Alternatively, circular flow can cause solids to behave in a particular way, and this has been exploited in a number of configurations of 'vortex overflow'. If the first foul flush can be stored in a tank until the storm flow subsides (and then returned to the continuation pipe) pollution will be reduced ('an overflow with storage'). It is also common to incorporate screens within CSO structures (when the main focus is solids retention).

12.4 Control of pollution from combined sewer systems

Many of the landmark developments in urban drainage in the UK in the last 50 years have had combined sewers as their focus, and have established principles of good practice for combined sewer systems, particularly for control of pollution. Some of these developments were referred to briefly in Chapter 1, and are considered in more detail now. It will quickly become clear that the crucial parameter in a combined sewer system is the 'setting' of the CSOs – that is, the flow they retain in the system for treatment. The setting influences the flow-rate expected in the system and, therefore, the capacity of the pipes and, more importantly, determines the flow-rate (and therefore the amount of pollutants) expected to leave the system at CSOs and enter the environment.

The latest advice on improving combined systems recommends choosing from a range of techniques to find the most suitable solution to each problem. The methods below – even the simplest – are still in use, or are available for use when appropriate.

'Technical Committee Formula A'

Until 1970 the traditional CSO setting had been 6 × dry weather flow (6DWF). A very extensive study in the 1950s and 1960s of the effects of CSOs was carried out by the government-appointed 'Technical Committee on Storm Overflows and the Disposal of Storm Sewage' whose Final Report was published in 1970 (Ministry of Housing and Local Government, 1970). Among the conclusions was that it was illogical to base a CSO setting merely on a multiple of DWF, and that, because of the harmful effects of CSO pollution, the new standard setting should give a 'modest improvement' (that is, divert less pollution to watercourses).

The setting of 6DWF had allowed for diurnal variations in wastewater flow, plus some stormwater (so that for low intensity rainfall there would be no overflow). If people in one area happened to use more water than they did in another, that was no reason for more stormwater to be retained in the sewer system. It was considered appropriate for a CSO setting to be based on DWF plus some storm allowance (related to population, but not to water consumption). The Committee also felt there were ambiguities about inclusion of infiltration and industrial flows in traditional practice. The proposed new standard CSO setting was given by their 'Formula A':

$$\text{setting} = DWF + 1360P + 2E \text{ (litres/day)} \qquad \text{(Formula A)}$$

for which $DWF = PG + I + E$

P population
G water consumption per person (litres/day)
I pipe infiltration rate (litres/day)
E average industrial effluent (litres/day)

In a catchment where G is 200 litres/head.day and there is no infiltration or industrial flow, Formula A gives a setting of 7.8DWF (see Example 12.2).

The report recommended that the coefficients could be treated with flexibility. $1360P$ could be decreased where discharge was to a large river, or increased if to a small stream; and industrial effluent of high strength might require an increase in the term $2E$.

The report also contained many other detailed recommendations about CSO design.

Scottish Development Department (SDD)

The report of the Working Party on Storm Sewage (Scotland), *Storm Sewage: Separation and Disposal* (Scottish Development Department, 1977), gave details of another significant study of CSOs. Guidelines for CSO setting and storage volume were related to the amount of dilution when combined sewer flow was overflowed to a watercourse (a factor not covered explicitly in Formula A). Dilution was defined as the 'minimum flow' in the stream (the flow-rate exceeded 95% of the time) compared with the sewer dry weather flow. For dilution of more than 7 to 1, Formula A was considered satisfactory. For dilution of 6 to 1, it was recommended that either the setting should be enhanced to Formula A + 455 P, or that Formula A should be used in its original form and storage should be provided. For any lower dilution, increasing amounts of storage were recommended (Table 12.1), in conjunction with a Formula A setting.

QUALSOC

This emphasis on the capacity of the receiving water to cope with discharges from CSOs was taken further by a procedure developed in the 1980s called QUALSOC (QUALity impacts of Storm Overflows: Consent procedure). It is a calculation based on pollutant mass balances that considers the CSO setting, the concentration of pollutants in the wastewater, the storm flow, the river flow and background pollutant levels, and compares the likely concentrations of pollutants in the river during CSO discharge with acceptable limits.

SRM

The *Sewerage Rehabilitation Manual* will be considered in detail in Chapter 18. The second edition (1986) contained a method of estimating impacts of CSO discharges on watercourses based on use of a sewer system model of flow, but not of quality (since such models did not exist at the time). The *Manual* gave factors which could be multiplied by pollutant concentrations in dry weather flow to give estimated average concentrations in storm flow. These were later revised (Threlfall *et al.*, 1991) and the

Table 12.1 Scottish Development Department (1977) storage recommendations

Dilution	Storage tank capacity (litres)
6 to 1	40 P
4 to 1	40 P
2 to 1	80 P
1 to 1	120 P

P = population

more recent figures are presented in Table 12.2. These can be used, in conjunction with output from a flow model, to give total pollutant loads to watercourses resulting from CSO discharge.

CARP

A weakness in the available methods was still seen to be the inability to take into account the complex way in which pollutant discharges to a river affect its water quality. CARP (Comparative Acceptable River Pollution procedure, introduced in 1988) was promoted by the Water Research Centre for use in conjunction with the method within SRM2 of estimating storm flow pollutant concentrations (above). It was envisaged that both would be replaced by detailed quality models for sewer system and river when these became established.

In CARP, time-series rainfall (see Section 5.5) and a sewer flow model are applied, and storm discharge concentrations are determined, to estimate the loading on a particular river length. This is then compared with the loading, calculated the same way, for a comparable river reach in which water quality is known to be satisfactory. If the loading exceeds that of the acceptable river, further improvements are needed.

AMP

The background to the AMP2 Guidelines (NRA, 1993) has been described in Chapter 3. A major part of AMP2 and subsequent AMP programmes has been to identify and set out plans to improve unsatisfactory CSOs. Discharges were identified as having low, medium or high significance, and this determined how the impact should be assessed and discharge consents specified (see Table 12.3).

In addition, the documentation proposed standards for bathing waters. Bathing standards have been presented in Section 3.5.2. In order to avoid detailed modelling for each relevant CSO, a surrogate emission standard was proposed. Thus, CSOs should not spill more frequently than 3 times per bathing season. Compliance with this requirement, it is said, should ensure conformance with the standard given in Chapter 3.

Table 12.2 Factors to convert DWF concentrations to average storm concentrations

Determinand	Multiplying factor[1]
BOD	0.5
COD	1.0
Ammonia	0.3
Suspended solids	2.0
Total dissolved solids	0.4

[1] for systems in which average sewer gradient is no steeper than 1 in 50

Table 12.3 Criteria to assess significance of CSOs on freshwaters

Significance	Dilution	Interactions	Population equivalent	Fisheries	Assessment criteria
Low	>8:1	None	–	–	Emission control, e.g. Formula A
Medium	<8:1	Limited	>2000	Cyprinid	Simple models, e.g. SDD, QUALSOC, CARP + sewer hydraulic model
High	<2:1	Significant	>10 000	Cyprinid or salmonid	Complex models e.g. sewer and river quality models

Standards for amenity areas were also specified in terms of importance of the amenity area defined in Chapter 3. The standards required for control of gross solids are chosen from:

- *6 mm solids separation* – separation from the effluent of a 'significant quantity' of solids greater than 6 mm in any two dimensions (excluding high flows, as defined in the guidelines)
- *10 mm solids separation* – separation of solids, giving a performance equivalent to that of a 10 mm bar screen
- *Good engineering design.*

Specification of a particular standard depends on the amenity-use category and the expected frequency of operation of the CSO as shown in Table 12.4.

UPM

The Urban Pollution Management Manual (Foundation for Water Research, 1998) set out procedures for management of wet weather discharges from urban drainage systems. The 1st edition (1994) formalised the basic procedures laid down in the AMP2 documents. The Manual provides quantitative standards on intermittent discharges (given in Chapter 3), and has also shown how, in some situations, CSO discharges

Table 12.4 Amenity area emission standards

Amenity category	Expected frequency of spills per year	Standard
High	>1	6 mm solids separation
	≤1	10 mm solids separation
Medium	>30	6 mm solids separation
	≤30	10 mm solids separation
Low	–	Good engineering design

Example 12.2

A combined sewer catchment serves a population of 50 000 and has an impervious area (A_i) of 18 ha. Determine the overflow setting required upstream of the main outfall sewer using the following approaches:
a) 6DWF
b) Formula A
c) River water quality adjacent to the overflow limited to a BOD_5 of 10 mg/l for the 1 year return period, 20 minute duration event.

Additional information:
Wastewater flow = 250 l/hd.d
Rainfall intensity for the 1 year, 20 minute event, $i = 20$ mm/h
River flow upstream of CSO = 1 m³/s, with BOD_5 of 2 mg/l
Overflow BOD_5 = 500 mg/l
Pipe infiltration and industrial flows are negligible.

Solution

a) DWF $= 50\,000 \times 250 = 12.5 \times 10^6$ l/d or 145 l/s
    setting $= 6 \times 145 = 870$ l/s
b) setting $=$ DWF $+ 1360P + 2E$
    $= 12.5 \times 10^6 + 1360 \times 50\,000$
    $= 932$ l/s
c) Runoff flow-rate $= iA_i = 18 \times 10^4 \times 20/3600 = 1000$ l/s
    BOD_5 load-rate: river upstream + overflow = river downstream

    $$1000 \times 2 + Q_{overflow} \times 500 = (1000 + Q_{overflow}) \times 10$$
    $$Q_{overflow} = 16 \text{ l/s}$$

    DWF + runoff = setting + $Q_{overflow}$
    $145 + 1000 =$ setting $+ 16$
    setting $= 1129$ l/s

should not be seen in isolation, but as part of the whole urban water system. More detail is given in Chapter 21, together with discussion of integrated system modelling.

12.5 Approaches to CSO design

12.5.1 Stilling pond

Principles

The main principles of a stilling pond CSO are illustrated in Fig. 12.4. In dry weather and low intensity rain, the flow enters via the inlet pipe,

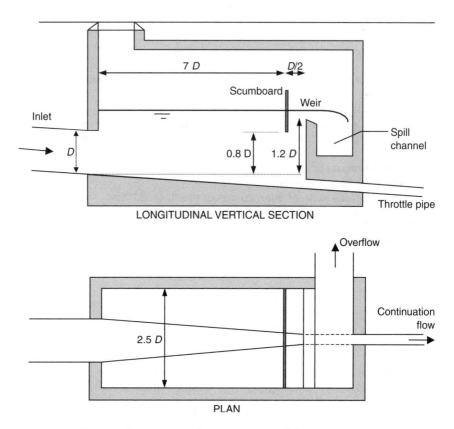

Fig. 12.4 Stilling pond CSO: general arrangement and dimensions

passes along a channel through the overflow and leaves via the throttle pipe. In heavier rainfall, as the inflow increases, the capacity of the throttle pipe is exceeded and flow backs up inside the chamber. The level usually has to rise above the top of the inlet pipe before it reaches the crest of the weir. This causes the inflow to become stilled, which helps to ensure that sinking solids are not carried over the weir. When the water level is above the weir crest, water spills over the weir and out via the spill channel and pipe. The scumboard is positioned to limit the passage of floating solids over the weir. In fact it does more: it sets up a pattern of circulation in the chamber which brings many floating solids back to the upstream end of the chamber – making it even less likely that they will flow over the weir.

So, the stilling pond functions hydraulically by means of a throttle pipe and a weir, and it limits pollution in two ways: by stilling the flow so that sinking solids pass out with the retained flow, and by using a scumboard to discourage floating solids from passing over the weir.

Development

The first major investigation was by Sharpe and Kirkbride (1959). The results are much-quoted and had a genuine impact on engineering practice. They concluded that the best conditions were achieved when the inlet velocity was low and the upstream sewer was well flooded in order to create stilling conditions in the chamber. The scumboard created a reverse surface flow which took the floating solids away from the weir. Their recommendation was that the distance from the inlet to the scumboard be at least 4.2 times the diameter of the inlet pipe (*D*). Other recommended dimensions included a chamber width of 2.5 *D*, a distance from the scumboard to the weir of 0.5 *D*, and a weir level similar to the soffit of the incoming pipe.

Frederick and Markland (1967) carried out laboratory studies of model stilling pond arrangements. Many of their conclusions confirmed those of Sharpe and Kirkbride (1959), 'in particular, the incoming sewer needs to be surcharged in order to produce a favourable reverse current near the surface'. The main difference in their conclusions was in terms of length, which they recommended should be as great as possible, with overall length no less than 7 times inlet diameter.

Balmforth (1982) studied the separation of a wide variety of solids in a model stilling pond. A particular aim was to resolve differences in the recommendations for chamber dimensions between Sharpe and Kirkbride and Frederick and Markland. Balmforth confirmed that there were significant advantages in the longer length recommended by Frederick and Markland.

Dimensions and layout

Recommendations for chamber dimensions, based on the development described above and best knowledge of CSO operation, are given in the *Guide to the Design of Combined Sewer Overflow Structures* by Balmforth, Saul and Clifforde (1994). Their recommended dimensions for a stilling pond are given in Fig. 12.4. They are based on the diameter of the incoming pipe. A method of determining this diameter (common to a number of different CSO types) is given in Section 12.7.

A dry weather flow channel runs along the centre of the chamber, contracting in area from the inlet to the throttle pipe. It should have sufficient size and longitudinal slope to carry flow-rate equal to the capacity of the throttle pipe and avoid sediment deposition. The base on either side of this channel slopes towards the centre to drain liquid to the dry weather flow channel. The capacity of the throttle pipe is crucial in determining the setting of the CSO. The upstream head is a function of the crest level and characteristic of the weir.

An example of stilling pond sizing is given as Example 12.6 in Section 12.7.

12.5.2 Hydrodynamic vortex separator

Principles

Several types of overflow arrangement exploit the separation of solids that occurs in the circular motion of a liquid.

When such a flow is considered in two dimensions, by studying a horizontal section, theory suggests that heavier solids will follow a path towards the outside of the circle. This leads to a design of 'vortex overflow' in which the overflow weir is placed on the inside of the chamber and the continuation pipe on the outside (where heavier solids tend to congregate). Floating solids are prevented from flowing over the weir by a baffle.

Studies in three dimensions, with particular chamber shapes, have suggested that heavy solids collect at the bottom of the chamber, in the centre. These have led to designs with the opposite arrangement: the weir on the outside, and the continuation pipe at the centre. Other, more elaborate chamber arrangements, have led to flow patterns with even more complex properties.

Development

Smisson (1967) carried out extensive work on models and full-scale vortex overflows, giving detailed descriptions of flow patterns, and design recommendations. The weir was positioned on the inside of the vortex and the continuation pipe on the outside.

A different type of vortex chamber was proposed by Balmforth *et al.* (1984), called a 'vortex overflow with peripheral spill'. This design 'makes use of the known ability of vortex motion to separate settleable solids, but differs from earlier designs in that the foul outlet pipe is set in the centre of the chamber floor, and the overflow occurs over a weir formed in the peripheral (outer) wall'.

Modern descendants of the vortex overflow are called hydrodynamic separators. A patented design, the Storm King® Overflow, in which separation of solids takes place within a complex flow pattern of upward and downward helical flow, is common in the UK. The arrangement is shown in Fig. 12.5. The internal hydraulics of this device have been modelled using computational fluid dynamics software by Harwood and Saul (1996), and detailed representations of liquid movement within the chamber have been simulated (Fig. 12.6).

Similar use of these principles has been made in other countries, the US EPA 'swirl regulator' (Field, 1974), and the German 'Fluidsep' vortex separator (Brombach, 1987, 1992) for example.

Much of the development is related to specific patented devices, but research is continuing into the more general principles of devices of this

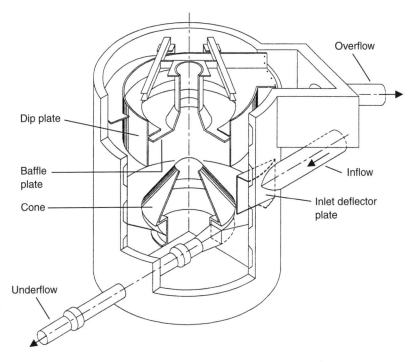

Overflow

Dip plate

Baffle plate

Cone

Inflow

Inlet deflector plate

Underflow

Fig. 12.5 Storm King® Overflow (courtesy of Hydro International)

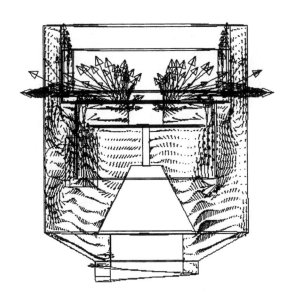

Fig. 12.6 CFD simulation of flow in a hydrodynamic separator (reproduced from Harwood 1999 with permission of the author)

type (for example, Huebner and Geiger, 1996). Fenner and Tyack (1997, 1998) have proposed scaling protocols for physical models. Saul and Harwood (1998) have studied retention efficiency for full-scale sanitary gross solids. A review of the various types of hydrodynamic separators in use has been given by Andoh (1998).

Dimensions and layout

Hydrodynamic separators are designed and fabricated by their manufacturers based on performance specifications.

12.5.3 High side weir

Principles

The high side weir overflow is an advanced development of the crude side weir that we considered in Section 12.3.3. An overflow with high weirs, scumboards and a stilling zone upstream, can provide good retention of both floating and sinking solids. Double side weirs can provide good hydraulic control. High side weirs are often associated with storage. This can be a storage zone downstream of the weir for retention of floating solids, or a large storage volume for retention of the first flush.

Development

Saul and Delo (1982) worked with a laboratory model of a high side weir with stilling and detention zones. A computer-controlled valve on the inlet allowed the study of unsteady inflow using realistic hydrograph shapes. Delo and Saul (1989) proposed a design method for the weirs themselves, which has been described in Section 9.2.2.

Dimensions and layout

Recommendations are given by Balmforth, Saul and Clifforde (1994) (Fig. 12.7). Flow in the chamber must be subcritical. Design of a high side weir CSO and calculation of flow depths over the weirs (as presented in Section 9.2.2) are illustrated in Example 12.3.

12.5.4 Storage

Principles

The aim of providing storage at a CSO is to retain pollutants in the sewer system rather than allowing them to be overflowed to a watercourse, even after a weir on the main sewer has come into operation during a storm.

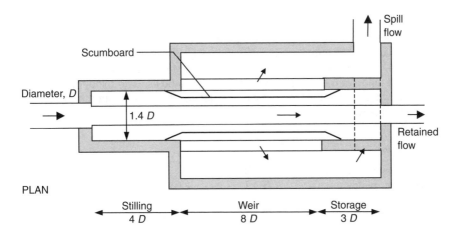

PLAN

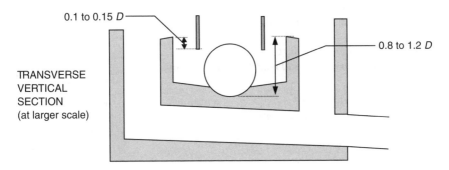

Fig. 12.7 High side weir CSO: general arrangement and dimensions

When flows in the system have subsided after the storm, the polluted flow retained in the storage can be passed onward to treatment. Clearly the larger (and more expensive) the storage, the lower the amount of pollution reaching the watercourse. Optimum sizing of storage needs to take into account the fact that polluting loads during storm flow vary with time. It is common (but not universal) that early flows are particularly polluted as a result of a first foul flush, as considered in Section 12.3.2.

Storage has tended to be provided in conjunction with a high side weir arrangement, but storage can be used to supplement any CSO configuration; it is becoming increasingly common, for example, to provide storage in conjunction with a hydrodynamic separator.

Storage can be provided on-line or off-line. In an *on-line* arrangement, the flow passes through the tank even in dry weather when the capacity of

Example 12.3

A high side weir CSO is being designed. The diameter of the inlet pipe has been fixed at 750 mm. Propose dimensions for the chamber.

The invert in the chamber will be level, and Manning's *n* will be taken as 0.01. If inflow is 600 l/s and continuation flow is 60 l/s, determine the depth of the water above the weir crest at the upstream and downstream ends of the double weirs.

Solution

Dimensions (Fig. 12.7):

chamber width	1.4 D	1.05 m
stilling length	4 D	3.0 m
weir length	8 D	6.0 m
storage length	3 D	2.25 m
weir crest above invert	0.8 D	0.6 m
bottom of baffle below crest	0.15 D	0.11 m

Width of chamber = 1.05 m = B_u = B_d
Weir height = 0.6 m = P_u = P_d

Q_d/Q_u = 60/600 = 0.1, B_d/B_u = 1
P_u/B_u = $P_d/B_u \approx 0.6$ ... so Fig. 9.11 is appropriate.

$$\text{Inlet flow ratio} = \frac{Q_u^2}{gB_u^5} = \frac{0.6^2}{g1.05^5} = 0.029$$

Length of double weirs = 6 m, so L/B_u = 5.7

From Fig. 9.11:

$$\frac{Y_u - P_u}{B_u} = 0.057 \text{ so } Y_u - P_u = 0.06 \text{ m or 60 mm}$$

(depth above crest, upstream)

$$\frac{Y_d - P_d}{B_u} = 0.095 \text{ m so } Y_d - P_d = 0.10 \text{ m or 100 mm}$$

(depth above crest, downstream)

i.e. depth increasing, as shown for Type II on Fig. 9.9.

the tank is not being utilised. When flow-rate increases during a storm, a downstream control will cause the level to rise, to fill up the storage volume, and eventually overflow at the weir (Fig. 12.8(a)). After the storm, the tank empties by gravity into the continuation pipe. In an *off-line* arrangement, flow is diverted to the tank via a weir as the level begins to rise (Fig. 12.8(b)). When the tank is full, a higher weir comes into operation and diverts further flows to the watercourse. After the storm, the tank is emptied into the continuation pipe, by gravity or by pumping. The rate at which the storage tank can be emptied is governed by the amount of spare capacity in the pipe and/or treatment plant.

Storage at a CSO can be provided in a number of forms: rectangular chamber, circular vertical shaft, or oversized pipe or tunnel.

Development

A method of designing CSOs incorporating storage was proposed by Ackers, Harrison and Brewer (1968). It was used well into the 1980s, and many operating cases exist (for example, Murrel *et al.*, 1983). The method aims to retain the first foul flush of pollutants by determining the volume of overtaken baseflow from the catchment. The method is approximate, as the authors admit; other causes of first flushes are not considered and the sewer system in all cases is assumed to be unbranched.

The volume of flow that is too polluted to spill, V_o, is assumed to be related to the volume of wastewater flow in the system when the storm

(a)

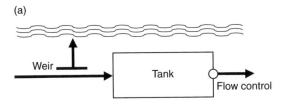

(b)

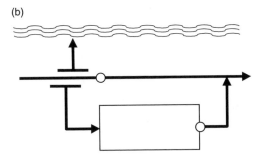

Fig. 12.8 Storage tank: (a) On-line; (b) Off-line

starts. The volume of the first foul flush, V_f (for which storage is desirable), is less than V_o by the volume of wastewater that passes the overflow while the storm wave is arriving. So:

$$V_f = V_o - Q_o T_a$$

Q_o wastewater baseflow
T_a time between rain first entering sewer upstream and storm wave reaching CSO

Hydraulic design of the tank is then based on consideration of flows entering and leaving the tank while it fills. A calculation procedure and design charts are presented for this.

The Ackers, Harrison and Brewer method was not really superseded until the development of the UPM procedures. In these, the size of storage is optimised by application of sewer quality modelling (either simple or complex). Because of the UPM emphasis on consideration of the system as a whole, design rules for sizing individual tanks are not proposed. Example 12.4 is an illustration of the way in which model simulations can be used to investigate possible storage proposals. A decision is not made

---

Example 12.4

Fig. 12.9 gives a simulated flow hydrograph and COD pollutograph (concentration and load-rate) for a catchment, in response to a particular rainfall pattern. Rainfall in this case started at a low intensity, causing a slight increase in flow-rate and dilution of COD concentration (from the dry weather level of 470 mg/l). During the early period, load-rate is constant. After 45 minutes, rain became more intense, causing a significant first flush, apparent from both the concentration and load-rate graphs.

Determine the approximate size of storage that would be needed (i) to retain pollutants until COD concentrations no longer exceed dry weather level, and (ii) to retain pollutants until COD load-rate no longer exceeds dry weather level. Assume that the detention tank will have outflow via a control that limits flow-rate to 100 l/s, and via an overflow that operates when the storage is full.

Solution

i)   The volume retained would equal the area under the hydrograph (above the 100 l/s line) up to a.
From the graph, volume $\approx$ 440 m³
ii)   The volume retained would equal the area under the hydrograph (above the 100 l/s line) up to b.
From the graph, volume $\approx$ 580 m³

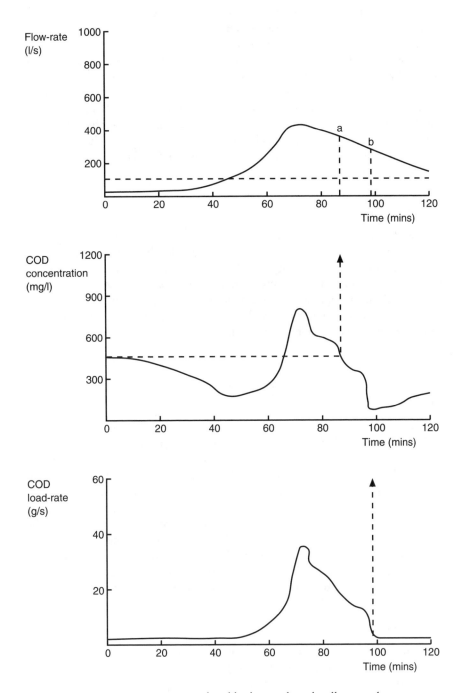

Fig. 12.9 Storage: interpreting simulated hydrograph and pollutograph

until the full range of design rainfall patterns have been considered, the costs of alternative storage strategies have been determined, and the effects on the rest of the system have been assessed.

As well as optimising the size of storage, designers need to pay attention to the layout of the chambers themselves to avoid excessive sedimentation. When a tank is full of virtually stationary liquid, conditions are ideal for deposition of suspended solids (and concentrations are likely to be high in the first flush that the tank will have been designed to retain). Work by Saul and Ellis (1990, 1992) has been aimed at creating self-cleansing conditions in storage tanks. Other work has involved CFD modelling (Stovin and Saul, 2000).

Dimensions and layout

Saul and Ellis (1992) found that a dry weather flow channel with a steep longitudinal gradient was helpful by creating suitably high velocities. Long narrow tanks with a single dry weather flow channel had better self-cleansing properties than wider tanks with multiple channels. Length to width ratio should be as high as possible, and width should not exceed 4 m. Guidance is also given in Water Research Centre (1997). CFD modelling has confirmed the importance of length to width ratio (Stovin and Saul, 2000).

12.5.5 Screens

Principles

Screens (traditionally uniformly spaced bars) have occasionally been included in CSO designs for many years as a direct attempt to remove gross solids, but they have not been common in the past partly because of the disadvantages of extra energy and maintenance costs. The more recent focus on aesthetic pollution has brought screens to the forefront.

The aim of the screen is to prevent solids from entering the overflow pipe. For example, stilling ponds have been designed with a screen attached to the weir, spanning the overflow channel. It is raked mechanically so that the screenings fall into a small screenings chamber beyond the overflow channel and are washed out through the throttle pipe. A similar arrangement has been used with side weirs, with the screenings retained in the continuation flow.

Development

Mechanically-raked bar screens at CSOs have not always been considered to be successful in operation. A field study of screens at CSOs (Meeds and Balmforth, 1995) concluded that mechanically-raked bar screens are unlikely to achieve retention efficiencies of greater than 50% (of all gross solids). Recent developments have tended to favour screens which consist of a mesh rather than parallel bars, partly as a result of solids separation requirements. A laboratory study (Saul *et al.*, 1993) demonstrated a mesh screen to be more effective at retaining solids than a bar screen with the same spacing, though a field study using actual wastewater concluded that 6 mm mesh screens are unlikely to achieve retention efficiencies of greater than 60% (Balmforth, Meeds and Thompson, 1996). Mesh screens cannot be raked in the same way as bar screens, and cleaning is usually by brushes or by liquid flushing. The mesh may also be in the form of a rotating drum; an innovative rotating drum 'sieve-filter' is described by Brombach and Pisano (1996). Another goal of recent developments is a self-cleansing screen – one with no requirement for power supply or maintenance of machinery (Faram *et al.*, 2001). There is much development work in this area, especially on devices that use the energy of the liquid itself to carry out the cleaning. Fig. 12.10 shows an example.

In Section 12.4 we have referred to the AMP standards for control of gross solids as being either '6 mm separation', '10 mm separation', or 'good engineering design'. In order to give these standards more precise meaning, the UPM manual (FWR, 1998, Appendix C) contains the results of a testing programme using a dedicated CSO test facility at Wigan Wastewater Treatment Plant. The tests used CSOs fitted with:

- 6 mm mesh screens
- 10 mm bar screens
- no screens, but designed according to the principles of the *Guide to the Design of Combined Sewer Overflow Structures* (Balmforth, Saul and Clifforde, 1994).

The resulting plot of solids separation efficiency are presented as standards against which any alternative or novel approach can be compared using a defined test procedure. The plot for 'good engineering design' is given as

Fig. 12.10 Self-cleansing screen: Swirl-Cleanse™ Screen (courtesy of Hydro International)

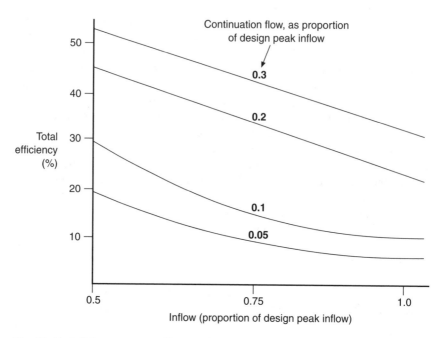

Fig. 12.11 Solids separation efficiency for 'good engineering design' (based on
UPM *Manual* (FWR, 1998) with permission of Foundation for Water
Research, Marlow)

Fig. 12.11, and the plot for 6 mm solids separation is given as Fig. 12.12.
'Total efficiency' is the percentage of the total mass of solids entering the
CSO that is retained within the sewer system (in the continuation flow or in
the chamber itself). The 6 mm separation does not come near to retaining *all*
solids. A wide range of proprietary screen arrangements have since been
tested at the same facility and compared with these standards in order to
help engineers choose suitable devices (Saul, 2000).

Dimensions and layout

It is possible to fit screens to any of the CSO types described in this
chapter. For example, there is a version of the Storm King® Overflow
which is supplied with a self-cleansing screen ready fitted. However the
standard UK recommendation (WaPUG, 2001) is for CSOs incorporating
screens to be of the high-sided weir type.

In reviewing research on the performance of CSOs with screens, Saul
(2002) states that 'the performance of the screen has been shown to dominate the overall performance of the screened CSO'. This may not seem surprising but it has significant implications. If a CSO has a screen, do
existing guidelines apply?

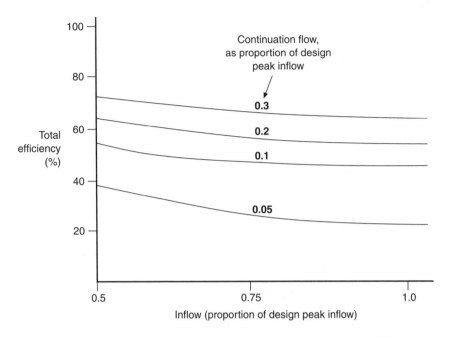

Fig. 12.12 Efficiency for 6 mm solids separation (based on *UPM Manual* (FWR, 1998) with permission of Foundation for Water Research, Marlow)

The recommended dimensions and layout in the WaPUG (Wastewater Planning User Group) guide to *The Design of CSO Chambers to Incorporate Screens* (WaPUG, 2001) are based on physical tests and on CFD modelling. The recommended chamber width is 1.4 *D* (where *D* is a diameter of incoming pipe), which is the same as the recommendation by Balmforth, Saul and Clifforde (1994) for a high-sided weir CSO without a screen (Fig. 12.7). However, other recommended dimensions are smaller. The weir length need be no more than 6 *D* (compared with 8 *D* on Fig. 12.7), and the stilling and storage lengths are not needed, though shorter inlet and outlet lengths are still recommended.

The guide refers to three possible positions for the screen in a high-sided weir CSO:

- mounted vertically on the weir, so that flow over the weir passes horizontally through it
- horizontally mounted above the main channel, so that flow over the weir must first pass upwards through the screen
- mounted on the downstream face of the weir, so flow that has passed over the weir then passes down through the screen.

12.6 Effectiveness of CSOs

12.6.1 *Performance measures*

Hydraulic performance can be expressed in terms of liquid volumes, using the term *flow split*.

$$\text{Flow split} = \frac{\text{storm volume retained in the sewer system}}{\text{total storm inflow volume}}$$

The split can also be expressed in terms of pollutant loads (cumulative mass of pollutant), using the term *total efficiency*.

$$\text{Total efficiency} = \frac{\text{storm load retained in the system}}{\text{total storm inflow load}}$$

Both terms are needed to judge the success of a CSO design, combined in the term *treatment factor*.

$$\text{Treatment factor} = \frac{\text{total efficiency}}{\text{flow split}}$$

The success of a CSO design in separation of pollutants is indicated by the amount by which the treatment factor exceeds 1.0. A treatment factor less than 1.0 indicates that a design is unsuccessful in this respect. This is illustrated in Example 12.5.

Example 12.5

For a particular storm, a CSO gives a flow split of 20% and a total efficiency of 33%. Comment on its effectiveness. How effective would it have been if the total efficiency had been 15%?

Solution

A flow split of 20% for a particular storm indicates that one-fifth of the total inflow volume was retained in the system, and four-fifths was overflowed. If the total efficiency was 33% (one-third of pollutants retained in the system, two-thirds overflowed), we can deduce that in these conditions the design has some qualities in retaining pollutants, over and above the straightforward split in flow. The treatment factor is 1.65 (33% divided by 20%).

If the total efficiency was only 15%, this would suggest that instead of the desired effect of retaining pollutants, the CSO was giving the opposite effect. The resulting treatment factor would be 0.75.

12.6.2 *Comparative research*

There have been a number of comparative studies – mostly of models. The studies have contributed significantly to knowledge of CSO performance, but the results have not produced a clear 'winner'. This is partly because the performances of the different types have in many cases been genuinely similar, and partly because the comparisons have been of specific examples of each type, leaving researchers unable to generalise.

More recently, comparison work has been based on larger scale testing. In the UK, this has been carried out at the CSO test facility at Wigan Wastewater Treatment Plant (Saul, 1998).

The results of practical research have been incorporated into a piece of software for CSO design, 'Aesthetisizer', by UK Water Industry Research (UKWIR, 1998).

A further tool for assessing and comparing CSO configurations, as we have seen, is CFD modelling. Advances in hardware and software mean that this tool can be used in place of, or in conjunction with, physical modelling (Harwood and Saul, 2001).

12.6.3 *Gross solids*

A laboratory comparison (Saul *et al.*, 1993) of the ability of large-scale model CSOs (stilling pond, vortex with peripheral spill, hydrodynamic separator, high side weir) to retain sanitary gross solids in steady and unsteady flow came to the disappointing conclusion that the performance of all types of chamber was relatively poor at design flow-rate, with treatment factor rarely much above unity. This was further confirmed by studies at the CSO test facility at Wigan (Saul, 1999). This is a result of the fact that the solids – those most likely to cause problems at actual CSOs – have a density close to that of water and, therefore, have low terminal velocities. Solids retention efficiencies plotted against terminal velocity for many studies have consistently demonstrated a characteristic cusp or 'gull's wing' shape (for example Fig. 12.13). Often the range of terminal velocities of particles studied has been wide, giving quite good efficiencies for the clear 'floaters' and 'sinkers'. But the reality is that the most common, and most aesthetically sensitive solids, with their close-to-neutral buoyancy, show the CSO designs at their worst.

Approaches to removing gross solids from the spill flow are:

- use of screens
- good design of stilling pond, vortex or high side-weir overflows, using the principles discussed, and, in particular, increasing the inlet diameter and chamber dimensions to improve solids retention efficiency
- provision of storage.

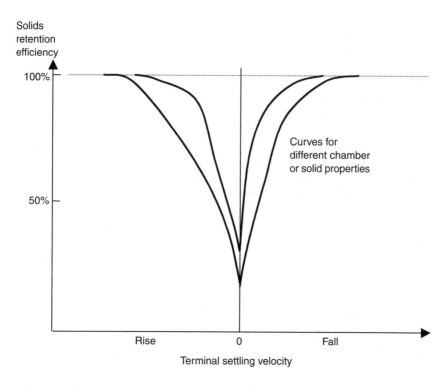

Fig. 12.13 Plots of CSO solids separating efficiency (conceptual)

A study involving simulation of alternatives to screens for a wide range of conditions has concluded that these other methods of enhancing CSO performance can be more cost-effective in certain circumstances (Balmforth and Blanksby, 1996).

12.6.4 *Choice of CSO design*

All the main CSO types described in this chapter are potentially the best choice in appropriate circumstances. For example, hydrodynamic separators are usually installed in circular shafts constructed off the line of the main sewer, which reduces the problems of construction on a live sewer. The stilling pond and high side weir incorporate the principle of stilling, and are therefore more suited to sewers with mild slopes. If construction can only take place on the line of the existing sewer, a stilling pond or high side-weir is most suitable, and if space is limited to a small width either side of the sewer, only a high side-weir is likely to be feasible. The main issues are summarised in Table 12.5. Other considerations may be local experience or uniformity of design on a particular scheme.

Table 12.5 Issues in choosing CSO type

		Stilling pond	Hydrodynamic separator	High side-weir
Surcharge of the upstream sewer permissible?	Yes	✓	✓	–
	No	✗	✗	–
Incoming sewer gradient	Mild*	✓	–	✓
	Steep*	✗	✓	✗
Construction on-line of existing sewer		✓	–	✓
Construction off-line of existing sewer		–	✓	–

* see Section 8.5.4   ✓ definitely appropriate   – neutral   ✗ definitely inappropriate

It should be added that we have only been considering the conventional types of CSO in this chapter, and have tended to concentrate on the control of larger solids. Some studies have concentrated on fine suspended particles and dissolved pollutants, and on advanced devices which include high-rate treatment processes (for example, Vetter *et al.*, 2001).

12.7 CSO design details

Diameter of inflow pipe

Dimensions of CSO chambers are based on the diameter of the inflow pipe. The minimum diameter of this pipe is determined from:

$$D_{min} = KQ^{0.4} \tag{12.1}$$

Q peak inflow with a return period of one year (m³/s)
K constant taken from Table 12.6

The general form of equation 12.1 has its origin in the Darcy-Weisbach equation 8.8, which can be rearranged to:

$$h_f = \frac{\lambda L Q^2}{12.1\, D^5} \quad \text{or} \quad D = \left(\frac{\lambda L}{12.1 h_f} \right)^{0.2} Q^{0.4}$$

D diameter (m)
λ friction factor (–)
L length (m)
h_f friction head loss (m)
Q flow-rate (m³/s)

In Table 12.6, flow ratio is the overflow setting divided by Q above, and total efficiency is a composite for storms over a particular 12-month period.

Table 12.6 Value of K in equation 12.1 (after Balmforth, Saul and Clifforde, 1994)

Flow ratio %	Total efficiency %				
	20	40	60	80	90
5	1.27	1.47	1.60	1.72	–
10	1.13	1.37	1.52	1.66	1.83
20	0.825	1.19	1.34	1.50	1.65
30	0.815	1.02	1.18	1.33	1.43

Use of equation 12.1 and Table 12.6 is demonstrated in Example 12.6.

Increasing K improves performance, but also increases cost by requiring a larger inlet pipe diameter and chamber dimensions. The incoming sewer should have the diameter D for a length of at least $25\,D$ upstream of the CSO. Where the pipe is an existing sewer, and the diameter is less than D, it should be replaced for this length.

Example 12.6

Determine the basic dimensions of a stilling pond CSO for the following conditions. Peak inflow with a return period of 1 year is estimated as 600 l/s. Overflow setting is to be 80 l/s. 60% total efficiency over a period of a year is sought.

Solution

Flow ratio (for use in Table 12.6) is 80/600 = 13.3%
For total efficiency of 60%, K should be 1.46.
From equation (12.1), $D_{min} = 1.46 \times 0.6^{0.4} = 1.19$ m. Nearest pipe size is 1.2 m, so fix D at 1.2 m.
With reference to Fig. 12.4, proposed dimensions are given in Table 12.7.

Table 12.7 Proposed dimensions, Example 12.6

Dimension	Recommended, Fig. 12.4	Distance (m)
Length, inlet to scumboard	7 D	8.4
Distance of scumboard from weir	D/2	0.6
Height of weir crest above inlet invert	1.2 D	1.44
Height of bottom of scumboard above inlet invert	0.8 D	0.96
Width of chamber	2.5 D	3.0

Control of outflow

Control of the continuation flow is an important part of the hydraulic design of a CSO. The setting of the overflow is normally defined as the continuation flow when spill starts – that is, when liquid level reaches the weir crest. As flow-rate over the weir increases, so will depth. It is best if the retained outflow does not vary greatly as a consequence. The common methods of control are:

- fixed orifice
- adjustable penstock
- vortex flow regulator
- throttle pipe.

These have been described in Section 9.1.

Weirs

The hydraulic characteristics of weirs have been considered in Section 9.2.

Chamber invert

Chambers should be as self-cleansing as possible. Deposition of solids can be minimised by suitable longitudinal and lateral slopes. The dry weather flow channel should ensure a velocity of 1 m/s at 2DWF, and lateral benching should slope at between 1:4 and 1:6.

Design return period

As stated above, design of inlet pipe and determination of the main chamber dimensions is based on the peak flow-rate with a one-year return period. A check should also be made to see how a proposed chamber would respond to more extreme events – including a once in 20 years storm. This is particularly true for the spill channel and outlet pipe, which is the route taken by most of the flow in extreme events.

Top water level

The top water level in the chamber can be determined from the design maximum inflow and the hydraulic properties of both outflow pipes. If the spill flow pipe could be drowned at the downstream end or if it discharges to tidal water, this will also need to be considered. The TWL will be one consideration in deciding the level of the roof of the structure.

Access

Human access is normally via manhole covers at ground level. Where screens are included, there will need to be appropriately sized and positioned access for vertical installation and removal of any machinery. There should be access to clear potential blockages, especially in throttle pipes. Thorough safety precautions are required during maintenance (considered in Section 16.5).

Problems

12.1   A combined sewer with diameter 750 mm, slope 0.002, and k_s 1.5 mm, drains a catchment with a DWF of 15 l/s. For a particular rainfall, the flow of stormwater is 750 l/s. Does the sewer have sufficient capacity to carry stormwater + DWF? Would the daily maximum flow in dry weather provide self-cleansing conditions? What are the likely consequences of this? How could the design of this pipe have been improved?   [yes, no, v = 0.44 m/s]

12.2   What is meant by the 'first foul flush'. What may cause it, and what are its implications?

12.3   What are the main functions of a combined sewer overflow? Explain the common alternative overflow configurations.

12.4   Define the term 'CSO setting'. Describe the importance of a CSO setting, and ways in which it can be fixed.

12.5   The population of a catchment is 5000, average wastewater flow is 180 l/hd.d. Infiltration is 10% of the domestic wastewater flow-rate, and average industrial flow is 2 l/s. Determine the DWF and the CSO setting according to 'Formula A'. Express the CSO setting as a multiple of DWF.   [13.5 l/s, 96.2 l/s, 7.1]

12.6   If the receiving water for the CSO in Problem 12.5 offers dilution of 2:1, how much storage should be provided in conjunction with the setting determined in 12.5 (on the basis of the recommendations of the Scottish Development Department, 1977)? If the overflow is operating at this setting and all overflow is diverted to storage, how long would the storage take to fill if the inflow was constant at 500 l/s?   [400 m³, 16.5 minutes]

12.7   In the case considered in Problem 12.6, assume that average dry weather concentration of suspended solids is 400 mg/l. If, for this case, storm inflow continues at the same rate (500 l/s) for 10 minutes after the storage is full, what will be the total mass of suspended solids discharged to the receiving water? (Use the data in Table 12.2, and assume that the continuation flow is equal to the setting throughout.)   [194 kg]

12.8   Increasingly stringent standards are being set to limit discharge of gross solids to the environment. Explain approaches to CSO design by which solids can be reduced in CSO spills.

12.9 Propose dimensions for a high side-weir CSO using the data on Fig. 12.7, for a case where the inlet diameter has already been fixed at 600 mm. If inflow is at the design maximum of 350 l/s, and the continuation flow is 35 l/s, determine the depth of water relative to the weir crest at the upstream and downstream ends of the double weirs. (Assume that the channel invert is level, and Manning's *n* is 0.01.)   [width 0.84 m, length of weirs 4.8 m, etc, 50 mm, 80 mm]

12.10 A stilling pond CSO is being designed. The peak inflow with a return period of 1 year is 380 l/s, and the setting is 57 l/s. The designers require 40% total efficiency over a 12-month period. Select a suitable inlet pipe diameter from the following available: 750, 825, 900, 975, 1050 mm. Propose the following dimensions for the chamber: length inlet to scumboard, width of chamber, height of weir crest above inlet invert.

[900 mm, 6.3 m, 2.25 m, 1.08 m]

12.11 Select an appropriate CSO type for the following conditions: construction will be on the line of the existing sewer; the incoming sewer has a mild slope; and surcharge of the upstream sewer is not permissible.

Key sources

Balmforth, D.J., Saul, A.J. and Clifforde, I.T. (1994) *Guide to the Design of Combined Sewer Overflow Structures*, Report FR 0488, Foundation for Water Research.

Foundation for Water Research (1998) *Urban Pollution Management Manual*, 2nd edn, FR/CL 0002.

References

Ackers, P., Harrison, A.J.M. and Brewer, A.J. (1968) The hydraulic design of overflows incorporating storage. *Journal of the Institution of Municipal Engineers*, 95, January, 31–37.

Andoh, R. (1998) Improving environmental quality using hydrodynamic separators. *Water Quality International*, January/February, 47–51.

Balmforth, D.J. (1982) Improving the performance of stilling pond storm sewage overflows. *Proceedings of the 1st International Seminar on Urban Drainage Systems*, Southampton, September, 5.33–5.46.

Balmforth, D.J. and Blanksby, J. (1996) Alternatives to screens in controlling aesthetic pollutants. *Proceedings of the 7th International Conference on Urban Storm Drainage*, 2, Hannover, September, 911–916.

Balmforth, D.J., Lea, S.J. and Sarginson, E.J. (1984) Development of a vortex storm sewage overflow with peripheral spill. *Proceedings of the 3rd International Conference on Urban Storm Drainage*, Gotenborg, June, 107–116.

Balmforth, D.J., Meeds, E. and Thompson, B. (1996) Performance of screens in controlling aesthetic pollutants. *Proceedings of the 7th International Conference on Urban Storm Drainage*, 2, Hannover, September, 989–994.

Brombach, H. (1987) Liquid–solid separation at vortex storm overflows. *Proceedings of the 4th International Conference on Urban Storm Drainage, Topics in Urban Storm Water Quality, Planning and Management,* Lausanne, September, 103–108.

Brombach, H. (1992) Solids removal from CSOs with vortex separators. *Novatech 92, International Conference on Innovative Technologies in the Domain of Urban Water Drainage,* Lyon, November, 447–459.

Brombach, H. and Pisano, W. (1996) Operational experience with CSO sieving treatment. *Proceedings of the 7th International Conference on Urban Storm Drainage,* 2, Hannover, September, 1007–1012.

Delo, E.A. and Saul, A.J. (1989) Charts for the hydraulic design of high side-weirs in storm sewage overflows. *Proceedings of the Institution of Civil Engineers,* Part 2, 87, June, 175–193.

Faram, M.G., Andoh, R.Y.G. and Smith, B.P. (2001) Optimised CSO screening: a UK perspective. *Novatech 2001, Proceedings of the 4th International Conference on Innovative Technologies in Urban Drainage,* Lyon, France, 1031–1034.

Fenner, R. and Tyack, J.N. (1997) Scaling laws for hydrodynamic separators. *American Society of Civil Engineers, Journal of Environmental Engineering,* 123(10), October, 1019–1026.

Fenner, R. and Tyack, J.N. (1998) Physical modeling of hydrodynamic separators operating with underflow. *American Society of Civil Engineers, Journal of Environmental Engineering,* 124(9), September, 881–886.

Field, R. (1974) Design of a combined sewer overflow regulator/concentrator. *Journal of WPCF,* 46(7), 1722–1741.

Frederick, M.R. and Markland, E. (1967) The performance of stilling ponds in handling solids. Paper No 5, *Symposium on storm sewage overflows,* Institution of Civil Engineers, May, 51–61.

Harwood, R. (1999) *Modelling combined sewer overflow chambers using comutational fluid dynamics.* Unpublished PhD Thesis, University of Sheffield.

Harwood, R. and Saul, A.J. (1996) CFD and novel technology in combined sewer overflow. *Proceedings of the 7th International Conference on Urban Storm Drainage,* 2, Hannover, September, 1025–1030.

Harwood, R. and Saul, A.J. (2001) Modelling the performance of combined-sewer overflow chambers. *Journal of the Chartered Institution of Water and Environmental Management,* 15(4), 300–304.

Huebner, M. and Geiger, W. (1996) Influencing factors on hydrodynamic separator performance. *Proceedings of the 7th International Conference on Urban Storm Drainage,* 2, Hannover, September, 899–904.

Meeds, B. and Balmforth, D.J. (1995) Full-scale testing of mechanically raked bar screens. *Journal of the Chartered Institution of Water and Environmental Management,* 9(6), 614–620.

Ministry of Housing and Local Government (1970) *Technical Committee on Storm Overflows and the Disposal of Storm Sewage, Final Report,* HMSO, London.

Murrel, M.D., Daws, G. and White, T.E. (1983) Design and construction of Accrington's storm sewage overflow tank. *Tunnels and Tunnelling,* 15(7), July, 28–29.

NRA (1993) *General Guidance Note for Preparatory Work for AMP2 (Version 2),* Oct.

Saul, A.J. (1998) CSO state of the art review: a UK perspective. *4th International Conference on Developments in Urban Drainage Modelling,* 2, London, September, 617–626.

Saul, A.J. (1999) CSO performance evaluation: results of a field programme to assess the solids retention performance of side weir and stilling pond chambers. UKWIR Report 97/WW/08/01.

Saul, A.J. (2000) *Screen efficiency (proprietary designs).* UKWIR Report 99/WW/08/5.

Saul, A.J. (2002) CSO: state of the art review. *Global Solutions for Urban Drainage: Proceedings of the 9th International Conference on Urban Drainage,* Portland, Oregon, September, on CD-ROM.

Saul, A.J. and Delo, E.A. (1982) Laboratory tests on a storm overflow chamber with unsteady flow. *Proceedings of the 1st International Seminar on Urban Drainage Systems,* Southampton, September, 5.23–5.32.

Saul, A.J. and Ellis, D.R. (1990) Storage tank design in sewerage systems. *Proceedings of the 5th International Conference on Urban Storm Drainage,* Osaka, Japan, July, 713–718.

Saul, A.J. and Ellis, D.R. (1992) Sediment deposition in storage tanks. *Water Science and Technology,* **25**(8), 189–198.

Saul, A.J. and Harwood, R. (1998) Gross solid retention efficiency of hydrodynamic separator CSOs. *Proceedings of the Institution of Civil Engineers, Water, Maritime and Energy,* **130**, June, 70–83.

Saul, A.J., Ruff, S.J., Walsh, A.M. and Green, M.J. (1993) *Laboratory studies of CSO performance,* Report UM 1421, Water Research Centre.

Scottish Development Department (1977) *Storm sewage: separation and disposal.* Report of the Working Party on Storm Sewage (Scotland), HMSO, Edinburgh.

Sharpe, D.E. and Kirkbride, T.W. (1959) Storm-water overflows: the operation and design of a stilling pond. *Proceedings of the Institution of Civil Engineers,* **13**, August, 445–466.

Smisson, B. (1967) Design, construction and performance of vortex overflows. Paper No 8, *Symposium on storm sewage overflows,* Institution of Civil Engineers, May, 99–110.

Stovin, V.R. and Saul, A.J. (2000) Computational fluid dynamics and the design of sewage storage chambers. *Journal of the Chartered Institution of Water and Environmental Management,* **14**(2), 103–110.

Threlfall, J.L., Crabtree, R.W. and Hyde, J. (1991) *Sewer quality archive data analysis,* Report FR 0203, Foundation for Water Research.

UKWIR (1998) *Aesthetisizer 97 – Development of software for the aesthetic design of combined sewer overflows,* Report No 97/WW/08/4, UK Water Industry Research.

Vetter, O., Stotz, G. and Krauth, K. (2001) Advanced stormwater treatment by coagulation process in flow-through tanks. *Novatech 2001, Proceedings of the 4th International Conference on Innovative Technologies in Urban Drainage,* Lyon, France, 1089–1092.

WaPUG (2001) *The design of CSO chambers to incorporate screens.* WaPUG Guide. www.wapug.org.uk.

Water Research Centre (1997) *Sewerage Detention Tanks – a Design Guide,* WRc, Swindon.

13 Storage

13.1 Function of storage

In an urban drainage system, storage can have the functions of

- limiting flooding
- reducing the amount of polluted storm flow discharged to a water-course.

Storage can be provided by construction of detention tanks and other devices, or may exist within the system without being deliberately provided, especially in pipes with spare capacity. Storage for stormwater can also be created outside the system, as mentioned in Section 9.1, and is an integral element in SUDS (Chapter 21).

Storage, in the context of combined sewer overflows, has been considered in Section 12.5.4. The current chapter is concerned with storage at other locations within sewer systems: both combined and separate stormwater systems. An example is a new development to be drained by a conventional separate sewer system discharging stormwater to a small stream. To reduce the risk of flooding in the stream, the maximum discharge from the new development must be restricted to a low value. If a detention tank is provided to achieve this, the outflow is likely to be via a flow control (as considered in Section 9.1), often in conjunction with a weir (see Section 9.2) to operate at higher flow-rates. The typical relationship between inflow and outflow for a case where outflow is controlled, and does not vary significantly with water level, is shown in Fig. 13.1(a). The volume of water stored for the case illustrated is given by the shaded area. When outflow exceeds inflow, the tank empties.

It is also useful to consider the hydraulic role of storage in more general cases (beyond specific application to detention tanks) where outflow may vary significantly, for example in reservoirs, or where conceptual 'reservoirs' are used to represent more complex systems (as in Sections 6.3.4 and 11.6.2). A general relationship between inflow and outflow is shown in Fig. 13.1(b). At any value of time, the difference between the inflow and outflow ordinates $(I_t - O_t)$ gives the overall rate at which water in the storage is increasing (if inflow exceeds outflow) or decreasing (if outflow

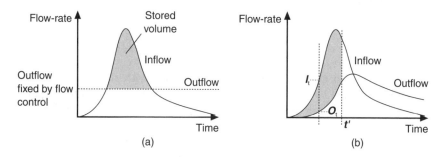

Fig. 13.1 Storage: inflow and outflow hydrographs

exceeds inflow). The total volume of water entering the storage up to any given time, say t' on Fig. 13.1(b), is given by the shaded area between the curves.

13.2 Overall design

Storage devices come in a number of shapes, sizes and configurations. Small volumes can be provided in manholes or in oversized pipes. Proprietary concrete or GRP tanks are also available. An alternative to a conventional tank (Andoh *et al.*, 2001) is a system such as Stormcell® by Hydro International, based on a three-dimensional plastic matrix with a high void ratio within which the water is stored, removing the need for the structural function of a tank (Fig. 13.2). Larger systems include purpose-built rein-

Fig. 13.2 Installation of a Stormcell® storage device (courtesy of Hydro International)

forced concrete tanks or multiple-barrelled tank sewers. An important distinction is whether they operate on- or off-line (see also Chapter 12).

On-line

On-line detention tanks are constructed in series with the sewer network and are controlled by a flow control at their outlet. Flow passes through the tank unimpeded until the inflow exceeds the capacity of the outlet. The excess flow is then stored in the tank, causing the water level to rise. An emergency overflow is provided to cater for high flows (an on-line storage tank is depicted in Fig. 12.8(a) in Chapter 12). As the inflow subsides at the end of the storm event, the tank begins to drain down, typically by gravity.

The flow control is normally one of those described in Chapter 9: an orifice, weir, vortex regulator or throttle pipe. An electrically-actuated gate linked to a downstream sensor may also be fitted. This will provide more precise control and also enable tank size to be minimised. Details of sewer system control are given in Chapter 22.

A common arrangement for an on-line tank is an oversized pipe. These *tank sewers* are provided with a dry weather (in combined systems) or low flow channel to minimise sediment deposition (Fig. 13.3). Benching with a positive gradient is also provided. Another arrangement uses smaller multiple-barrelled sewers operating in parallel. These provide the necessary storage, and have better self-cleansing characteristics.

Off-line

Off-line tanks are built in parallel with the drainage system as shown in Fig. 12.8(b). These types of tank are generally designed to operate at a pre-determined flow rate, controlled at the tank inlet. An emergency overflow is provided, as for the on-line tank. Flow is returned to the system

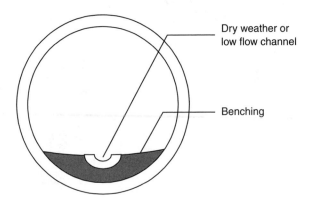

Fig. 13.3 Tank sewer

either by gravity or by pumping, depending on the system configuration and levels. A flap valve is normally used for gravity returns.

Off-line tanks require less volume than on-line tanks for equivalent performance and hence less space, but the overflow and throttling devices necessary to divert, regulate and return flows tend to be more complicated. Maintaining self-cleansing is also more difficult for this type of tank. Regular maintenance is therefore important.

Flow control

The points at which flow control is required for both on-line and off-line tanks are marked on Fig. 12.8. Table 13.1 presents a summary of the flow control requirements. The common devices have been described in Chapter 9.

More information on the use of flow control devices in conjunction with storage tanks is given by WRc (1997). Further details on a variety of flow control devices and their application within larger storage facilities are presented by Hall *et al.* (1993).

Table 13.1 Flow control for tanks

Type of storage	Type of control	Purpose	Common devices
On-line	Outlet restriction	To match continuation flow to the capacity of the downstream sewer	Orifice Penstock Vortex regulator Throttle pipe
	Relief	To divert excess flow when capacity of storage (and downstream sewer) has been exceeded	High side weir
Off-line	Continuation restriction	To match continuation flow to the capacity of the downstream sewer	Orifice Penstock Vortex regulator Throttle pipe
	Tank inlet	To pass flow into the tank when the downstream capacity has been exceeded	Orifice Penstock Side weir High side weir
	Relief	To divert excess flow when capacity of storage (and downstream sewer) has been exceeded	High side weir
	Tank outlet	To return stored flow to the sewer once the storm has passed	Orifice Penstock Vortex regulator Throttle pipe Pump

13.3 Sizing

The hydraulic design of a tank or pond serving a new development usually entails limiting the outflow for a specific storm event. So, typical design criteria are:

- Rate of outflow – this can be fixed by one of a number of approaches:
  – no greater than estimated values from the undeveloped site
  – a value linked to the area of the site (e.g. 8–12 l/s.ha)
  – the capacity of the downstream sewer or watercourse.
  The first of these approaches is particularly problematic as it is difficult to accurately predict runoff flows from small undeveloped catchments. The last approach is preferred.
- Design storm – small tanks are typically designed for 1- to 2-year storms and possibly up to 5 years. For large lakes, much higher return periods may be specified.

The question in design is, what active storage volume is required to achieve the outflow limitation and which is the critical storm that produces the worst case? It is not simply a case of using the Rational Method, as the critical storm is usually of longer duration than the one giving maximum instantaneous flow.

Preliminary storage sizing

A preliminary estimate of storage volume requirements for peak flow attenuation (in on-line tanks) can be obtained by using:

$$S = V_I - V_O \tag{13.1}$$

S storage volume (m³)
V_I total inflow volume (m³)
V_O total outflow volume (m³)

In this case, outflow is via an outlet restriction using one of the devices in Table 13.1.

Fig. 13.4 shows a plot of inflow volume, V_I, versus storm duration, D, for a particular return period. Outflow volume, V_O, has been also plotted, assuming a constant discharge. The difference in the ordinates of the two curves gives the storage, S, required for any duration storm. The design storage (S_{max}) is the maximum difference between the curves (Davis, 1963). Example 13.1 shows how S_{max} can be identified using a tabular approach.

Storage routing

A more accurate assessment of the effect of the storage can be obtained by routing an inflow hydrograph through the tank/pond. This can be done

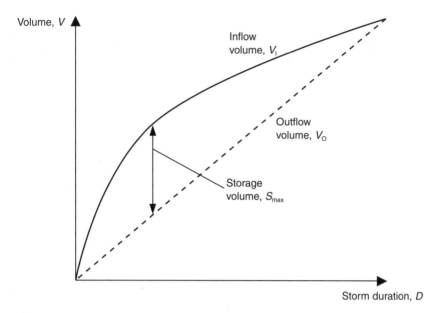

Fig. 13.4 Storage volume as a function of storm duration

using the level-pool routing technique described in the next section. In practice, most engineers will establish storage volume using one of the proprietary models discussed in Chapter 19.

13.4 Level pool (or reservoir) routing

Calculating the relationship between inflow and outflow as flow passes through storage (for example, as shown on Fig. 13.1(b)) is called 'routing'. A standard calculation method with a wide range of applications is now given.

The difference between inflow and outflow equals the rate at which the volume of water in the storage changes with time, or:

$$I - O = \frac{dS}{dt} \tag{13.2}$$

I	inflow rate (m³/s)
O	outflow rate (m³/s)
S	stored volume (m³)
t	time (s)

The simplest application is shown in Fig. 13.5. Here, there is one outflow controlled by an arrangement such as a weir, giving a simple relationship

Example 13.1

A housing development has an impermeable area of 25 ha. Determine approximately the volume of storage required to balance the 10-year return period storm event. Downstream capacity is limited, and a maximum outflow of 100 l/s has been specified. Rainfall statistics are the same as those derived in Example 5.2.

Solution

In the table below, column (3) is the inflow volume which is derived from the product of (1), (2) and the impermeable area (25 ha). Column (4) is the outflow volume; the product of column (1) and the outflow rate (100 l/s). The storage is the difference between (3) and (4).

(1) Storm duration, D (h)	(2) Intensity, i (mm/h)	(3) $V_I = iA_iD$ (m^3)	(4) $V_O = Q_OD$ (m^3)	(5) $S = V_I - V_O$ (m^3)
0.083	112.8	2350	30	2320
0.167	80.4	3350	60	3290
0.25	62.0	3875	90	3785
0.5	38.2	4775	180	4595
1	24.8	6200	360	5840
2	14.9	7450	720	6730
4	8.6	8600	1440	7160
6	6.1	9150	2160	6990
10	4.0	10 000	3600	6400
24	2.0	12 000	8640	3360

The maximum storage, S_{max} is 7160 m³

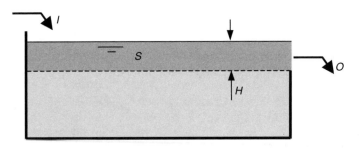

Fig. 13.5 Simple application of level pool routing

between O and H (height of water above the weir crest). S in this case is the 'temporary storage', the volume created when there is outflow. The key to the method is that both O and S are functions of H.

We will solve equation 13.2 for fixed time steps, and consider 'average' conditions over the period of each time step. Therefore, the average inflow during a time step minus the average outflow equals the change in stored volume during the step:

$$\frac{I_1 + I_2}{2} - \frac{O_1 + O_2}{2} = \frac{S_2 - S_1}{\Delta t} \tag{13.3}$$

I_1, O_1, S_1 inflow, outflow, stored volume at the start of the time step
I_2, O_2, S_2 inflow, outflow, stored volume at the end of the time step
Δt time step

A typical application is to calculate outflow for known values of inflow. In each time step, the unknown will be O_2. Since O and S are related via H, we put S_2 with O_2 on the left-hand side of the equation:

$$\frac{S_2}{\Delta t} + \frac{O_2}{2} = \frac{S_1}{\Delta t} - \frac{O_1}{2} + \frac{I_1 + I_2}{2}$$

It is convenient to have the term $\left[\dfrac{S}{\Delta t} + \dfrac{O}{2}\right]$ on both sides, so we rearrange to:

$$\left[\frac{S_2}{\Delta t} + \frac{O_2}{2}\right] = \left[\frac{S_1}{\Delta t} + \frac{O_1}{2}\right] - O_1 + \frac{I_1 + I_2}{2} \tag{13.4}$$

Now we need to incorporate the way both O and S vary with H. The neatest way of doing this is to create a relationship between $\dfrac{S}{\Delta t} + \dfrac{O}{2}$ and O (based on the variations of O and S with H). This is demonstrated by Example 13.2.

13.5 Alternative routing procedure

One disadvantage of the routing method just described is that it is difficult to implement using widely available computational tools such as spreadsheets. However, this can be overcome by transforming the rate of change of storage into the rate of change of head, as follows:

$$\frac{dS}{dt} = A\frac{dH}{dt} \tag{13.5}$$

Example 13.2

Outflow from a detention tank is given by $O = 3.5\,H^{1.5}$. The tank has vertical sides and a plan area of 300 m². Inflow and outflow are initially 0.6 m³/s, then inflow increases to 1.8 m³/s at a uniform rate over 6 minutes. Inflow then decreases at the same rate (over the next 6 minutes) back to a constant value of 0.6 m³/s. Using a time step of 1 minute, determine the outflow hydrograph.

Solution

We first use the way O and S vary with H to create a relationship between

$\dfrac{S}{\Delta t} + \dfrac{O}{2}$ and O, as on Table 13.2.

Table 13.2 Variation with H

H	O $3.5\,H^{1.5}$	S $300\,H$	$\dfrac{S}{\Delta t} + \dfrac{O}{2}$
(m)	(m^3/s)	(m^3)	(m^3/s)
0	0	0	0
0.2	0.31	60	1.16
0.4	0.89	120	2.44
0.6	1.63	180	3.81
0.8	2.50	240	5.25

We can use the data in Table 13.2 to plot $\dfrac{S}{\Delta t} + \dfrac{O}{2}$ against O (Fig. 13.6).

The calculation now progresses as in Table 13.3. The values of I, and therefore $\dfrac{I_1 + I_2}{2}$, are known. The first value of O is 0.6 m³/s, and from this the first value of $\dfrac{S}{\Delta t} + \dfrac{O}{2}$ can be determined via the graph in Fig. 13.6 (giving a value of 1.8 m³/s, as indicated). So for the first time step, $\dfrac{S_2}{\Delta t} + \dfrac{O_2}{2}$ is calculated from equation 13.4 giving $1.8 - 0.6 + 0.7 = 1.9$ (circled on Table 13.3). The corresponding value of O is determined from Fig. 13.6, working this time from the y axis to the x axis, giving 0.65 m³/s (boxed on Table 13.3).

Table 13.3 Routing calculation (extract)

Time	I	O	$\dfrac{S}{\Delta t} + \dfrac{O}{2}$	$\dfrac{I_1 + I_2}{2}$
(minutes)	(m³/s)	(m³/s)	(m³/s)	(m³/s)
0	0.6	0.6	1.8	
				0.7
1	0.8	0.65	1.9	
				0.9
2	1.0	0.76	2.15	

So now we know O after the first time step. For the next time step, $\dfrac{S_1}{\Delta t} + \dfrac{O_1}{2}$ is 1.9, and $\dfrac{S_2}{\Delta t} + \dfrac{O_2}{2}$ is calculated again from equation 13.4: $1.9 - 0.65 +$

$0.9 = 2.15$. O_2 is again determined from Fig. 13.6, giving 0.76 m³/s – the outflow at 2 minutes. The calculation proceeds in this way until all the values of O have been determined. The resulting outflow hydrograph is given in Fig. 13.7.

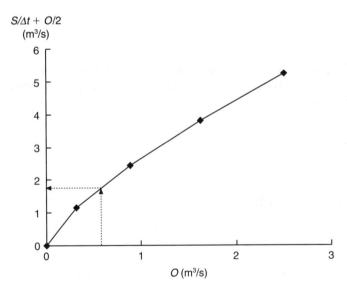

Fig. 13.6 Graph of $S/\Delta t + O/2$ against O (Example 13.2)

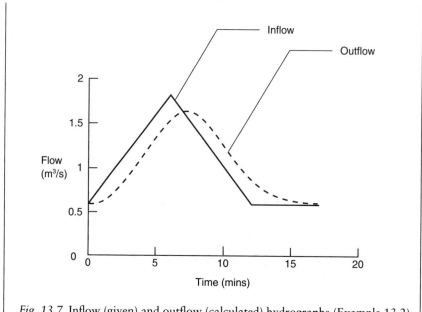

Fig. 13.7 Inflow (given) and outflow (calculated) hydrographs (Example 13.2)

A	plan area of the storage
H	head over the downstream control device

Typically, there is a relationship between *A* and *H*. For storage ponds with vertical sides, *A* is constant, but for more complex shapes a function could be used such as:

$$A = \alpha H^{\beta}$$

13.6

α	constant
β	constant

So, from equation 13.2:

$$I = O + \frac{dS}{dt}$$

and if the outflow regulator is an orifice outlet (equation 9.1):

$$O = C_d A_o \sqrt{2gH}$$

then:

$$I = C_d A_o \sqrt{2gH} + A\frac{dH}{dt}$$

and therefore:

$$\frac{dH}{dt} = \frac{I - C_d A_o \sqrt{2gH}}{A} = f(H, t) \qquad 13.7$$

The derivative can be simply represented (using Euler's approximation) as:

$$\frac{dH}{dt} \approx \frac{H(t + \Delta t) - H(t)}{\Delta t}$$

So, substituting into equation 13.7 and solving for $H(t+\Delta t)$ gives:

$$H(t + \Delta t) = H(t) + \Delta t. f(H, t) \qquad 13.8$$

which can be solved iteratively as shown in example 13.3. As Euler's approximation assumes linear change of head over time, it is most accurate when small time increments are used. It is recommended that

$$\Delta t < 0.1 \ T_p.$$

Example 13.3

An on-line balancing pond is needed to limit the peak storm runoff from the site to 100 l/s. Design a suitable vertically sided storage tank using an orifice plate ($C_d = 0.6$) as the outflow regulator. The maximum head available on the site is 1.5 m. The inflow hydrograph is given below.

Time (h)	0	0.25	0.50	0.75	1.00	1.25	1.50	1.75
Flow (l/s)	0	75	150	225	300	375	450	525

Time (h)	2.00	2.25	2.50	2.75	3.00	3.25	3.50	3.75
Flow (l/s)	600	525	450	375	300	150	75	0

Solution

Determine the required orifice diameter D for the maximum outflow (100 l/s) at the available height (1.5 m):

$$A_o = \frac{O}{C_d \sqrt{2gH}} = \frac{0.100}{0.6 \sqrt{2g1.5}} = 0.031 \ \text{m}^2$$

where A_o is the orifice cross-sectional area.

$$D = \sqrt{\frac{4A_o}{\pi}} = 0.197 \ \text{m}$$

Thus use a $D = 200$ mm orifice:

$$O = C_d A_o \sqrt{2gH} = 0.6 \times 0.031\sqrt{2g}\ H^{0.5} = 0.082H^{0.5}$$

What plan area A is needed? Equation 13.7 gives:

$$f(H, t) = \frac{I - O}{A}$$

$\Delta t = 0.25\ \text{h} = 900\ \text{s}$

In the table below, the shaded portion refers to data known initially. Column (4) is calculated from the orifice equation. Column (5) is column (2) minus column (4) divided by the plan area of the storage (equation 13.7). The 'new' head in column (6) is the sum of column (3) and column (5) times the time increment (equation 13.8). Column (3) takes the head from column (6) at the previous time step.

(1) T (h)	(2) I (m^3/s)	(3) $H(t)$ (m)	(4) O (m^3/s)	(5) $f(H,t)$ (m/s)	(6) $H(t+dt)$ (m)
0.00	0.000	0.00	0.000	0.000000	0.00
0.25	0.075	0.00	0.000	0.000030	0.03
0.50	0.150	0.03	0.013	0.000055	0.08
0.75	0.225	0.08	0.023	0.000081	0.15
1.00	0.300	0.15	0.032	0.000107	0.25
1.25	0.375	0.25	0.041	0.000134	0.37
1.50	0.450	0.37	0.050	0.000160	0.51
1.75	0.525	0.51	0.059	0.000187	0.68
2.00	0.600	0.68	0.068	0.000213	0.87
2.25	0.525	0.87	0.076	0.000179	1.03
2.50	0.450	1.03	0.083	0.000147	1.16
2.75	0.375	1.16	0.088	0.000115	1.27
3.00	0.300	1.27	0.092	0.000083	1.34
3.25	0.225	1.34	0.095	0.000052	1.39
3.50	0.150	1.39	0.097	0.000021	1.41
3.75	0.075	1.41	0.097	−0.000009	1.40
4.00	0.000	1.40	0.097	−0.000039	1.36
4.25	0.000	1.36	0.096	−0.000038	1.33
4.50	0.000	1.33	0.095	−0.000038	1.30

A number of areas were tried iteratively. The above refers to: $A = 2500\ \text{m}^2$.
  At this area (volume), $H_{max} = 1.41$ m (< 1.5 m) and $O_{max} = 97$ l/s (< 100 l/s) which is acceptable.

13.6 Storage in context

When stormwater from a new development is drained by a conventional pipe system to a river, the peak outflow is much greater and is reached much more quickly than when the land was in its natural state. The effect is usually an increase in the risk of flooding and pollution. These issues have been considered in Chapters 1 and 2.

The idea of moving away from conventional pipe systems and relying more on semi-natural drainage techniques (sustainable drainage systems, 'SUDS') has been introduced in Chapter 2 and will be considered in detail in Chapter 21. These techniques reduce the peak discharge and slow down the run-off, and so reduce the risk of flooding and pollution. They allow areas to be developed without the stormwater runoff having an undesirable impact on the river.

As we have seen in this chapter, a storage tank can also have this effect. But whereas the use of SUDS is to be seen as trying to achieve a more natural state in the engineered urban environment, providing a large concrete storage tank at the downstream end of a drainage system is at the opposite end of the hard/soft engineering spectrum: it is the ultimate in 'end-of-pipe solutions'.

Of course there are many engineering approaches spread over this spectrum. Most SUDS devices themselves include some element of storage. And storage within a sewer system does not have to be provided by large downstream tanks. Small storage devices can be distributed within a catchment close to individual properties, with flow controls set to make optimum use of the storage volumes created. Andoh and Declerk (1999) show that distributed storage can lead to significant cost savings by reducing the capacity needed downstream in a sewer system. (Though the same study suggests that SUDS devices can lead to even greater savings.)

Problems

13.1 A development has an impermeable area of 1.8 ha. Basing rainfall estimation on the formula $i = 750/(t + 10)$, where i is rainfall intensity in mm/h and t is duration in minutes, determine the volume of storage needed to limit outfall to 70 l/s. (Try storm durations of 8, 12 and 16 minutes.) [72.4 m³]

13.2 A detention tank on a sewerage scheme is rectangular in plan: 25 m × 4 m. It is being operated in such a way that the only outflow is over a weir. The flow-rate over the weir is given by the standard expression for a rectangular weir, in which $C_D = 0.63$. The width of the weir is 1.5 m. Consider the following case. Initially inflow is zero, and the water surface is at the level of the weir crest. Then inflow increases at a uniform rate over 12 minutes to 0.9 m³/s, and reduces immediately at the same rate back to zero.

Determine the peak outflow, using a time step of 2 minutes.

[0.8 m³/s]

13.3   How much did the tank in Problem 13.2 attenuate or delay the hydrograph peak? How would normal operation of the tank differ from that described in Problem 13.2?

References

Andoh, R.Y.G. and Declerk, C. (1999) Source control and distributed storage: a cost-effective approach to urban drainage for the new millennium? *Proceedings of the 8th International Conference on Urban Storm Drainage*, Sydney, August/September, 1997–2005.

Andoh, R.Y.G., Faram, M.G., Stephenson, A. and Kane, A. (2001) A novel integrated system for stormwater management. *Novatech 2001, Proceedings of the 4th International Conference on Innovative Technologies in Urban Drainage*, Lyon, France, 433–440.

Hall, M.J., Hockin, D.L. and Ellis, J.B. (1993) *Design of Flood Storage Reservoirs*, CIRIA/Butterworth-Heinemann.

Water Research Centre (1997) *Sewerage Detention Tanks – a Design Guide*, WRc, Swindon.

14 Pumped systems

14.1 Why use a pumping system?

As indicated in Chapter 1, the need for urban drainage arises from human interaction with the natural water cycle. Sewers usually drain in the same direction that nature does: by gravity. Gravity systems tend to be seen as requiring little maintenance, certainly when compared with systems involving a significant amount of mechanical equipment or the need to maintain fixed pressures. And while neglect is undesirable, so is unnecessary maintenance, and so gravity sewer systems prevail. This can be seen as the result of an implicit decision to accept high capital costs (for deep, large and expensive sewers) if they result in low operating costs.

But in some cases gravity is not enough, usually when it is not cost-effective to provide treatment facilities for each natural sub-catchment. In these circumstances, it is appropriate to pump, and the overall methods and technology used are considered in this chapter. Some engineers have favoured non-gravity systems for more general application, and these approaches are discussed at the end of the chapter.

14.2 General arrangement of a pumping system

Sewer pumping systems have a number of general features.

- In sewer systems based mostly on gravity flow, pumped sections require comparatively high levels of maintenance. Engineers, therefore, prefer to keep the pumped lengths to a minimum: to lift the flow as required and then the system can revert to gravity flow as soon as possible. (Fig. 14.1 gives a simple section of a typical arrangement.)
- The liquid being pumped contains solids, and therefore pumps must be designed with the risk of clogging in mind. The nature of the liquid also creates risks of septicity, corrosion of equipment and production of explosive gases (as will be considered in Chapter 17).
- It is common for pumps to deliver flow at a fairly constant rate or, where there are a number of pumps which may work in combination,

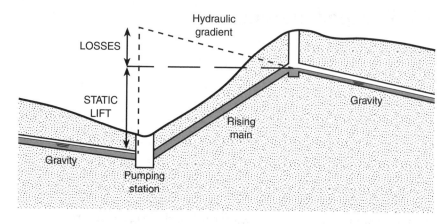

Fig. 14.1 Typical sewer pumping arrangement

there may be a number of alternative fixed rates. Whatever the flow-rate handled by the pumping system, it must generally exceed the rate of flow arriving from the gravity system, otherwise there would be a risk of flooding. So pumping systems tend to work on a stop–start basis, with flow arriving at a reception storage (a 'wet well'), as shown in the simplified pumping station arrangement in Fig. 14.2. When the pumps are operating, the wet well empties; and when the pumps are not operating the wet well fills. The water level in the sump is used to trigger the stop and start of the pumps.

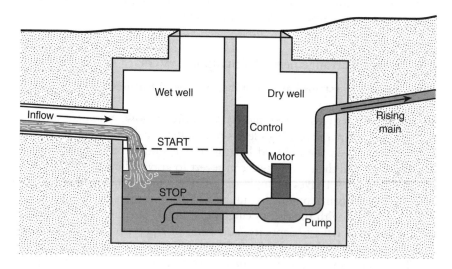

Fig. 14.2 Simplified pumping station arrangement

14.3 Hydraulic design

14.3.1 Pump characteristics

Hydraulically, the function of a pump is to add energy, usually expressed as head (energy per unit weight) to a liquid. The hydraulic performance of a pump can be summed up by the 'pump characteristic curve', a graph of the head added to the liquid, plotted against flow-rate. A typical pump characteristic is given in Fig. 14.3(a); this shows generally-reducing head for increasing flow-rate, but it is not a simple relationship – what goes on inside a pump is complex in hydraulic terms. The different types of pumps available are considered in Section 14.5. The characteristics for each type of pump are derived from tests carried out by the manufacturer, and are available in the manufacturer's literature.

But at what values of flow-rate and head will the pump operate when connected to a particular pipe system? The engineer answers that question at the design stage in the following way.

14.3.2 System characteristics

The pipe system to which the pump will be connected will have a characteristic curve of its own: the 'system characteristic'. Water must be given head in order to:

- be lifted physically (the 'static lift' – see Fig. 14.1)
- overcome energy losses due to pipe friction and local losses at bends, valves etc. As flow-rate increases, energy losses increase in proportion to the square of velocity (as set out in Section 8.3)
- provide velocity head if the water is discharged to atmosphere at a significant velocity.

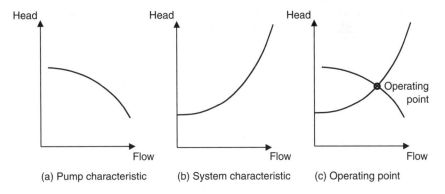

Fig. 14.3 Pump and system characteristic curves

So the system characteristic can be determined from:

head = static lift + losses and velocity head

Losses and velocity head are proportional to velocity squared. A typical system characteristic is shown in Fig. 14.3(b).

So there are two characteristics: the pump characteristic, which gives the head that a pump is capable of producing while delivering a particular flow-rate, and the system characteristic, which gives the head that would be required for the system to carry a particular flow-rate. If a specific pump is going to be connected to a specific system, there is only one set of conditions where what the pump has to offer can satisfy what the system requires: it is the point where the pump characteristic and the system characteristic cross, as shown on Fig. 14.3(c). This is called the *operating point* or *duty point*.

14.3.3 Power

The power required at the operating point can be derived from the operating flow-rate and head in conjunction with the pump's efficiency.

Power (P), *energy per time*, is the product of *weight per time* $\rho g Q$ and head, *energy per weight*. So:

$$P = \rho g Q H \qquad (14.1)$$

ρ density (taken as 1000 kg/m³ for water)
g gravitational acceleration, 9.81 m/s²
Q operating flow-rate (m³/s)
H operating head (m)

A pump gives power to the water, and it receives power ('power supply'), usually in the form of electrical power. The pump and motor are not 100% efficient at converting the power supply into power given to water. Efficiency (power given to the water divided by power supplied to the pump) varies with flow-rate, and can be taken from the manufacturer's plot (for example, Fig. 14.4). Therefore:

$$\text{Power supplied} = \frac{\rho g Q H}{\eta} \qquad (14.2)$$

where η = efficiency (–).

Example 14.1 demonstrates these calculations. Note:

- Water level in the sump is not constant because, as the pump drains the sump, the level goes down (and therefore the static head increases). This may be significant and, if it is, must be taken into account in design.

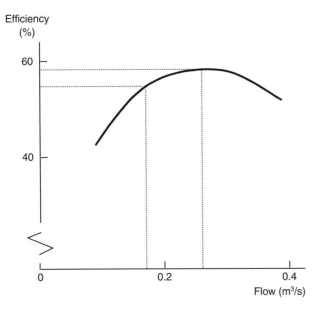

Fig. 14.4 Pump efficiency against flow

Example 14.1

A pump in a sewer system is connected to a rising main with a diameter of 0.3 m and a length of 105 m. The rising main discharges to a manhole at a level 20 m above the water level in the sump. The roughness k_s of the rising main is 0.3 mm, and local losses total $0.8 \times v^2/2g$.

The pump has the following characteristics:

Q (m³/s)	0	0.1	0.2	0.3	0.4
H (m)	33	32	29	24	16
efficiency (%)		42	56	57	49

Determine the flow-rate, the head and the power supplied at the operating point.

Solution

The system characteristic is given by:

total head required = static lift + friction losses + local losses + velocity head

This can be expressed as:

$$H = 20 + \frac{\lambda L}{D}\frac{v^2}{2g} + 0.8\frac{v^2}{2g} + \frac{v^2}{2g}$$

We can find λ from the Moody diagram (Fig. 8.4), and its value will be constant, provided flow is rough turbulent. Assuming that it is,

$$\frac{k_s}{D} = \frac{0.3}{300} = 0.001, \text{ giving } \lambda = 0.02$$

So $\qquad H \quad = 20 + \dfrac{v^2}{2g}\left[\dfrac{0.02 \times 105}{0.3} + 0.8 + 1\right]$

$$= 20 + 8.8\frac{v^2}{2g}$$

Of course: $Q \quad = vA$

So $\qquad v \quad = \dfrac{4Q}{\pi 0.3^2}$

From this we can determine the relationship between H and Q for the pipe system (the system characteristic).

Alternatively we can use Wallingford charts or tables (see Section 8.3.4) to determine the system characteristic. Local losses + velocity head $= (0.8 + 1)\dfrac{v^2}{2g}$, and this can be expressed as an equivalent length using equation 8.15.

$$\text{So } L_E = D\frac{1.8}{\lambda} = 27 \text{ m}$$

For the system, for any value of Q: $\quad H = 20 + 132 \times S_f$ (from chart or table).

The system characteristic (from either method) is plotted (as on Fig. 14.3(c)) together with the pump characteristic. The operating point is where the lines cross; and, at this point, flow-rate is 0.26 m³/s. This gives a velocity of 3.7 m/s, giving R_e of about 10^6 – in the rough turbulent zone, so the assumption about constant λ is valid.

At the operating point, head is 26 m.

Pump efficiency has been plotted on Fig. 14.4. At a flow-rate of 0.26 m³/s, efficiency is 57%.

So: power supplied $= \dfrac{\rho g Q H}{\eta} = \dfrac{\rho g \times 0.26 \times 26}{0.57} = 116 \text{ kW}$

- The velocity head is sometimes insignificant in relation to the losses and is ignored. In this example, velocity head was not insignificant and was rightly included.
- In another case, instead of discharging to atmosphere at a manhole, the rising main outlet might be 'drowned', for example, submerged in a tank into which the liquid is being pumped. In this case, the static lift must be measured up to the liquid surface in the tank. This surface

is unlikely to have a velocity and, therefore, velocity head will not be included. However, there will be exit losses at the point where the rising main discharges to the tank.

14.3.4 *Pumps in parallel*

A common arrangement is for two (or more) pumps to be placed in parallel (Fig. 14.12). One pump may act as a standby to replace others when there is a fault, or reinforce them when high discharges are needed. When two identical pumps are operating in parallel, each delivers flow-rate Q and raises the head by H (Fig. 14.5), so overall the flow-rate is $2Q$, all experiencing an increase in head of H. The characteristic for two pumps in parallel is given in Fig. 14.6. For each value of H, the flow-rate is doubled

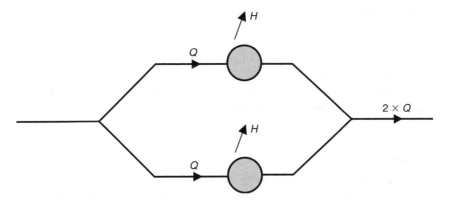

Fig. 14.5 Pumps in parallel (schematic)

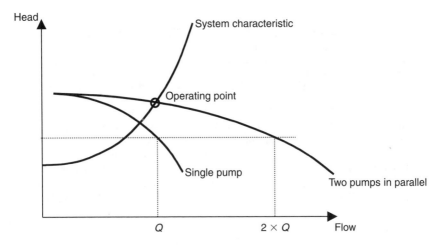

Fig. 14.6 Operating point: pumps in parallel

to 2Q. The new operating point is given by the intersection with the system characteristic (Fig. 14.6).

For pumps in parallel, care is needed when determining the efficiency. The operating flow-rate on Fig. 14.6 is for both pumps together. Half that value gives the flow-rate in each pump, and this should be used in determining efficiency from Fig. 14.4, as this gives the efficiency for a single pump. Example 14.2 demonstrates this.

14.3.5 Suction and delivery pipes

The pipe on the upstream, or inlet side of a pump is referred to as the *suction pipe*, and the pipe on the downstream, or outlet side, is referred to as the *delivery pipe*. In Examples 14.1 and 14.2, the suction pipe was short and was not considered separately. This arrangement is common in drainage applications. It is also common for pumps to be below the level of liquid in the sump, as in Fig. 14.2, to ensure that the pumps remain 'primed' (full of liquid). However, where this is not the case and the suction pipe is long or the pump is at a higher level than the sump level, it is important to ensure that pressure on the suction side of the pump stays well above the vapour pressure of the liquid. This is to avoid cavitation – explained in more detail by Chadwick and Morfett (1998).

Example 14.2

For Example 14.1, what would be the flow-rate, head and power supplied at the operating point if an additional pump, identical to the first, was arranged in parallel?

Solution

The pump characteristic for the pumps in parallel is determined by doubling the flow-rate for each value of H:

For one pump, Q (m³/s)	0	0.1	0.2	0.3	0.4
Two pumps in parallel, Q (m³/s)	0	0.2	0.4	0.6	0.8
Head, H	33	32	29	24	16

The characteristic for two pumps in parallel can be plotted, together with the system characteristic (and, for the purposes of illustration, the characteristic for one pump) as on Fig. 14.6. At the operating point, flow-rate is 0.34 m³/s, and head is 30 m. As has already been pointed out, care is needed when handling efficiency for pumps in parallel. The flow-rate in each pump is 0.17 m³/s, therefore the efficiency of each pump (Fig. 14.4) is 54%.

So, power supplied $= 2 \times \dfrac{\rho g \times 0.17 \times 30}{0.54} = 185\,\text{kW}$

14.4 Rising mains

14.4.1 *Differences from gravity sewers*

It is useful to consider the ways in which rising mains are different from gravity sewers.

Hydraulic gradient

Gravity pipes are designed assuming that the hydraulic gradient is numerically equal to the pipe slope. As shown in Section 8.4, this is because, in part-full pipe flow, the hydraulic gradient coincides with the water surface and therefore, in uniform flow, is parallel to the invert of the pipe. In a rising main, of course, none of this applies. At the pumps, the flow is given a sudden increase in head and this is 'used' to achieve the static lift and overcome losses along the pipe (Fig. 14.1). The slope of the hydraulic gradient is the natural one: downwards in the direction of flow, while the rising main does its job: it rises. The pipe can be laid at a constant depth and follow the profile of the ground.

Flow is not continuous

At times there may be no flow in the main, and at others there may be a number of alternative flows, depending on how many pumps are operating. When the pumps are not operating, wastewater stands in the rising main. Therefore, it is important that when pumping resumes, the velocities are sufficient to scour deposited solids.

A standard value for minimum (scour) velocity is 0.75 m/s (considered further in Chapter 15), but if a velocity of 1.2 m/s is achieved for several hours a day, the minimum could be as low as 0.5 m/s. The minimum suitable diameter is usually considered to be 100 mm.

To avoid septicity, wastewater should not be retained in a rising main for more than 12 hours. It is sometimes necessary to arrange for addition of oxygen or oxidising chemicals to control septicity (considered in Sections 17.6 and 22.5).

When the range of flows is high, dual rising mains can be used to maintain velocities high enough to prevent deposition. One can also act as standby; but, in this case, both mains must be used regularly to avoid septicity.

One possible consequence of starting and stopping the flow is extremely high or low pressures resulting from surge waves, considered in Section 14.4.3.

Power input

We must provide power to create flow in the system. The power is needed year after year for as long as the system operates. This creates trade-offs in selection of an economic design.

A smaller diameter pipe will be cheaper, and the resulting higher velocities will help to ensure scouring of deposits, but the higher velocities will also create higher head losses (proportion to velocity squared) and, therefore, higher power costs. The economic decision will need to consider design life, and time-related comparisons of capital and operating costs.

The pipes are under pressure

There is, of course, no open access to rising mains in the way that there is for gravity sewers.

There are some economic advantages of rising mains in comparison with gravity sewers. Because rising mains are under pressure, and always full, the diameter tends to be smaller and the depth of excavation less than a gravity pipe (which is usually not full and must slope downwards).

14.4.2 Design features

Common materials for rising mains are ductile iron, steel and some plastics. Flexible joints are preferable to allow for differential settlement and other causes of underground stress. (There is more detail on pipe material and construction in Chapter 15.)

Valves and other hydraulic features need to be included at key points in a rising main. There should be an isolating valve, normally a 'sluice' (or gate) valve, near the start of the rising main, so that the pumping station pipework can be worked on without emptying the main. There must be a non-return (or *reflux*) valve which prevents back-flow when pumping stops. Summits (local high points) in the rising main should be avoided, but where unavoidable, should be provided with air release valves. Washout facilities (for emptying the main) should be provided at low points in the main.

Thrust blocks may be needed to withstand the forces created when water is forced to change direction. Their design is beyond the scope of this text but is covered by Thorley and Atkinson (1994).

14.4.3 Surge

One possible risk in rising mains – flowing under pressure with potentially rapid changes of flow – is *surge*.

A change in flow in a liquid is always associated with a change in pressure. If flow changes rapidly, for example as a result of a pump suddenly stopping, these changes in pressure can be high. The effect is known as surge, and the consequences of ignoring it at the design stage can be disastrous, with the creation of pressures high or low enough to cause damage to pipes. Not all pumping systems are likely to suffer from serious surge problems, and many devices for overcoming the problems can be very simple, but surge must be considered when a pumping system is being designed.

A crucial factor is the rate at which the flow changes. If the flow changes gradually, the normal assumption that water (or wastewater) is incompressible can be maintained, and the changes in pressure are not high. If the flow changes rapidly, the compressibility of water must be considered, and the changes of pressure can be great. Methods of predicting the changes in pressure are considered by Chadwick and Morfett (1998), and in greater detail by Creasey (1977) and Swaffield and Boldy (1993).

Most standard (clean water) surge protection devices (e.g. air vessels or surge tanks) are inappropriate for wastewater application because of the problems of blockage or stagnation of the stored liquid. A simple way of ensuring a gradual cut-out is to add a *flywheel* to the pump (though this, of course, adds a load when the pump is started). Perhaps the best way is the judicious use of motor controls. Fortunately, most rising mains have relatively low pumping head.

14.5 Types of pump

As has been stated, the function of a pump is to add energy to a liquid. There are a number of ways in which this energy can be transferred, but the most common is by a rotating 'impeller' driven by a motor (a *rotodynamic pump*).

The most common rotodynamic pump for use with wastewater is a centrifugal pump in which the impeller forces the liquid radially into an outer chamber called a 'volute' (Fig. 14.7). In effect, the volute converts velocity head into pressure head. The impeller has a special design to avoid clogging by solids, and this feature means that centrifugal pumps for wastewater tend to have lower efficiencies (about 50 to 60%) than centrifugal pumps for clean water (up to 90%). A common requirement is for these pumps to be capable of handling a 100 mm diameter sphere. They are suitable for a wide range of conditions – flow-rate: 7 to 700 l/s; and head: 3 to 45 m and typically operate at low speeds of around 900 rpm. Centrifugal pumps require priming (filling with water before pumping can begin) and so must normally be installed below the lowest level of wastewater to be pumped.

Axial-flow pumps are simpler than centrifugal pumps and have impellers (acting like a propeller) that force the liquid in the direction of the longitudinal axis (Fig. 14.8). Axial flow pumps are suited to relatively high flow-rates and low heads with efficiencies of 75–90%. Unlike centrifugal pumps, axial-flow pumps suffer a rapid decrease in head with

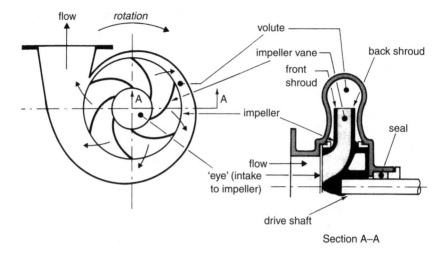

Fig. 14.7 Centrifugal pump (reproduced from Chadwick and Morfett [1998] with permission of E & FN Spon)

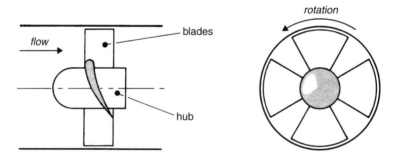

Fig. 14.8 Axial flow pump (reproduced from Chadwick and Morfett [1998] with permission of E & FN Spon)

increased discharge. In *mixed-flow pumps*, the direction in which the water is forced by the impeller is at an intermediate angle, so flow is partially radial and partially axial. Mixed-flow pumps are recommended for medium heads between 6 and 18 m. Axial and mixed-flow pumps are most appropriate for pumping stormwater. Fig. 14.9 illustrates the shape of pump characteristics for the main types of pumps. In practice, pump selection for a particular application is made by matching requirements to manufacturer's data.

The pump and the motor that drives it are often kept in a 'dry well', separate from the wastewater (Fig. 14.2). But an alternative is a *submersible pump* in which the pump and motor are encased in waterproof protection and submerged in the wastewater which is to be pumped

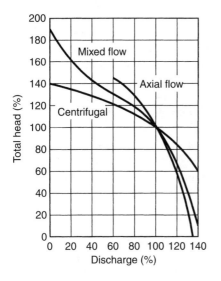

(a) Discharge–head

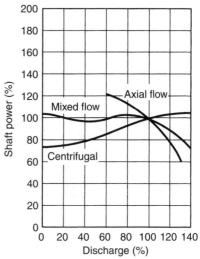

(b) Discharge–power

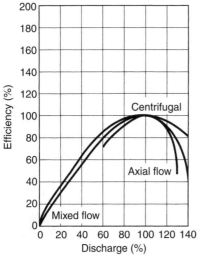

(c) Discharge–efficiency

Fig. 14.9 Characteristics of various pump types (reproduced from Kay [1998] with permission of E & FN Spon)

(Fig. 14.10). This greatly simplifies the design of the pumping station, and is common for fairly small installations.

For very small installations, a rotodynamic pump may not be suitable, because the risk of clogging places a limit on the smallness of a pump. An alternative system is a *pneumatic ejector*, in which the wastewater flows by gravity into a sealed unit and is then pushed out using compressed air. These require little maintenance and are not easily blocked by solids. However, they are of low efficiency and limited capacity (1 to 10 l/s).

While many pumps operate at a single fixed speed, some types switch between two or more speeds ('multi-speed'), and others can run at continuously-variable speeds. The benefit of variable speed pumps is that the pumped outflow from the pumping station can be more closely matched to the inflow (from the system), and therefore less storage volume is required. Also, pumps do not have to be stopped and started so frequently, deposition resulting from liquid lying still in the rising main is reduced, and flow-rates, velocities and therefore losses tend to be lower. However, variable speed pumps are more expensive, require more complex control arrangements, and may be very inefficient at some speeds.

14.6 Pumping station design

14.6.1 *Main elements*

The design of pumping stations usually involves the integration of a number of branches of engineering, including civil, structural, mechanical, electrical and electronic. Also, a pumping station is one of the few elements of an urban drainage system that can be seen above ground, so there may also be significant architectural aspects (Fig. 14.11). They will require a planning application, in which noise and odour, as well as appearance, may be issues. The extent of all these aspects will, of course, depend on size – pumping stations in urban drainage schemes may be very small, serving just a few people, or they may be large and complex engineering structures serving large populations.

The basic components of a pumping station have already been described in this chapter. Pumps (nearly always more than one) take sewer flow from a reception volume, a sump, and deliver it with increased head into a rising main. The pumps, most commonly centrifugal, are driven by motors, which must be provided with a supply of electrical power. There must be arrangements for controlling the pumps, usually related to liquid level in the sump.

Wet well–dry well

When the pumps and motors are kept completely separate from the liquid, the sump is referred to as the 'wet well', and the chamber containing

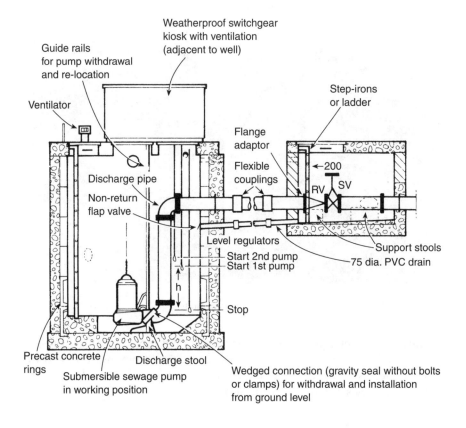

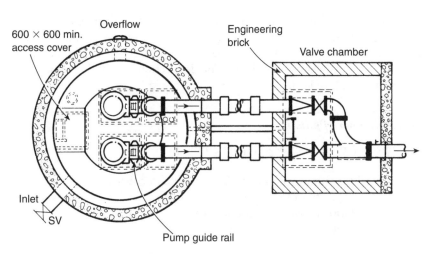

Fig. 14.10 Submersible pump (reproduced from Woolley [1988] with permission of E & FN Spon)

Fig. 14.11 Pumping station: architectural treatment

pumps, etc. is referred to as the 'dry well'. The motors may be directly beside the pumps or, to provide further remoteness from moisture and for ease of access, may be at a higher level, connected by a long shaft. Fig. 14.12 shows a typical configuration.

Wet well only

The wet well–dry well separation is not needed when submersible pumps are used (as already illustrated in Fig. 14.10). The single wet well in which submersible pumps are placed can be of a simple construction, based on precast concrete segmental rings. For inspection or maintenance, the pumps must be lifted out.

14.6.2 Number of pumps

The appropriate number of pumps is a function of

- the need for standby pumps to be available to cover for faults
- the flow capacity of the pumps, alone and in parallel, determined from the calculations covered in Section 14.3
- the variation in inflow.

The simplest pumping station consists of a duty/standby arrangement. It is common, however, in larger installations to have a number of pumps,

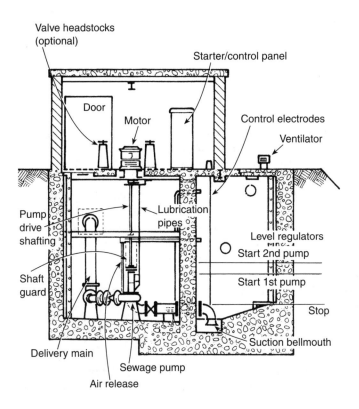

Valve headstocks
(optional)

Starter/control panel

Door

Motor

Control electrodes

Ventilator

Pump
drive
shafting

Lubrication
pipes

Level regulators
Start 2nd pump

Shaft
guard

Start 1st pump

Stop

Suction bellmouth

Delivery main

Sewage pump

Air release

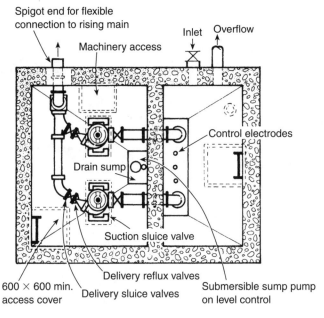

Spigot end for flexible
connection to rising main

Inlet   Overflow

Machinery access

Control electrodes

Drain sump

Suction sluice valve

600 × 600 min.
access cover

Delivery reflux valves

Delivery sluice valves

Submersible sump pump
on level control

Fig. 14.12 Pumping station arrangement (reproduced from Woolley [1988] with permission of E & FN Spon)

arranged in parallel, which are brought into use successively as inflow increases.

14.6.3 Control

In most systems, while the pumps are running, the level in the sump is falling. At a fixed level, the pumps are turned off and the level starts to rise. Subsequently, the level reaches the point at which pumping is resumed.

All pumping stations require some control system. The basic requirement is sensing of upper and lower sump level, and the consequent starting and stopping of the pumps. With more pumps, and more complex starting and stopping procedures, the complexity of the control system increases. Common methods of sensing water level are by ultrasonic detector, float-switches and electrodes. The safe frequency of operation of the electric motor starter is limited; it is typical to design for between 6 and 12 starts/hour.

14.6.4 Sump volume

To determine the required sump volume (V) between 'stop' and 'start' levels, the time taken to fill the sump while the pump is idle (t_1) is given by:

$$t_1 = \frac{V}{Q_I}$$

where Q_I is the inflow rate. The time taken to pump out the sump (t_2) is:

$$t_2 = \frac{V}{(Q_o - Q_I)}$$

where Q_o is the outflow (pump) rate. Thus, the time between successive starts of the pump, the pump cycle (T), is:

$$T = \frac{V}{Q_I} + \frac{V}{(Q_o - Q_I)} = \frac{VQ_o}{Q_I(Q_o - Q_I)} \tag{14.3}$$

Now, to find the minimum sump volume required, V is differentiated with respect to Q_I and equated to zero:

$$\frac{dV}{dQ_I} = \frac{T(Q_o - 2Q_I)}{Q_o} = 0$$

$$Q_o = 2Q_I \tag{14.4}$$

So, the pump sump should be sized for a pumped outflow rate that is double the inflow rate. Substituting equation 14.4 into equation 14.3 gives:

$$V = \frac{TQ_o}{4}$$

If $n = 3600/T$ is the number of motor starts per hour:

$$V = \frac{900Q_o}{n} \tag{14.5}$$

Thus, the required sump volume is determined from the rate of outflow in conjunction with the allowable frequency of motor starts.

Example 14.3

The peak inflow to a sewerage pumping station is 50 l/s. What capacity sump and duty/standby pumps will be required if the number of starts is limited to 10 per hour. How long will the pump operate during each cycle?

Solution

For minimum sump volume, $Q_o = 2Q_I$ so:
Capacity of duty pump, $Q_o = 100$ l/s
Capacity of standby pump, $Q_o = 100$ l/s
Pump sump volume (equation 13.5),

$$V = \frac{900 \times 0.1}{10} = 9 \text{ m}^3$$

Time taken to empty sump,

$$t_2 = \frac{9}{(0.1 - 0.05)} = 180 \text{ s} = 3 \text{ min}$$

14.6.5 Flow arrangements

Within a pumping station, pipework is usually ductile iron with flanged joints. Flexible joints should be placed outside walls to allow for differential settlement. For each pump there should be a sluice valve on the suction and delivery sides for isolating the pump (Fig. 14.12).

The base of the sump is usually given quite steep slopes to limit deposition of solids (Fig. 14.12). More detail on sump arrangements is given by Prosser (1977).

In large pumping stations, some form of preliminary treatment to remove large solids, most commonly by means of screens, may be necessary.

All pumping stations should have an emergency overflow in case of complete failure of the pumps, with storage for wastewater inflow during emergency repairs. In combined systems, it may be necessary to provide an overflow for storm flows. This would be based on the same principles as other CSOs, described in Chapter 12.

14.6.6 Maintenance

A pumping station, with mechanical, electrical and control equipment, is one part of a sewer network that has obvious maintenance needs. And while it is true of any part of a sewer system, it is particularly important that the maintenance needs of pumping systems are taken into account at the design stage. Care and expense in design may reduce the cost of maintenance, and care and expense in maintenance may reduce the cost of replacement.

Priorities in taking maintenance requirements into account in design are to ensure that:

- it is possible to isolate and remove the main elements of pipework and equipment. There must be access to allow the pumps to be lifted out vertically; this is especially true for submersible pumps which it must be possible to lift out with ease
- problems caused by solids can be overcome (suitable pumps, pipe sizes, access to clear blockages)
- emergencies caused by breakdown, power failure, etc. can be coped with.

The possibility of power failure needs to be taken into account. Provision of a standby generator, or a diesel-powered pump in larger stations, is a common precaution.

Maintenance procedures for pumping installations are covered by Wharton *et al.* (1998) and Sewers and Water Mains Committee (1991).

An important element in maintenance is monitoring performance. Small to medium-sized pumping stations are usually controlled from the wastewater treatment plant that they serve, by telemetry (considered further in Chapter 22). The types of information likely to be communicated are:

- failure in the electricity supply
- pump failure
- unusually high levels in the wet well

- flooding of the dry well
- operation of the overflow.

The information is needed for effective operation of the system, and in particular to aid decisions about when to attend to operational problems. Pumping stations may also be fitted with flow measurement devices, used to monitor performance, and (potentially) as part of a management system for the catchment (Chapter 22).

More detailed guidance on practical aspects of pumping station design is given by Prosser (1992), Wharton *et al.* (1998), BS EN 752–6, and for smaller installations in *Sewers for Adoption* (WRc, 2001).

14.7 Vacuum systems

Where the ground surface is very flat, or where ground conditions make construction of deep pipes difficult, an alternative to gravity drainage is 'vacuum sewerage'. Wastewater is drained from properties by gravity to collection sumps. When the liquid surface in the sump rises to a particular level, an *interface valve* opens and wastewater is drawn into a pipe in which low pressure (in the order of −0.6 bar) has been created by a pump. After the collection sump has been emptied, the interface valve remains open for a short time to allow a volume of air at atmospheric pressure to enter the pipe. The mixture travels at high velocity (5–6 m/s) towards the vacuum source. The wastewater is then retained in a collection vessel for subsequent pumped removal.

Vacuum systems consist of small, shallow pipes with relatively high running costs, compared with the large, deep pipes and low running costs of gravity systems. In the right circumstances, vacuum systems show overall cost advantages (Consterdine, 1995).

There are a few vacuum systems in the UK (see, for example, Stanley and Mills, 1984; Ashlin *et al.*, 1991).

Problems

14.1 A pumping system has a static lift of 15 m. The pump characteristics are below, together with the total losses in the rising main (velocity head can be neglected).

	Q (m^3/s)	0	0.05	0.1	0.15	0.2
Pump:	*H (m)*	25	24	20	14	7
	efficiency (%)		45	55	55	50
Rising main:	*losses (m)*	0	1	4	9	16

Determine the flow-rate, head delivered and power supplied to the pump at the operating point. If the diameter of the rising main is 250 mm, are conditions suitable for scouring of deposited solids? If the rising main became rougher with age, would the flow-rate, and head, increase or decrease?

[0.105 m³/s, 19 m, 36 kW, v = 2.1 m/s OK, Q decrease, H increase]

14.2 For the same system as Problem 14.1, if an additional identical pump is operating in parallel, determine the total flow-rate, head delivered and power supplied to the pumps at the operating point. Which arrangement – one pump or two in parallel – uses power more efficiently?   [0.14 m³/s, 23 m, 64 kW, one pump]

14.3 There are three main categories of rotodynamic pumps. Describe for each category (a) their basic mode of operation (b) their advantages and disadvantages and (c) their application in urban drainage.

14.4 A pumping station sump is being designed to suit an inflow of 30 l/s. What rate of pumped outflow would give the minimum sump volume? What sump volume would then be required if the pump was to operate at (i) 6 starts/hour and (ii) 12 starts/hour? At 12 starts/hour there is 5 minutes between each start. For how much of that time is the pump operating, and for how long is it idle?

[60 l/s, 9 m³, 4.5 m³, 2.5 minutes]

14.5 Designing a pumping system presents a different set of challenges from designing a gravity system. Explain why.

Key sources

Prosser, M.J. (1992) *Design of low-lift pumping stations*. Report 121, CIRIA, London.

References

Ashlin, D.E., Bentley, S.E. and Consterdine, J.P. (1991) Vacuum sewerage – the Four Crosses experience. *Journal of the Institution of Water and Environmental Management*, 5(6), December, 631–640.

BS EN 752–6: 1998 Drain and sewer systems outside buildings – Part 6: Pumping installations.

Chadwick, A. and Morfett, J. (1998) *Hydraulics in Civil and Environmental Engineering*, 3rd edn, E & FN Spon.

Consterdine, J.P. (1995) Maintenance and operational costs of vacuum sewerage systems in East Anglia. *Journal of the Chartered Institution of Water and Environmental Management*, 9(6), December, 591–597.

Creasey, J.D. (1977) *Surge in water and sewage pipelines. Measurement, analysis and control of surge in pressurized pipelines*, Technical Report TR51, Water Research Centre.

Kay, M. (1998) *Practical Hydraulics*, E & FN Spon.

Prosser, M.J. (1977) *The hydraulic design of pump sumps and intakes*, CIRIA with BHRA (Fluid Engineering).

Sewers and Water Mains Committee (1991) *A guide to sewerage operational practices*, WSA/FWR.

Stanley, G.J. and Mills, D. (1984) Vacuum sewerage. *International Conference on the Planning, Construction, Maintenance and Operation of Sewerage Systems*, Reading, 317–326.

Swaffield, J.A. and Boldy, A.P. (1993) *Pressure surge in pipe and duct systems*, Avebury Technical.

Thorley, A.R.D. and Atkinson, J.H. (1994) *Guide to the design of thrust blocks for buried pressure pipelines*, Report 128, CIRIA, London.

Wharton, S.T., Martin, P. and Watson, T.J. (1998) *Pumping stations, design for improved buildability and maintenance*, Report 182, CIRIA, London.

Woolley, L. (1988) *Drainage Details*, 2nd edn, E & FN Spon.

WRc (2001) *Sewers for Adoption – a Design and Construction Guide for Developers*, 5th edn, Water UK.

15 Structural design and construction

15.1 Types of construction

Most sewers are constructed underground. This is achieved by one of three general methods:

- open-trench construction
- tunnelling
- trenchless construction.

Open-trench construction consists of excavating vertically along the line of the sewer to form a trench, laying pipes in the trench, and backfilling – see Fig. 15.1. It is suitable for a wide range of pipe sizes and depths, and is the common method for small- to medium-scale sewer construction.

Both *tunnelling* and *trenchless construction* involve excavating vertically at a particular location for access and then excavating outwards at an appropriate gradient to form the space for the sewer to be constructed.

Tunnelling generally involves sizes large enough for human access in which a lining (eventually part of the sewer fabric) is constructed from inside the excavation. Tunnelling tends to be associated with large-scale projects like interceptor sewers.

Underground construction techniques that involve inserting pipes in the ground without a trench are called *trenchless* or 'no-dig' methods. They avoid disruption on the surface, and have become increasingly popular as the technology has developed and engineers have become more aware of the costs to business and society of conventional trench construction. As well as its use in construction of new sewers, this type of technology is widely used in sewer rehabilitation, as will be described in Chapter 18.

This chapter describes these three methods of construction. Before doing so, we will consider other physical aspects of sewer pipes and their design.

Pipelines must possess a number of physical properties (see Section 15.2, below). They must also give satisfactory performance hydraulically

Fig. 15.1 Open-trench construction (courtesy of Clay Pipe Development
   Association)

and structurally. Hydraulic performance has been considered in Chapter 8. The important issue of structural performance is considered in Section 15.3.

15.2 Pipes

15.2.1 General

The nominal size (DN) of a pipe is the diameter of the pipe in mm rounded up or down to a convenient figure for reference. In some materials (including clay and concrete) the DN refers to the inside diameter, and in some (including plastics), it refers to the outside diameter. The actual diameter may be slightly different from the DN. A concrete pipe with an actual inside diameter of 305 mm is referred to as 'DN 300'. Of course, the precise diameter must be clear in the manufacturer's product data, and must be used in accurate calculations of hydraulic or structural properties.

General requirements for all materials used in gravity sewer systems are given in BS EN 476: 1998 *General requirements for components used in discharge pipes, drains and sewers for gravity systems.*

15.2.2 Materials

The main characteristics and applications of the common sewer pipe materials will now be considered. Relevant British/European Standards are listed at the end of this chapter; these provide more detailed guidance on properties, specification and structural behaviour. A detailed survey of pipe materials is also given in the *Materials Selection Manual for Sewers, Pumping Mains and Manholes* (Sewers and Water Mains Committee, 1993).

In general, the most important physical characteristics of a sewer pipe material are:

- durability
- abrasion-resistance
- corrosion-resistance
- imperviousness
- strength.

Clay

Vitrified clay is a commonly-used material for small- to medium-sized pipes. Its major advantages are its strength and its resistance to corrosion, making it particularly suitable for foul sewers. However, clay is both heavy and brittle and, therefore, susceptible to damage during handling, storage and installation.

Concrete

Plain, reinforced and prestressed concrete pipe is generally used for medium- to large-sized pipes. It is particularly suited to use in storm sewers because of its size, abrasion resistance, strength and cost. There is potential for corrosion (see Chapter 17), though generally domestic wastewater is not harmful to concrete pipes. Non-circular cross-section pipes are also available. A specific type of prestressed concrete pipe is made for pressure applications.

Ductile iron

Ductile (centrifugally spun) iron pipes are used where significant pressures are expected, such as in pumping mains (Chapter 14) and inverted siphons (Chapter 9), or where high strength is required, such as in onerous underground loading cases and above-ground sewers. Ductile iron is susceptible to corrosion, and needs protection such as zinc coating, bitumen paint and polyethylene sleeving.

Steel

Steel pipes tend to be used in specialist applications where high strength is required. These include sea outfalls, above-ground sewers and pipe bridges. Steel pipes require protection from corrosion – by internal lining and external coating, often supplemented by cathodic protection.

Unplasticised PVC (PVC-U)

PVC-U pipe has found general application in small size pipes. It is lightweight, making installation straightforward, and is corrosion resistant. However, as the pipe is flexible, strength relies on support from the bedding and good construction practice is therefore critical. Smaller sizes are routinely used in building drainage applications, but are also used to some extent for public sewers. External rib reinforced PVC-U pipes utilise the special shape of the pipe wall to increase stiffness for the same volume of material.

Other pipe materials in use include: medium density polyethylene (MDPE), glass reinforced plastic (GRP) and fibre cement. Many existing sewers are made of brick.

Sizes

The range of sizes, and increments in size, depend on the pipe material. For example, clay pipes are available at a number of smaller diameters, with larger diameters at multiples of 100 mm. Traditional sizes for concrete pipes start at 150 mm and increase at 75 mm increments over a wide range. Table 15.1 gives size ranges, together with British/European standards.

15.2.3 Pipe joints

Sewer pipes are usually supplied and laid in the trench in standard straight lengths, and jointed *in situ*. There are several alternative jointing methods, providing either rigid or flexible joints. In most cases, flexible joints are preferred to allow for differential settlement, nonuniform support, drying or other effects, without introducing unacceptable bending moments and stresses in the pipe.

The standard jointing methods are as follows.

Spigot and socket

This joint is illustrated in Fig. 15.2(a). It is the normal jointing method for concrete pipes, larger clay pipes and most ductile iron pipes. The spigot is inserted into the socket. A rigid joint can be created using a material such as cement mortar; however, it is much more common for flexibility and watertightness to be provided by an O-ring of rubber, or equivalent material, placed in a groove at the end of the spigot before insertion (Fig. 15.2(a)). Insertion causes suitable compression of the O-ring. In some arrangements, a ring, gland or gasket is compressed by tightening bolts.

Sleeve

An alternative is to use a separate sleeve, as shown in Fig. 15.2(b). Clay pipes of smaller diameters are commonly jointed using flexible polypropylene sleeves. Plastic pipes use plastic sleeves, including angled flexible strips or O-rings.

Bolted flange joints

Simple bolted flanges do not provide flexibility. They are used in rigid installations where exposed pipework (usually ductile iron) needs to be readily dismantled, as in a pumping station (see Section 13.6).

Table 15.1 Pipe materials: size ranges, and British/European standards

Material	Normal size range (mm)	Principal British/European Standards (full titles are given in References)
Clay	75–1000	BS EN 295, BS 65
Concrete	150–3000	BS 5911
Ductile iron	80–1600	BS EN 598
Steel	60–2235	BS 534
PVC-U	17–630	BS 4660, BS 5481

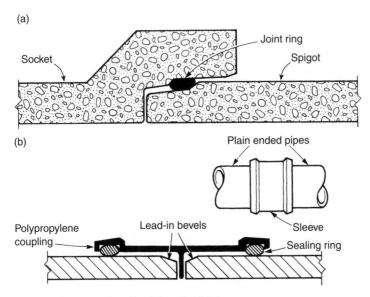

Fig. 15.2 (a) Spigot and socket joint (flexible);
(b) Sleeve joint (reproduced from Woolley [1988] with permission of
E & FN Spon)

15.3 Structural design

15.3.1 Introduction

This section deals with structural design of open-trench sewers. The main
components of an open-trench arrangement are shown in Fig. 15.3. A pipe
laid in a trench has to be strong enough to withstand loads from the soil
above it, from traffic and other imposed loads, and from the weight of
liquid it carries. With increasing depth of cover, the load from the soil
increases and the load from traffic decreases. The ability to withstand the
loads is derived from the strength of the pipe itself and from the nature of
the *bedding* on which it is laid. A number of standard 'classes' of bedding
are defined in Fig. 15.4. The material used to support the pipe, together
with the depth of support, and the material used to back-fill the excavation, are specified. The contribution of the bedding to the overall strength
of the system, is characterised by a 'bedding factor' (given in Fig. 15.4, and
demonstrated in use in the next section).

Different calculation procedures are used depending on whether the
pipe is considered to be rigid (clay, concrete, fibre cement), semi-rigid
(ductile iron) or flexible (plastic). (Steel pipe can be semi-rigid or flexible
depending on its dimensions.) There are also differences between trenches

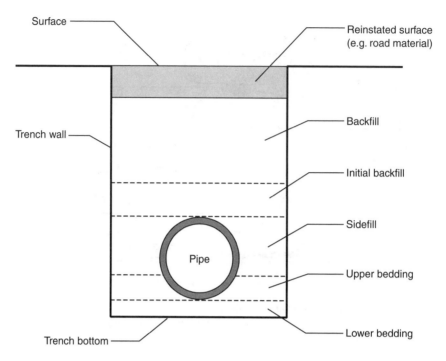

Surface

Reinstated surface
(e.g. road material)

Backfill

Trench wall

Initial backfill

Sidefill

Pipe

Upper bedding

Lower bedding

Trench bottom

Fig. 15.3 Open trench arrangement

considered 'narrow' and those considered 'wide', and between pipes in trenches and pipes under embankments.

Practising engineers tend to carry out structural design of pipes using standard charts, tables or software, which relate loads to the properties of the soil and fill material, and the pipe diameter, width of trench and height of cover. An example is the publication *Simplified tables of external loads on buried pipelines*, Young *et al.* (1986). Engineering firms also use their own in-house reference material.

The basis of the procedures is summarised in BS EN 1295: 1998 *Structural design of buried pipelines under various conditions of loading*. Other useful sources are: Young and O'Reilly (1983), and, specifically for clay pipes, Clay Pipe Development Association (1999). More detailed treatment can be found in two thorough textbooks on the subject: Young and Trott (1984) and Moser (1990).

The procedure most commonly appropriate for sewer design is for a *rigid pipe* carrying gravity flow, that is considered in some detail here. Design of semi-rigid, flexible, and pressure pipes is not considered in detail, but more information can be found in BS EN 1295: 1998, Compston *et al.* (1978) and Moser (1990).

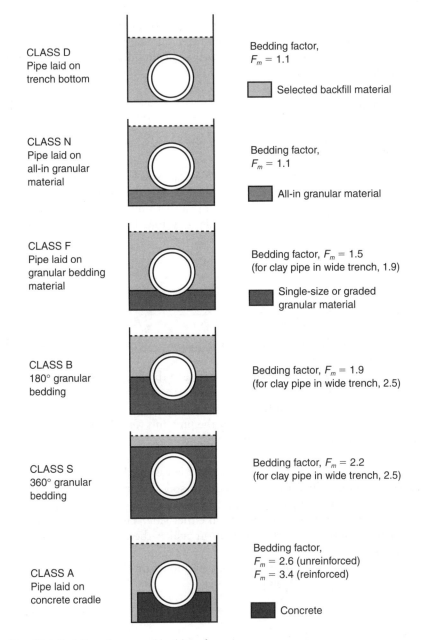

CLASS D
Pipe laid on
trench bottom

Bedding factor,
$F_m = 1.1$

Selected backfill material

CLASS N
Pipe laid on
all-in granular
material

Bedding factor,
$F_m = 1.1$

All-in granular material

CLASS F
Pipe laid on
granular bedding
material

Bedding factor, $F_m = 1.5$
(for clay pipe in wide trench, 1.9)

Single-size or graded
granular material

CLASS B
180° granular
bedding

Bedding factor, $F_m = 1.9$
(for clay pipe in wide trench, 2.5)

CLASS S
360° granular
bedding

Bedding factor, $F_m = 2.2$
(for clay pipe in wide trench, 2.5)

CLASS A
Pipe laid on
concrete cradle

Bedding factor,
$F_m = 2.6$ (unreinforced)
$F_m = 3.4$ (reinforced)

Concrete

Fig. 15.4 Bedding classes and bedding factors

15.3.2 Rigid pipe

The total design external load per unit length of pipe (W_e) is given by the sum of the soil load (W_c), the concentrated surcharge load (W_{csu}) and the equivalent external load due to the weight of liquid in the pipe (W_w) (per unit length in each case):

$$W_e = W_c + W_{csu} + W_w \tag{15.1}$$

Soil load, W_c

In analysis of a *narrow trench* (Fig. 15.5), the soil load is considered to be due to the weight of material in the trench minus the shear between the fill material and the trench sides. W_c – soil load per unit length – is determined from Marston's narrow trench formula, developed using the principles of soil mechanics:

$$W_c = C_d \gamma B_d^2 \tag{15.2}$$

where $C_d = \dfrac{1 - e^{-2K\mu' \, H/B_d}}{2K\mu'}$ (15.3)

K Rankine's coefficient: ratio of active lateral pressure to vertical pressure (–)

μ' coefficient of sliding friction between the fill material and the sides of the trench (–)

γ unit weight of soil (typically 19.6 kN/m³)

B_d width of trench at the top of the pipe (m) (Fig. 15.5)

H depth of cover to crown of pipe (m)

In analysis of a *wide trench* (Fig. 15.5), it is assumed that the soil directly above the pipe will settle less than the soil beside it. The soil load is considered to be due to the weight of soil directly above the pipe plus the shear between this and the soil on either side. The effect is considered to reach up only to a certain height, at which there is a 'plane of equal settlement'.

    W_c – soil load per unit length – is determined on the basis of Marston's theory as developed by Spangler, giving:

$$W_c = C_c \gamma B_c^2 \tag{15.4}$$

where B_c is the outside diameter of pipe (m).

There are two possible cases, illustrated in Fig. 15.5: (1) where the vertical shear planes extend to the top of the cover – '*complete projection*', for which:

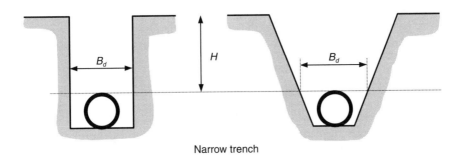

Narrow trench

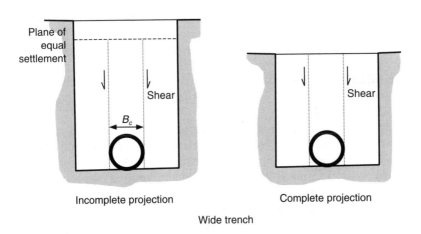

Incomplete projection   Complete projection

Wide trench

Fig. 15.5 Narrow and wide trench

$$C_c = \frac{e^{2K\mu \; H/B_c} - 1}{2K\mu} \tag{15.5}$$

where μ is the coefficient of friction within the soil mass (–).

or (2) where the top of the cover is above the plane of equal settlement – '*incomplete projection*' – for which C_c is determined from H, B_c, and the product of two terms: r_{sd}, the 'settlement deflection ratio', and p, the 'projection ratio'. Expressions for C_c are given in Table 15.2. The term r_{sd} is related to the firmness of the foundation of the trench as given in Table 15.3. The term p is the proportion of the external diameter of the pipe that is above firm bedding.

Table 15.2 C_c for incomplete projection

$r_{sd}p$	Expression for C_c
0.3	$1.39H/B_c - 0.05$
0.5	$1.50H/B_c - 0.07$
0.7	$1.59H/B_c - 0.09$
1.0	$1.69H/B_c - 0.12$

Table 15.3 Values of r_{sd}

Foundation	r_{sd}
Unyielding (e.g. rock)	1.0
Normal	0.5 to 0.8
Yielding (e.g. soft ground)	less than 0.5

K, μ', μ and γ are properties of the fill material and soil. Values of the products $K\mu'$ and $K\mu$ for specific soil types are given in Table 15.4. For a narrow trench, the value of $K\mu'$ used in the calculation should be the lower of the values for the backfill material and for the existing soil in the trench sides. When the soil type is not known, $K\mu'$ is usually taken as 0.13, and $K\mu$ as 0.19. The value of γ is that of the fill material, or a standard value of 19.6 kN/m^3.

In design, it is not known in advance whether the case will be one of complete or incomplete projection. Therefore, both cases should be determined and the lower value of C_c used in equation 15.4.

Similarly, to determine W_c for a trench, the *lower* of the values determined from equations 15.2 and 15.4 should be used. When the value from equation 15.2 is used, the trench is defined as narrow, and the specified width must not be exceeded during construction.

Concentrated surcharge load

The method of determining of this term originates from Boussinesq's equation for distribution of stress resulting from a point load at the surface. Some simplification allows W_{csu} – concentrated surcharge load – to be derived from:

$$W_{csu} = P_s B_c \tag{15.6}$$

P_s surcharge pressure (N/m^2)
B_c outside diameter of pipe (m)

P_s is a function of the depth of cover and the type of loading (for example, light road, main road, or different types of railway) as shown in Fig. 15.6. Whatever the type of loading when the pipe is in use, it is important to check at the design stage that loadings from construction vehicles will not exceed the concentrated surcharge load predicted.

Table 15.4 Values of $K\mu'$ and $K\mu$

Soil	$K\mu'$ or $K\mu$
Granular, cohesionless materials	0.19
Maximum for sand and gravel	0.165
Maximum for saturated top soil	0.15
Ordinary maximum for clay	0.13
Maximum for saturated clay	0.11

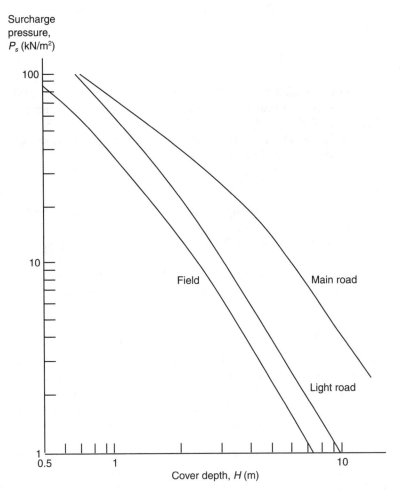

Fig. 15.6 Graph of surcharge pressure against cover depth

Liquid load

The weight of liquid in the pipe is not strictly an external load, so W_w is the *equivalent* external load due to weight of liquid per unit length of pipe. It is taken as a certain proportion, C_w ('water load coefficient'), of the weight of liquid held when the pipe is full.

$$W_w = C_w \rho g \pi D^2/4 \qquad (15.7)$$

ρ density of liquid, in a sewer (1000 kg/m³)
D internal diameter of pipe (m)

C_w depends on the type of bedding, with a general range of values between 0.5 and 0.8; but for simplicity, the relatively conservative value of 0.75 is often used. W_w does not tend to make a significant contribution to the overall loading for pipes under DN 600.

Strength

The strength provided by the chosen combination of pipe material and bedding class is determined by multiplying the strength of the pipe by the factor that indicates the additional strength provided by the bedding, the bedding factor. This overall strength must be sufficient to withstand the total load with an applied factor of safety:

$$W_t F_m \geq W_e F_{se} \qquad (15.8)$$

W_t crushing strength of pipe, provided by manufacturer (N/m²)
F_m the bedding factor (–)
F_{se} factor of safety, normally 1.25 for clay and concrete pipes

The normal alternatives for bedding, and the bedding factor for each, have been given on Fig. 15.4. Example 15.1 shows a full calculation.

15.4 Site investigation

Site investigation identifies problems with ground conditions and special hazards that may have a significant effect on planning, choice of pipe material, structural design and construction. The objectives are to provide information useful in the selection of a scheme from a number of alternatives, to inform the detailed design, to estimate costs and foresee difficulties.

    Site investigation is of great importance for all types of sewer construction, but may have particular significance for schemes involving construction of sewers in tunnel. In this case, the main areas of interest are geological structure, groundwater, existing services and structures, and

Example 15.1

A sewer pipe has an internal diameter of 300 mm and external diameter 400 mm. It is to be laid under a light road in a trench 0.9 m wide, with a cover depth of 2 m. The original ground is unsaturated clay, and the fill is granular and cohesionless. Assume a value for $r_{sd}\,p$ of 0.7, use unit weight of soil $\gamma = 19.6$ kN/m³, and density of liquid $\rho = 1000$ kg/m³.

The pipe is available with a crushing strength of either 36 or 48 kN/m. Select a suitable class of bedding for each of these pipe strengths.

Solution

First assume wide trench:

(1) complete projection, equation 14.5 $C_c = \dfrac{e^{2K\mu H/B_c} - 1}{2K\mu}$

Fill is granular cohesionless, so from Table 14.4, $K\mu = 0.19$, so

$$C_c = \frac{e^{2\times0.19\times2/0.4} - 1}{2 \times 0.19} = 15.0$$

(2) incomplete projection

From Table 15.2, for $r_{sd}p = 0.7$, $C_c = 1.59H/B_c - 0.09 = 7.86$

Choose lower value, $C_c = 7.86$, so this is a case of incomplete projection.

Equation 15.4: $W_c = C_c\gamma B_c^2 = 7.86 \times 19.6 \times 0.4^2 = 24.6$ kN/m

Now assume narrow trench:

Equation 15.3: $C_d = \dfrac{1 - e^{-2K\mu'H/B_d}}{2K\mu'}$

The value of $K\mu'$ used should be the lower of that for the backfill material (0.19) and that for the existing soil in the trench sides (unsaturated clay, from Table 15.4, 0.13).

So $C_d = \dfrac{1 - e^{-2\times0.13\times2/0.9}}{2 \times 0.13} = 1.69$

Equation 15.2: $W_c = C_d\gamma B_d^2 = 1.69 \times 19.6 \times 0.9^2 = 26.8$ kN/m

Choose lower value for wide and narrow trench,
so $W_c = 24.6$ kN/m – this is a wide trench case.

Equation 15.6: $W_{csu} = P_s B_c$

from Fig. 15.6 (light road) for $H = 2$ m, P_s is 22 kN/m

so $W_{csu} = 22 \times 0.4 = 8.8$ kN/m

Equation 14.7:

$$W_w = C_w \rho g \pi D^2/4 = 0.75 \times 1000 \times 9.81 \times \pi \times 0.3^2/4 = 0.5\,\text{kN/m}$$

This uses the usual value of 0.75 for C_w. (W_w is not very significant since the diameter is below 600 mm, as suggested.)

Equation 15.1: $W_e = W_c + W_{csu} + W_w = 24.6 + 8.8 + 0.5 = 33.9\,\text{kN/m}$

From equation 15.8, we require $W_t F_m \geq W_e F_{se}$

For pipe strength of 36 kN/m, $36 \times F_m \geq 33.9 \times 1.25$

Bedding factor, F_m , would need to exceed 1.18, so class D or N bedding would not be sufficient, but class F would be.

For 36 kN/m strength pipe, use class F bedding

For pipe strength of 48 kN/m, $48 \times F_m \geq 33.9 \times 1.25$

Bedding factor, F_m, would need to exceed 0.88, so class D bedding would be adequate.

For 48 kN/m strength pipe, use class D bedding

special hazards. Investigation of ground conditions for tunnel construction can typically be divided into three phases:

1   desk study: analysis of existing data, geological and other maps
2   site investigation: boreholes, trial excavations, analysis of samples, interpretation
3   during construction: observations and records, probing ahead, further trials and boreholes.

15.5 Open-trench construction

15.5.1 Excavation

In urban areas, all excavation must be carried out with great care so as not to damage existing services. In some areas these are densely packed. Their location may be indicated on plans held by the responsible authority and, where necessary, diversions may need to be arranged in advance. However, there are always problems with the precise location of services in the ground and with services that are unknown, omitted or wrongly located on the plans. Non-intrusive methods can be used to locate services from the surface, but these may need to be backed up by trial pits – small excavations dug by hand (since their purpose is to prevent machinery from causing damage).

Pipe trenches are generally excavated mechanically, though hand excavation is needed where access is limited and where existing services are a problem.

The minimum trench width specified in BS EN 1610 is given in Tables 15.5 and 15.6. The width must not exceed any maximum specified in the structural design, since this might affect the appropriateness of the structural design calculations. The normal maximum depth of a trench is 6 m, but this is less if there are extra surcharge loads on either side of the trench.

Where access is needed to the outside of structures like manholes, a working space of at least 0.5 m needs to be provided. Where more than one pipe is laid in the same trench, working space between the pipes should be 0.35 m for pipes up to 700 mm diameter, and 0.5 m between larger pipes.

The usual system for temporary support of the sides of the trench consists of vertical steel sheets supported by horizontal timber 'walings' and

Table 15.5 Minimum trench width related to pipe diameter, for supported trench with vertical sides (adapted from BS EN 1610)

DN	Minimum trench width (m) (OD = outside diameter)
Below 225	OD + 0.4
225 to 350	OD + 0.5
350 to 700	OD + 0.7
700 to 1200	OD + 0.85
Above 1200	OD + 1.0

Table 15.6 Minimum trench width related to trench depth (adapted from BS EN 1610)

Trench depth (m)	Minimum trench width (m)
Less than 1.0	no minimum
1.0 to 1.75	0.8
1.75 to 4.0	0.9
More than 4.0	1.0

adjustable steel struts (Fig. 15.1). Whether the sheets form a continuous wall or are placed at a particular separation depends on the condition of the ground and its need for support, as well as on possible inflow of ground-water. Alternatives are ready-made frames, boxes or trench shields, which are moved progressively as excavation, laying and backfilling proceed.

Where there is a significant problem of groundwater entering the trench during construction, dewatering may be necessary – either by pumping from the trench bottom or from remote points – to lower the water table.

15.5.2 Pipe laying

Pipes are delivered to site in bulk, and stored by stacking until needed. They must be stacked carefully to avoid damage, not so high as to cause excessive loads on the pipes at the bottom, and far enough from the trench to avoid any threat to the stability of the excavation.

The nature of the bedding will be specified in the design, and the alternatives have been considered in Section 15.3. Where a pipe is being laid directly on the bed, the trench should be carefully excavated to the correct gradient to ensure that the pipe is supported all along its length. Localised sections of poor ground at the base of the trench must be dug out and replaced by suitable material. Small extra excavations are needed to accommodate pipe sockets with some clearance, to ensure that the weight of the pipe is not bearing on the socket. Where granular bedding material has been specified, this must be similarly prepared to the correct gradient to give support all along the pipe.

It is most common to set out sewer pipes in open-trench using a laser, either set up inside the pipe or above the excavation. The pipe invert (defined in Section 7.3) should be used as the reference point, since this is the most important vertical position from a hydraulic point of view, and most pipes are not sufficiently round for setting out to be carried out by reference to the crown of the pipe.

Pipes are generally laid in the direction of the upwards gradient, so that water in the excavation drains away from the working area. Since the spigot is inserted into the socket, the normal orientation of a pipe is for the socket to be at the upstream end, 'pointing' upstream. The specification will indicate tolerances of line and level that must be complied with when the pipe is laid.

Methods of jointing pipes have been described in Section 15.2. Where a pipeline passes through a fixed structure, it is normal to include flexible joints and sometimes short pipe lengths ('rockers') close to the wall of the structure.

Completed sewer sections are tested for leaks by pressure tests using air or water. Criteria for pressure loss with time are given in BS EN 1610. Sewers that are subsequently to be adopted by the sewerage undertaker may be subject to CCTV inspection (considered further in Chapter 17).

The trench is backfilled in accordance with the design and specification, typically in carefully compacted layers of specified thickness. During back-filling, the trench support is removed progressively. When backfilling is complete, the surface is reinstated.

A good source of further information on open-trench construction is by Irvine and Smith (1983).

15.6 Tunnelling

15.6.1 Lining

Common methods of tunnel lining are capable of withstanding loads over a wide range of conditions. Linings of extra strength can be supplied where necessary. The loads experienced during construction may be more critical than those experienced by the completed sewer.

It is common for sewers in tunnel to have a primary lining, most commonly bolted concrete segments, to support construction and permanent loads, and a secondary lining, often *in situ* concrete, to provide suitably smooth hydraulic conditions.

Primary lining

A precast concrete lining consists of segments that are bolted *in situ* to form a ring, with a narrow key segment at or near the soffit. The excavated area will be slightly larger than the outside of the ring and the annular space between is filled with grout, injected through holes in the lining.

Standard segments are ribbed and are unsuitable for carrying flow as built. Special concrete blocks can be used to fill the panels between the ribs, but the most common method of achieving a smooth surface is by adding a secondary lining.

Secondary lining

In situ concrete can be injected behind circular travelling shutters. Alternatives are ready-made linings of glass reinforced plastics or fibre reinforced cement composites, with the annular space between the secondary and primary lining filled with grout.

15.6.2 Ground treatment and control of groundwater

Ground treatment for sewers in tunnels may be needed to control ground-water during construction or to stabilise the ground. The main methods are dewatering, ground freezing and injection of grouts or chemicals. Groundwater at the tunnel face can be controlled by compressed air.

Dewatering

Water is pumped from wells to lower the water table in the area of tunnel construction.

Ground freezing

The temperature of the ground is lowered to freeze the groundwater during construction. This is achieved by circulating refrigerated liquids – usually brine or liquid nitrogen – through pipes in the ground.

Injection of grouts or chemicals

This may be to reduce permeability, or improve strength of cohesionless soils or broken ground. Injection can be carried out through holes drilled from the tunnel face or from the ground surface. Less drilling may be needed from the tunnel face but this approach can hold up construction.

Compressed air

Groundwater can be held back by balancing the hydrostatic pressure with compressed air inside the tunnel. Pressures are commonly less than one atmosphere, but can be higher. The part of the tunnel under pressure is sealed off by air locks, through which personnel and materials pass. Air must be supplied continuously as some escapes through the ground. People working in compressed air must have regular health checks.

15.6.3 Excavation

Tunnels are driven from working shafts, usually supporting drives of roughly equal lengths both upstream and downstream.

Most ground can be excavated by a boring machine or by hand-held pneumatic tools. Hard rock may need to be drilled and blasted. If the ground cannot be left unsupported during erection of the primary lining, a tunnel shield, which pushes itself forward from the previously erected primary lining, is used to give continuous support. A tunnel boring machine may combine the functions of shield and mechanical excavator.

Shafts are excavated vertically, mechanically or by hand, and the ground is usually supported using precast concrete segmental rings similar to those used for tunnels. For sewers in tunnels, working shafts usually become manholes in the completed scheme.

15.7 Trenchless methods

The choice between open-trench construction and traditional tunnelling can be made on the basis of ease and cost of construction. Beyond a certain depth and diameter, it is simply cheaper – in purely construction terms – to tunnel, than to excavate and backfill a trench. In contrast, trenchless methods usually become an appropriate alternative to open-trench construction when an additional factor is taken into account: the disruption to business, traffic and everyday life caused by open-trench construction.

Trenchless methods are also applied commonly to other pipelaying fields, particularly gas and oil supply, for which some of the techniques were originally developed. Certain countries have been particularly active in development of the methods in the past, including Japan, the United States, Russia and Germany.

A brief introduction to some of the principal methods is given here; more information can be found in Thomson (1993), Watson (1987), Flaxman (1990) and Grimes and Martin (1998). Further important applications of trenchless technology – to sewer rehabilitation – are covered in Chapter 18.

Pipe jacking

Hydraulic jacks are used to push specially-made pipes (without protruding sockets, and strong enough to take the jacking forces) through an excavated space in the ground (Fig. 15.7(a)). Ahead of the pipes is a shield at which excavation takes places either mechanically or by hand-tool. The jacks push against a thrust wall located in a specially constructed thrust pit.

Microtunnelling

This is a form of pipe jacking for pipe diameters under 900 mm. Excavation is by unmanned, remotely-controlled equipment.

Auger boring

Soil is removed from the excavated face by an auger (Fig. 15.7(b)), and pipes are jacked into the excavated space. This is generally considered to be a rather inaccurate method, and is used for short drives only.

Impact moling

An earth displacement mole creates a hole in the ground by pushing the earth outwards. A pipe can then be pushed into the space. This method is used for small diameters only.

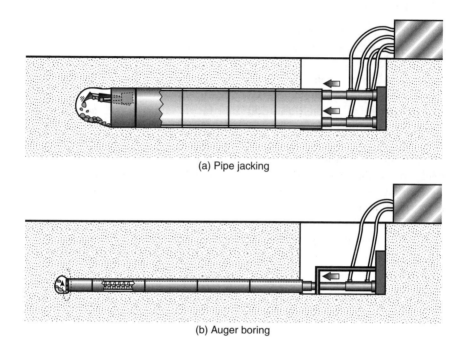

(a) Pipe jacking

(b) Auger boring

Fig. 15.7 Pipe jacking and auger boring

Problems

15.1 Describe the properties of the main sewer pipe materials. What factors affect the selection of a pipe material for a particular case?

15.2 A pipe has an internal diameter of 450 mm and a wall thickness of 50 mm. If it is laid in an open trench of depth 2 m, determine the minimum width of the trench. Similarly, determine the minimum trench width for a pipe of internal diameter 200 mm and wall thickness 22 mm (same depth of trench). [1.25 m, 0.9 m]

15.3 A pipe with internal diameter 225 mm and external diameter 280 mm is to be laid in a trench of minimum width. The cover depth will be 3 m; the ground is unsaturated clay and the backfill will be granular cohesionless material. Determine the soil load per m length (W_c) assuming a narrow trench condition. Use unit weight of soil $\gamma = 19.6$ kN/m^3. [35.4 kN/m]

15.4 For the case in Problem 15.3, determine the soil load assuming a wide trench condition. Use the data given in 15.3, and assume a value for $r_{sd} \, p$ of 0.7. Which assumption is appropriate (narrow or wide trench) for design in this case? [26 kN/m, wide]

15.5 For the case in Problems 15.3 and 15.4, determine the external load per unit length (W_e). The pipe will be under a light road. Use density of liquid $\rho = 1000$ kg/m^3. Pipe is available in this size with a crushing strength of 28, 36 or 48 kN/m. Determine the minimum bedding factor in this case for each of these pipe strengths. Propose two appropriate designs. [29.1 kN/m; 1.3, 1.0, 0.76; low strength pipe on Class *F* bedding, medium strength on Class *D*]

15.6 Describe the main methods of sewer construction. Discuss the factors that influence selection of the most appropriate method in particular cases.

15.7 Describe methods of ground treatment for sewer construction and the circumstances in which they might be used.

Key sources

BS EN 1295: 1998 *Structural design of buried pipelines under various conditions of loading.*

Read, G. (2004) *Sewers: replacement and new construction*, Butterworth-Heinemann

Sewers and Water Mains Committee (1993) *Materials selection manual for sewers, pumping mains, and manholes*, WSA/FWR.

References

Clay Pipe Development Association (1999) *The specification, design and construction of drainage and sewerage systems using vitrified clay pipes.* CPDA, Chesham.

Compston, D.G., Gray, P., Schofield, A.N. and Shann, C.D. (1978) *Design and construction of buried thin wall pipes*, Research Report No 78, CIRIA, London.

Flaxman, E.W. (1990) Trenchless Technology. *Journal of the Institution of Water and Environmental Management*, 4, April, 187–193.

Grimes, J.F. and Martin, P. (1998) *Trenchless and minimum excavation techniques, planning and selection*, Special Publication 147, CIRIA, London.

Irvine, D.J. and Smith, R.J.H. (1983) *Trenching practice*, Report 97, CIRIA, London.

Moser, A.P. (1990) *Buried pipe design*, McGraw-Hill.

Thomson, J.C. (1993) *Pipejacking and microtunnelling*, Blackie Academic and Professional.

Watson, T.J. (1987) *Trenchless construction for underground services*, Technical Note 127, CIRIA, London.

Woolley, L. (1988) *Drainage details*, 2nd edn, E & FN Spon.

Young, O.C. and O'Reilly, M.P. (1983) *A guide to design loadings for buried rigid pipes*, Transport and Road Research Laboratory, HMSO.

Young, O.C. and Trott, J.J. (1984) *Buried rigid pipes: structural design of pipelines*, Elsevier Applied Science Publishers.

Young, O.C., Brennan, G. and O'Reilly, M.P. (1986) *Simplified tables of external loads on buried pipelines*, Transport and Road Research Laboratory, HMSO.

British Standards

BS 65: 1991 *Specification for vitrified clay pipes, fittings and ducts, also flexible mechanical joints for use solely with surface water pipes and fittings.*

BS EN 295: 1991 *Vitrified clay pipes and fittings and pipe joints for drains and sewers,* Parts 1, 2 and 3.

BS EN 476: 1998 *General requirements for components used in discharge pipes, drains and sewers for gravity systems.*

BS 534: 1990 *Specification for steel pipes, joints and specials for water and sewage.*

BS EN 598: 1995 *Ductile iron pipes, fittings, accessories and their joints for sewerage applications. Requirements and test methods.*

BS EN 1295: 1998 *Structural design of buried pipelines under various conditions of loading.*

BS EN 1401–1: 1998 *Plastics piping systems for non-pressure underground drainage and sewerage, unplasticized polyvinyl chloride (PVC-U). Specifications for pipes, fittings and the system.*

BS EN 4660: 2000 *Thermoplastics ancillary fittings of nominal sizes 110 and 160 for below ground gravity drainage and sewerage.*

BS 5911–1: 2002 *Precast concrete pipes, fittings and ancillary products. Specification for unreinforced and reinforced concrete pipes (including jacking pipes) and fittings with flexible joints.*

BS EN 10224: 2002 *Non-alloy steel tubes and fittings for the conveyance of aqueous liquids including water for human consumption.*

16 Sediments

16.1 Introduction

Sediment is ubiquitously present in urban drainage systems. It is found deposited on catchment surfaces, in gully pots and in drains and sewers. Drainage engineers have long recognised its presence in stormwater and the problems it may cause. They have sought to exclude larger, heavier sizes from the piped system by the provision of gully pots and designed sewers to limit in-pipe deposition. The theory is that sediment that does enter the system is carried downstream, where it is eventually trapped and removed at the outlet of the system. This may be the case for newly-designed systems, but for older (especially combined) networks, sedimentation in sewers is commonplace. In fact, a review of sediment movement in sewers (Binnie and Partners and Hydraulics Research, 1987) concluded that 80% of UK urban drainage systems had at least some permanent sediment deposits.

The movement of sediment through a drainage catchment is a complex, multi-stage process. Sediment deposited on roads, for example, is initially freed from the surface and then washed transversely by overland flow to the channel, where it is transported parallel to the kerb-line under open channel flow. The sediment is discharged with the flow into the gully inlet under gravity, and is captured in the gully pot (in part) by sedimentation or transferred to the receiving sewer. Once in the sewer, material is transported under open channel flow as suspended or bed-load. Suspended sediment is carried along in the main body of the flow, whilst bed-load travels more slowly in contact with the invert of the pipe. Some material may be deposited and/or re-eroded as it progresses downstream.

During transport, sediment may be discharged to a watercourse via a CSO (if in operation) or settled in the WTP grit removal device. At points of deposition (surface, gully, sewer), sediment may be extracted from the system by cleaning. A representation of sediment inputs, outputs and movement through the system is given in Fig. 16.1. Further details on the removal of sediment from systems will be given in Chapter 17.

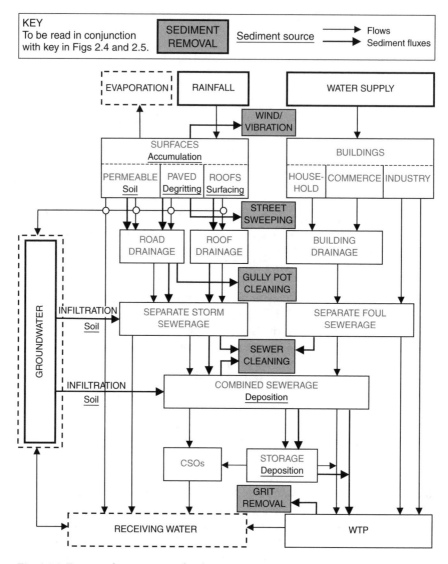

Fig. 16.1 Entry and exit points of sediment in drainage systems

The rate of progress of material through the system depends on factors such as:

*   the characteristics of the sediment (physical, chemical)
*   the characteristics of the flow (velocity, degree of unsteadiness)
*   the characteristics of the drainage network (layout, geometry).

For example, different types of sediment will move through the system in different ways. Particles of very small size or low density may remain in suspension under all normal flow conditions and be transported through the system without being deposited. Sediments with low settling velocities may only form deposits during periods of very low flow, and may readily be re-entrained when higher velocities occur in the pipes as a result of storms or diurnal variations in flow. By contrast, larger and denser particles may only be transported by peak flows that occur relatively infrequently, and in some cases they may form permanent stationary deposits near where they enter the sewer system.

Deposition commonly occurs during dry weather flow periods (particularly during low flow at night), and in decelerating flows during storm recession. Deposits also form at structural and hydraulic discontinuities such as at joints, changes in gradients and ancillary structures. Only the steepest sewers are immune from deposition.

This chapter reviews the origins, problems and effects of sediment. Sections look at how the sediment is transported through the system and at its detailed characteristics. A design method that takes sediment explicitly into account is presented.

16.2 Origins

16.2.1 Definition

Sewer sediments are defined as any settleable particulate material that is found in stormwater or wastewater and is able, under appropriate conditions, to form bed deposits in sewers and associated hydraulic structures. Using the basic solids classification presented in Table 3.1, this would include:

- grit
- suspended solids
  - sanitary
  - stormwater.

16.2.2 Sources

Sources of sediment entering sewers are quite diverse. Indeed, any particulate-generating material or activity in the urban environment is a potential source. Broadly, three categories can be established: sanitary, surface and sewer. These are defined in Table 16.1.

Table 16.1 Sources of sewer sediment

Source	Type
Sanitary	• Large faecal and organic matter with specific gravity close to unity.
	• Fine faecal and other organic particles.
	• Paper and miscellaneous materials flushed into sewers (sanitary refuse).
	• Vegetable matter and soil particles from the domestic processing of food.
	• Materials from industrial and commercial sources.
Surface	• Atmospheric fall-out (dry and wet).
	• Particles from erosion of roofing material.
	• Grit from abrasion of road surfaces or from re-surfacing works.
	• Grit from de-icing operations on roads.
	• Particulates from motor vehicles (e.g. vehicle exhausts, rubber from tyres, wear and tear, etc.).
	• Materials from construction works (e.g. building aggregates, concrete slurries, exposed soil, etc.) and other illegally-dumped materials.
	• Detritus and litter from roads and paved areas (e.g. paper, plastic, cans, etc.).
	• Silts, sands and gravels washed or blown from unpaved areas.
	• Vegetation (e.g. grass, leaves, wood, etc.).
Sewer	• Soil particles infiltrating due to leaks or pipe/manhole/gully failures.
	• Material from infrastructure fabric decay.

16.3 Effects

16.3.1 Problems

There are three major effects of sediment deposition leading to a number of serious consequences, and these are summarised in Table 16.2. The table also lists parts of the book where further information can be found on the consequences of sediment deposition.

The first effect of deposited sediment is its propensity to initiate block-

Table 16.2 Effects and consequences of sewer sediment deposition

Effect	Consequences	Further information (Section)
Blockage	• Surcharging.	8.4.5
	• Surface flooding.	
Loss of hydraulic capacity	• Surcharging.	
	• Surface flooding.	
	• Premature operation of CSOs.	12.3.1
Pollutant storage	• Washout to receiving waters during CSO operation.	3.4.1
	• Shock loading on treatment plants.	22
	• Gas and corrosive acid production.	17.6

age. Larger, gross solids and other matter may build up, leading to partial or total blockage of the pipe bore.

The second effect results from the fact that a deposited bed restricts the flow in the sewer, resulting in a loss of hydraulic capacity. The reasons for this effect are discussed in Section 16.3.2, but the result can be pipe or manhole surcharge. Another common effect is the premature operation of CSOs.

The third major effect results from the ability of the deposited sediment to act as a pollutant store or generator. The reasons and possible mechanisms for this are discussed later in the chapter. These pollutants are only stored temporarily, and can be released under flood flow conditions, probably contributing to the commonly observed first foul flush of heavy pollution (see Section 12.3.2). Biochemical changes in the bed of sediment can result in septic conditions, releasing gas that can be highly corrosive to the sewer fabric.

It should be clear from the previous discussion that excessive sediment deposition should be avoided if at all possible at the design stage. Extensive sediment removal is a difficult, recurrent and expensive process.

16.3.2 Hydraulic

The presence of sediment in sewer flows has three hydraulic effects of varying importance: dissipation of energy in keeping solids in suspension, reduction of flow cross-sectional area and increased frictional losses due to the texture of the bed.

Suspension

In the case of a sewer without deposits, the presence of sediment in the flow or moving along the invert causes a small increase in energy loss, and this is observed as a reduction in discharge capacity of about 1% for rough pipes (Ackers *et al.*, 1996a).

For flows in sewers in which there is a deposited bed of sediment, the energy losses associated with keeping the sediment in motion are relatively insignificant compared with the other effects.

Geometry

A deposited bed reduces the cross-sectional area available to convey flow and therefore increases the velocity and head loss for a given discharge and depth of flow. The loss of total area is relatively small (<2%) provided that the depth of sediment is less than 5% of the pipe diameter, but becomes important at sediment depths above about 10%.

Bed roughness

Usually of greatest significance is the increase in overall resistance caused by the rough texture of the deposited bed. Above the *threshold of movement*, sediment quickly forms into ripples and dunes and initially these grow in size as the flow velocity increases. The effective roughness value of dunes k_b (k_s in the Colebrook-White equation) can reach 10% or more of the pipe diameter, compared with typical values for the pipe walls that are in the range 0.15 mm to 6 mm (depending on the material and the degree of sliming). The value of k_b can be estimated (May, 1993) from:

$$k_b = 5.62R^{0.61}d_{50}^{0.39} \qquad (16.1)$$

R hydraulic radius (m)
d_{50} sediment particle size larger than 50% (by mass) of all particles in the bed (m)

Under these conditions, a 5% depth of deposited sediment with dunes could reduce the pipe-full capacity by 10–20%. However, at higher velocities the dunes tend to reduce in size until the bed again becomes flat with much lower roughness. The loss of hydraulic capacity due to a deposited bed can, therefore, vary considerably with the flow conditions. The approximate combined effects of shape and roughness are illustrated in Fig. 16.2.

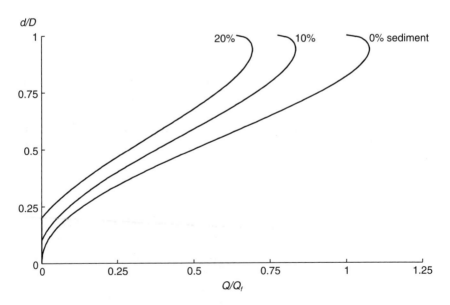

Fig. 16.2 Effect of sediment bed on flow capacity of sewers

16.3.3 *Pollutional*

Solids of sanitary origin are readily mixed, in combined sewers, with sediments entering from surface sources. The organic material tends to adhere to the heavier inorganic sediment and deposits with them. These deposits form a rough surface that encourages further adherence of organic matter. In such an environment, anaerobic conditions are likely to develop, resulting in partial digestion of the sediment. By-product fatty acids will be liberated into the interstitial liquor (as described further in Chapter 17) with the possibility of substantial BOD/COD loads being generated. Some evidence suggests that degradation processes in sediment can increase pollutant discharges by up to 400% (Binnie and Partners and HR Wallingford, 1987). Clearly, the presence of sediment deposits encourages the retention of solids and pollutants during low flows. This increases the potential for degradation before such material is flushed away.

The concept and importance of the first foul flush has been discussed in Chapter 12. Sediment deposits are commonly considered to be one of its major causes. Field evidence indicates that up to 90% of the pollution load discharged from CSOs can be derived from erosion of in-sewer deposits (Crabtree, 1989).

16.4 Transport

The movement of sediment has three broad phases: entrainment, transport and deposition.

16.4.1 *Entrainment*

As wastewater flows over a sediment bed in a sewer, hydrodynamic lift and drag forces are exerted on the bed particles. If these two combined forces do not exceed the restoring forces of sediment submerged weight, interlocking and cohesion (if applicable), then the particles remain stationary. If they exceed the restoring force, then entrainment occurs, resulting in movement of the particles at the flow/sediment boundary. Not all the particles of a given size at this boundary are dislodged and moved at the same time, as the flow is turbulent and contains short-term fluctuations in velocity. The limiting condition, below which sediment movement is negligible (the threshold of movement), is usually defined in terms of either a critical boundary shear stress (τ_o) or critical erosion velocity (v). The two are related as shown in Equation 8.22 reproduced below:

$$\tau_o = \frac{\rho \lambda v^2}{8}$$

ρ liquid density (kg/m³)
λ Darcy-Weisbach friction factor (−)

In storm sewers, sediments are mainly inorganic and non-cohesive, although some deposits may be cementitious and become permanent if undisturbed for long periods. Sediments in foul sewers generally have cohesive-like properties due to the nature of the particles and the presence of greases and biological slimes. In combined sewers, the sediments tend to be a combination of the first two types.

Cohesion will tend to increase the value of shear stress that the flow needs to exert on the deposited bed in order to initiate movement of particles in the surface layer. In laboratory tests, erosion of synthetic cohesive sediments has been observed to occur at bed shear stresses of 2.5 N/m^2 (surficial material) and 6–7 N/m^2 (granular, consolidated deposits). Field studies in large sewers, however, show that lower shear stresses of around 1 N/m^2 may initiate erosion (Ashley and Verbanck, 1996).

It is possible that cohesion may also alter the way in which the sediment then moves and thus affects the sediment-transporting capacity of the flow. However, experimental research (Nalluri and Alvarez, 1992) using synthetic cohesive sediments suggests that the second factor may not be very significant; once the structure of a cohesive bed is disrupted, the particles are stripped away and transported by the flow in a similar way to non-cohesive sediments.

16.4.2 Transport

Once sediment has been entrained into the flow, it travels, as mentioned in the introduction, in suspension or as bed-load. Finer, lighter material tends to travel in suspension and is primarily influenced by turbulent fluctuations in the flow, which in turn are influenced by bed shear. It is advected at mean flow velocity. Heavier material travels by rolling, sliding or saltating along the pipe invert (or deposited bed) as bed-load. This type of movement is affected by the local velocity distribution, and advection velocities in this mode are considerably lower than the flow mean velocity. In an urban drainage network, with graded materials of differing specific gravity, a combination of these modes exists.

Table 16.3 indicates that the mode of transport depends on the relative magnitude of the lifting effects due to turbulence, as measured by the shear velocity (U_*), and the settling velocity (W_s). Shear velocity is given by:

$$U_* = \sqrt{\frac{\tau_o}{\rho}} \qquad (16.2)$$

Table 16.3 Mode of sediment transport (after Raudkivi, 1998)

W_s/U_*	Mode
>0.6	Suspension
0.6–2	Saltation
2–6	Bed-load

Example 16.1

Analysis of the sediment in an urban catchment shows it to consist pre-dominantly of grit with a characteristic settling velocity of 750 m/h. Estimate the mode of transport of this sediment in a 0.15% gradient, 1.5 m diameter sewer flowing half-full.

Solution

$$R = \frac{D}{4} = 0.375 \text{ m}$$

For a pipe flowing half full, the wall shear stress is given by equation 8.21

$$\tau_0 = \rho g R S_O = 1000 \times 9.81 \times 0.375 \times 0.0015 = 5.5 \text{ N/m}^2$$

Shear velocity is given by equation 15.2.

$$U_* = \sqrt{\frac{5.5}{1000}} = 0.074 \text{ m/s}$$

So, as:

$$\frac{W_s}{U_*} = \frac{750/3600}{0.074} = 2.8 > 2$$

transport will be bed-load.

A large body of knowledge has been built-up, including many different predictive equations, for sediment transport, based particularly on loose-boundary channels including rivers (see, for example, Raudkivi, 1998). Equations are available, normally in terms of the volumetric sediment carrying capacity of the flow, for both suspended and bed-load transport.

Although these equations are useful in highlighting important prin-ciples, they should not be used uncritically for sewer design and analysis. Conditions in pipes are different to those in rivers: pipes have rigid bound-aries, significantly different and well-defined cross-sections, and transport different material. However, there have been a number of studies particu-larly focusing on pipes and sewers, both in the laboratory and in the field, and these have been comprehensively reviewed and compared by Ackers *et al.* (1996a). Recommended transport equations are given in Section 16.6.

16.4.3 Deposition

If the flow velocity or turbulence level decreases, there will be a net reduction in the amount of sediment held in suspension. The material accumulated at the bed may continue to be transported as a stream of particles without deposition. However, below a certain limit, the sediment will form a deposited bed, with transport occurring only in the top layer (the *limit of deposition*). If the flow velocity is further reduced, sediment transport will cease completely (the *threshold of movement*). The flow velocities at which deposition occurs tend to be lower than those required to entrain sediment particles.

16.4.4 Sediment beds and bed-load transport

If an initially clean sewer flowing part-full is subjected to a sediment-laden flow transported under bed-load, but conditions are not sufficient to prevent deposition, a sediment bed will develop. It will increase the bed resistance, causing the depth of flow to increase and the velocity to decrease.

Intuitively, it might be assumed that a reduction in velocity would cause a reduction in the sediment-transporting capacity of the flow, leading to further deposition and possibly blockage. In fact, laboratory evidence has shown (May, 1993) that the presence of the deposited bed actually allows the flow to acquire a *greater* capacity for transporting sediment as bed-load. This is because the mechanism of sediment transport is related to the width of the deposited bed, which can, of course, be much greater than the narrow stream of sediment which is present along the bed of the pipe at the limit of deposition. The effect more than compensates for the reduction in velocity caused by the roughness of the bed. Ultimately, the increased deposited bed depth (and width) and the associated increased sediment transport capacity may balance with the incoming sediment load and prevent further deposition. Thus, in principle, a small amount of deposition may be advantageous in terms of sediment mobility.

16.5 Characteristics

16.5.1 Deposited sediment

The characteristics of sewer sediment deposits vary widely according to the sewer type (foul, storm or combined), the geographical location, the nature of the catchment, local sewer operation practices, history and customs. Crabtree (1989) proposed that the origin, nature and location of deposits found within UK sewerage systems could be used to classify sediment under five categories A–E (see Fig. 16.3). The characteristics of these deposits are described below and summarised in Table 16.4.

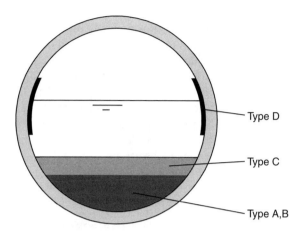

Fig. 16.3 Typical sediment deposits in a sewer

Table 16.4 Physical and chemical characteristics of sewer sediment type classes (adapted from Crabtree, 1989)

| | *Sediment type* | | | | |
	A	B	C	D	E
Description	Coarse, loose granular material	As A but concreted with fat, tars etc.	Fine-grained deposits	Organic slimes and biofilms	Fine-grained deposits
Location	Pipe inverts	As A	Quiescent zones, alone or above A material	Pipe wall around mean flow line	In CSO storage tanks
Saturated bulk density (kg/m³)	1720	N/A	1170	1210	1460
Total solids (%)	73.4	N/A	27.0	25.8	48.0
COD (g/kg)*	16.9	N/A	20.5	49.8	23.0
BOD₅ (g/kg)*	3.1	N/A	5.4	26.6	6.2
NH₄⁺–N (g/kg)*	0.1	N/A	0.1	0.1	0.1
Organic content (%)	7.0	N/A	50.0	61.0	22.0
FOG (%)	0.9	N/A	5.0	42.0	1.5

* Grams pollutant per kilogram of wet bulk sediment

Physical characteristics

Type A material is the coarsest material, found typically on sewer inverts. These deposits have a bulk density of up to 1800 kg/m³, organic content of 7%, with some 6% of particles typically smaller than 63 μm. The finer material (type C) is typically 50% organic, with a bulk density of approximately 1200 kg/m³ and some 45% of the particles are smaller than 63 μm. Type E is the finest material, although there is no definite boundary between any of the types A, C or E. This is perhaps to be expected, as the sediment actually deposited will depend on the material available for transport and the flow conditions in specific locations.

Chemical characteristics

Table 16.4 summarises the mean chemical characteristics of the deposited sediments. A high degree of variability is observed in practice (e.g. coefficient of variation 23–125%). On a mass for mass basis, wall slimes are the most polluting in terms of oxygen demand (49.8 gCOD/kg wet sediment). There is a broad decrease in strength among types, in the order D, E, C and A, with type A material having mean COD levels of just 16.9 g/kg.

However, this does not show the full significance of the relative polluting potential of each type of deposit. This is illustrated by Example 16.2.

Results from Example 16.2 show that although type D material is of higher unit strength, where small quantities are found in practice, it tends to be relatively insignificant. Type A material clearly shows up as having the bulk of the pollution potential (79% in this case) because of its large volume. The actual value will vary depending on location. It should be realised, too, that the total polluting load would only be released under extreme storm flow conditions that erode all the sediment deposits. More routine storm events will probably erode only a fraction of the type A deposits. It is also interesting to note that, in this case, the wastewater itself only represents 10% of the potential pollutant load.

Significance of deposits

Type A and B deposits are normally associated with loss of sewer capacity, and type A deposits are the most significant source of pollutants. The nature of the sediment appears to vary between areas, with large organic deposits being found nearer the heads of networks and with more granular material (type A) being found in trunk sewers. Larger interceptor sewers typically have more type C material intermixed with the type A (Ashley and Crabtree, 1992). Pipe wall slimes/biofilms (type D) are important because they are very common, highly concentrated, easily eroded and affect hydraulic roughness.

Example 16.2

A 1500 mm diameter sewer has a sediment bed (type A) of average thickness 300 mm. Above this is deposited a 20 mm type C layer and above that flows 350 mm of wastewater (BOD$_5$ = 350 mg/l). Along the walls of the sewer at the waterline are two 50 mm × 10 mm thick biofilm deposits (type D). Calculate the relative pollutant load associated with each element in the pipe.

Solution

For each of the sediment types, the cross-sectional area can be calculated from the pipe geometrical properties (Chapter 8) to give volume per unit length. Combining this information with the bulk density of the sediment and its pollutant strength enables unit pollutional load to be estimated as shown in Table 16.5.

Table 16.5

Type	Depth (mm)	Vol (m³/m)	Bulk density (kg/m³)	BOD$_5$ (g/kg)	Unit BOD$_5$ (g/m length)	% load
A	0–300	0.252	1720	3.1	1344	79
C	300–320	0.024	1170	5.4	152	9
Wastewater	320–670	0.488	1000	0.35	171	10
D	50 × 10 × 2	0.001	1210	26.6	32	2
Total					1699	100

16.5.2 Mobile sediment

Suspension

The predominant particles in suspension during both dry and wet weather flows are approximately 40 μm in size, and primarily attributed to sanitary solids. Most of the suspended material in combined sewer flow ($\approx$90%) is organic and is biochemically active with the capacity to absorb pollutants. Settling velocities are usually less than 10 mm/s (Crabtree *et al.*, 1991).

Near-bed

Under dry weather flow conditions, sediment particles can form a highly concentrated, mobile layer or 'dense undercurrent' just above the bed (see Fig. 16.4). Solids in this region are relatively large (>0.5 mm), organic (>90%) particles and are believed to be trapped in a matrix of suspended flow (Verbanck, 1995). Concentrations of solids of up to 3500 mg/l have

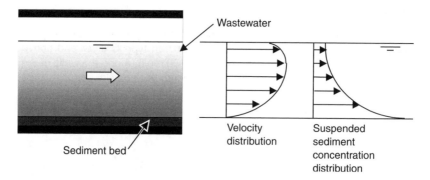

Fig. 16.4 Velocity and suspended sediment distributions in dry weather

been measured, and the corresponding biochemical pollutants are also particularly concentrated (Ashley and Crabtree, 1992). According to Ashley *et al.* (1994), typically, 12% of total solids are conveyed in the material moving near the bed. The rapid entrainment of near-bed solids is thought to make a significant contribution to first foul flushes (Chapter 12).

Granular bed-load

Granular particles (2–10 mm) are transported as 'pure' bed-load only in steeper sewers (>2%). In flatter sections (<0.1%), little granular material is observed in motion, presumably because it is deposited.

Particle size

In Chapter 6, it was shown how smaller particle sizes tend to be associated with a greater proportion of pollutant than might be expected from their mass. The same is true of mobile sewer sediment as shown in Table 16.6.

Table 16.6 Percentage of total pollution load associated with different particle size fractions (after Bertrand-Krajewski *et al.*, 1993)

Pollutant	Particle size fraction (µm)		
	<50	*50–250*	*>250*
BOD	52	20	28
COD	68	4	28
TKN	16	58	26
Hydrocarbons	69	4	27
Lead	53	34	13

16.6 Self-cleansing design

16.6.1 *Conventional methods*

The need for sewers to be designed to carry sediment has been recognised for many years. Conventionally, this has been done by specifying a minimum 'self-cleansing' flow velocity that should be achieved at a particular depth of flow or with a particular frequency of occurrence (see Chapters 10 and 11). Although this approach has apparently been successful in many cases, a single value of minimum velocity, unrelated to the characteristics and concentration of the sediment or to other aspects of the hydraulic behaviour of the sewer, does not properly represent the ability of sewer flows to transport sediment. In particular, it is known that a higher flow velocity is needed to transport a given concentration of sediment in large sewers than in small sewers.

It is also important to appreciate that conditions in gravity sewers are extremely variable. Flow rates and the sediment entering a system can vary considerably with time and location, so a sewer designed to be self-cleansing in normal conditions is still likely to suffer sediment deposition during periods of low flow and/or high sediment load.

16.6.2 *CIRIA method*

The CIRIA method (Ackers *et al.*, 1996a; Butler *et al.*, 1996a; 1996b) was developed in an attempt to relate minimum velocity to all the factors that affect it most, namely: pipe size and roughness, proportional flow depth, sediment size and specific gravity, degree of cohesion between the particles, sediment load or concentration and the presence of a deposited bed.

Self-cleansing

The method proposes the following definition of self-cleansing:

> An efficient self-cleansing sewer is one having a sediment-transporting capacity that is sufficient to maintain a balance between the amounts of deposition and erosion, with a time-averaged depth of sediment deposit that minimises the combined costs of construction, operation and maintenance.   (Ackers *et al.*, 1996a)

The important aspect of this definition is that it does not necessarily require sewers to be designed to operate completely free of sediment deposits if more economical overall solutions can be achieved by allowing some deposition to occur. This is a viable alternative because, as described in Section 16.4.4, the presence of the deposited bed can significantly increase the sediment transporting capacity of the pipe, despite the adverse effect of the deposits on the geometry and hydraulic roughness.

Movement criteria

The results of field and laboratory research indicate that, to achieve self-cleansing performance, sewers should be designed to:

1. transport a minimum concentration of fine-grain particles in suspension
2. transport coarser granular material as bed-load
3. erode cohesive particles from a deposited bed.

A thorough comparison of the application of mainly laboratory-based equations to typical sewer design situations was carried out, based on these three sediment movement criteria.

1.   SUSPENDED LOAD TRANSPORT

Macke's (1982) equation was found to provide a reasonable fit to laboratory data for suspended particles moving at the limit of deposition (no deposition) and was recommended for use as the normal design method.

$$C_v = \frac{\lambda^3 v_L^5}{30.4(S_G - 1)W_s^{1.5}A} \tag{16.4}$$

C_v volumetric sediment concentration (discharge rate of sediment/ discharge rate of water)
v_L limiting flow velocity without deposition (m/s)
S_G sediment specific gravity (–)
A flow cross-sectional area (m²)

The equation is valid beyond $\tau = 1.07 \text{ N/m}^2$.

Where sediment is to be transported over a sediment bed, the 1973 Ackers-White equation, originally developed for alluvial channels, is advocated. This has been applied to sewer design by Ackers (1991) with a reduced effective sediment bed width as follows:

$$C_v = J\left(W_e\frac{R}{A}\right)^\alpha \left(\frac{d'}{R}\right)^\beta \lambda_c^\gamma \left(\frac{v}{\sqrt{g(S_G - 1)R}} - K\lambda_c^\delta\left(\frac{d'}{R}\right)^\epsilon\right)^m \tag{16.5}$$

W_e effective width of sediment bed (m)
d' sediment particle size (m)
λ_c friction factor for pipe and sediment bed (–)

The various empirical coefficients (J, K, M, α, β, γ, δ and ϵ) depend on the dimensionless grain size D_{gr} and the mobility parameter at the threshold of movement A_{gr} (defined in more detail by Ackers *et al.*, 1996a).

2.   BED-LOAD TRANSPORT

For bed-load transport at the limit of deposition, no existing equation gave a good fit over a full range of the data and so a new equation was derived:

$$C_v = 3.03 \times 10^{-2} \left(\frac{D^2}{A}\right)\left(\frac{d'}{D}\right)^{0.6}\left(1 - \frac{v_t}{v_L}\right)^4\left(\frac{v_L^2}{g(S_G - 1)D}\right)^{1.5} \tag{16.6}$$

$$v_t = \sqrt{0.125g(S_G - 1)d'}\left(\frac{d}{d'}\right)^{0.47} \tag{16.7}$$

v_t threshold velocity required to initiate movement
d depth of flow

A similar procedure was followed to evaluate equations for bed-load transport in circular pipes with deposited beds. The best fit to the experimental data was provided by a slightly modified version of the method due to May (1993):

$$C_v = \eta\left(\frac{W_b}{D}\right)\left(\frac{D^2}{A}\right)\left(\frac{\theta\lambda_g v^2}{8g(S_G - 1)D}\right) \tag{16.8}$$

W_b sediment bed width (m)
λ_g friction factor corresponding to the grain shear factor (–)
θ transition coefficient for particle Reynolds number (–)
η sediment transport parameter (–)

Evaluation of this equation is more complicated because of the need to estimate the flow resistance produced by the deposited bed. Full details of the recommended method are given by Ackers *et al.* (1996a).

   Pipes designed in accordance with these equations should allow transport of sediment as bed-load at a rate sufficient to avoid deposition or limit it to a specified depth.

3.   BED EROSION

The effect of cohesion on the shear stress at the threshold of movement is specifically allowed for in criterion 3. Based on field evidence and experimental investigations into erosion shear stresses, various relationships between pipe diameter and minimum full-bore velocity were identified, depending on the chosen bed shear stress and bed roughness. It was recommended that the design flow conditions needed to erode cohesive particles from a deposited bed should have a minimum value of shear stress of 2 N/m² on a flat bed with a Colebrook-White roughness value of $k_b = 1.2$ mm (based on 1 mm cohesive sediment particles). The required full-bore velocity (v_f) is given by a modification of equation 8.21:

$$v_f = \sqrt{\frac{8\tau_b}{\rho\lambda_b}} \tag{16.9}$$

$$\lambda_b \approx \frac{1}{4\left[\log_{10}\left(\dfrac{k_b}{3.7D}\right)\right]^2} \tag{16.10}$$

τ_b critical bed shear stress (N/m^2)
λ_b friction factor corresponding to the sediment bed (–)

In many cases, the roughness of the deposited bed will be much higher resulting in bed shear stresses that are higher, making this approach conservative.

Design Procedure

The CIRIA method proposes a detailed procedure in which design tables are worked up from first principles using the equations and information supplied. As an alternative, a simplified procedure is provided in which selected standard values of sediment characteristics and other parameters have been adopted, allowing the presentation of ten simplified design tables. These cover foul, storm and combined sewers, medium and high sediment loads (Table 16.7), and criteria based on either no deposition (LOD) or an allowable average deposition of up to 2% of the pipe diameter. Fig. 16.5 gives the minimum design velocities for foul and storm sewers based on the design tables (Ackers *et al.*, 1996b).

Table 16.7 Typical sewer sediment characteristics and applicability

Sediment class	Normal transport mode	Types of sewer applicable	Parameter	Sediment load		
				Low (L)	Medium (M)	High (H)
Sanitary solids	Suspension	Foul (F) Combined (C)	X (mg/l)	100	350	500
			d_{50} (µm)	10	40	60
			S_G	1.01	1.4	1.6
Stormwater solids	Suspension	Foul (F) Storm (S) Combined (C)	X (mg/l)	50	350	1000
			d_{50} (µm)	20	60	100
			S_G	1.1	2.0	2.5
Grit	Bed-load	Storm (S) Combined (C)	X (mg/l)	10	50	200
			d_{50} (µm)	300	750	750
			S_G	2.3	2.6	2.6

Example 16.3

A 2100 m long storm sewer is to be designed to carry a discharge of 1500 l/s in a location with a maximum available fall of 7 m. There is no specific sediment data available, but the area will be subject to considerable development in the next 10 years. Select an appropriate diameter and gradient circular sewer. Take Manning's *n* as 0.012.

Solution

Solve Manning's equation with S_0 = 1:300 and assume the sewer will run full to obtain required pipe diameter:

$$D = \left(\frac{1.5 \times 0.012 \times 4^{5/3}}{\pi \times (1/300)^{1/2}} \right)^2 = 1.000 \text{ m}$$

Resulting velocity, v

$$v = \frac{1.5 \times 4}{\pi \times 1^2} = 1.9 \text{ m/s}$$

Assume the sediment loading category is high (H), due to anticipated development work, determine required velocity to avoid any permanent deposition (LOD).
   From curve S-H-LOD: The velocity required for this diameter is not shown, but it is clearly above 2.0 m/s, which is greater than that available. Within the available gradient of 1/300, the sewer cannot be designed to carry high sediment loading without deposition.
   Therefore, allow up to 2% deposition and determine the required velocity.
   From curve S-H-2%: required velocity = 1.35 m/s which is <1.9 m/s and thus achievable.
   Note: this velocity is in excess of typical minimum full-bore velocities (e.g. 1.0 m/s)

Limitations

Although the CIRIA method is a significant advance over conventional approaches, it should be noted that it does have limitations (Arthur *et al.*, 1999). The main one is that it is based mainly on data from laboratory pipe-flume experiments using single-sized, granular sediment. This is clearly a simplification of the conditions found in combined sewers (in particular), as described in the rest of this chapter. Some verification with field data has been attempted (May *et al.*, 1996), but comprehensive evaluation is still needed.

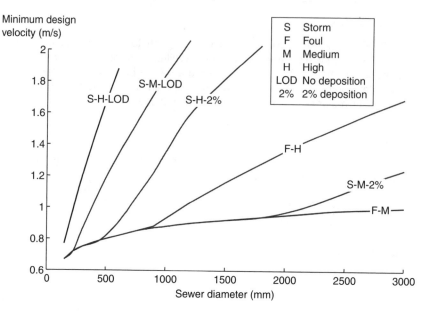

Fig. 16.5 Minimum design velocities using simplified procedure

16.7 Load estimation and application

16.7.1 Long-term washoff

The accurate determination of surface sediment washoff at any individual site is difficult, particularly because of the uncertainties in quantifying the following:

- sediment supply rates (related to land use)
- municipal cleansing practices (fixed by local authority practice)
- hydrology (linked to location).

An approximate approach is to assume that the supply rate of sediment to the catchment (κ) is constant and that its removal by rainfall and street sweeping is proportional to the amount on the surface (χ), so:

$$\frac{d\chi}{dt} = \kappa + b\chi \qquad\qquad 16.11$$

κ sediment supply rate (g/m².d)
b sediment removal constant (d⁻¹)
t time (d)

Integrating gives:

$$\chi = \frac{\kappa}{b}(1 - e^{-bt})$$

16.12

Thus in the long term (t is great), the amount of sediment on the catchment surface (χ_u) reaches equilibrium, such that:

$$\chi_u = \frac{\kappa}{b}$$

16.13

Table 16.8 gives some typical design sediment supply rates. The removal constant (b) represents the combined effect of natural washoff by rainfall (b_r) and maintenance by street sweeping (b_s), so:

$$b = b_s + b_r$$

16.14

The value of b_r, must be related to rainfall properties, but only limited information is available. Butler and Clark (1995) recommend a value of 0.05 per week, for southern England. Additionally, b_s is linked to sweeping frequency and efficiency such that:

$$b_s = f_s/10$$

16.15

f_s number of sweeps per week.

Thus the washoff rate (ζ) is:

$$\zeta = \chi_u b_r = \kappa \frac{b_r}{b_s + b_r}$$

16.16

So the sediment washoff rate is a function only of the sediment supply rate and the removal constants. Example 16.4 shows how this theory can be used to estimate the concentration of the sediment (grit and suspended solids) washoff.

Table 16.8 Design surface sediment supply rates

Land use	Sediment supply rate, a [g/m² imp area per day]
Suburban	0.25
Inner city	2
Industrial	10

Example 16.4

A catchment in a suburban area has a sediment supply rate of 0.2 g/m².d and an annual rainfall of 800 mm. Estimate the equilibrium surface sediment load and the average washoff concentration if there has been no regular street sweeping and the rainfall removal constant is 0.05 per week.

Solution

Equation 16.13:

$$X_u = \frac{0.2 \times 7}{0.05} = 28 \text{ g/m}^2$$

If there has been no street sweeping, $b_s = 0$, so from equation 16.16:

$$\zeta = \kappa = 2 \text{ g/m}^2.\text{day}$$

Assume the road is completely impervious,

Average runoff = 800/365 = 2.19 mm/day

Washoff concentration = 0.2/0.00219 = 91 mg/l

16.7.2 Gully pot sizing

Most gullies are provided with 'pots' which act as sediment intercepting traps. The size of pot required in any given location depends on a number of factors including the rainfall regime, gully spacing, land use and road sweeping/gully emptying frequency. Individual pots may be subject to localised conditions (e.g. heavy leaf fall, adjacent sand stockpiles) that are difficult to predict.

However, a general approach to sizing uses the theory developed in the previous section. If ε is the pot sediment retention efficiency, then the sediment accumulation rate in the gully (e) is:

$$e = \varepsilon\zeta = \frac{\varepsilon\kappa b_r}{b_s + b_r} \qquad\qquad 16.17$$

Thus, if ε' is the gully cleaning efficiency, the time between cleans (T_c) is given by:

$$T_c = \frac{h_{max}A_pS_d\varepsilon'}{A_r e} \qquad\qquad 16.18$$

h_{max} pot trap depth (m)
A_p pot cross-sectional area (m^2)
A_i impervious drainage area (m^2)
S_d sediment dry bulk density (kg/m^3).

Example 16.5 shows how the gully emptying frequency can be estimated.

Example 16.5

Determine the required pot size of gullies serving an average impervious area of 300 m^2 assuming that they are emptied twice a year. The sediment washoff rate is 8 g/m^2.wk, the sediment bulk density is 900 kg/m^3 and the pot retention and cleaning efficiencies are 50% and 75% respectively.

Solution

Sediment accumulation (equation 16.17):

$$e = \varepsilon\zeta = 4 \text{ g/m}^2.\text{wk}$$

Estimate pot dimensions from equation 16.18:

$$25 = \frac{h_{max}A_p \times 900 \times 10^3 \times 0.75}{300 \times 4}$$

$$h_{max}A_p = 0.044 \text{ m}^3$$

If $h_{max} = 0.5$ m, pot diameter is:

$$D = \sqrt{\frac{4 \times 0.044}{\pi \times 0.5}} = 0.37 \text{ m}$$

Thus, the required pot diameter is 370 mm or nearest commercially available.

16.7.3 Gully pot sediment retention efficiency

In the pot sizing procedure described in the previous section, the efficiency of the pot (ε) at retaining sediment has been assumed at a fixed value. By assuming the gully pot acts as a completely mixed reactor and that particle settling is determined by Stoke's law, the efficiency can be related to the key determining variables as follows:

$$\varepsilon = \frac{1}{1 + \dfrac{72Qv}{\alpha\pi gd'^2 \, D_p^2(S_G-1)}} \qquad\qquad 16.19$$

Q flow-rate (m³/s)
v kinematic viscosity (m²/s)
α turbulence correction factor (–), recommended as 0.60
g acceleration due to gravity (m/s²)
d' solid particle diameter (m)
D_p gully pot diameter (m)
S_G particle specific gravity (–)

The equation has been validated in the laboratory (Butler and Karunaratne, 1995) over a range of realistic flow-rates and particle sizes. Application of the equation indicates that particles of 500 μm diameter or greater are greatly retained (90% or more) at normal flow rates, but efficiency falls off rapidly for sub-200 μm sediment. Overall performance for these smaller particles may be even lower due to the possibility of re-erosion of previously settled material.

16.7.4 Sediment management

As discussed earlier in the chapter, most sediment is generated on the catchment surface and is washed from the surface to downpipes and gully pots, through drains or sewers to the WTP. Sediment is also removed from the catchment as a whole in a number of places by a number of means: street sweeping of surfaces, cleaning of gully pots, sewer cleaning and grit removal at WTPs.

In principle, sediment should be managed on a catchment-wide basis (Butler and Clark, 1995) rather than being specifically associated with the drainage network. However, in the UK, responsibility for sediment removal and cleaning rests with different authorities, and is rarely considered on a catchment-wide basis. Cleaning authorities perform their sediment removal operations for different reasons, and are often unaware of the effect of their activities on other parts of the drainage network. Street cleaning, for example, is primarily carried out to remove litter and improve the general appearance of the above-ground area, and gully cleaning is undertaken to avoid blockage and consequent ponding of water on roads.

The aim of considering the whole catchment when planning cleaning operations is to achieve more efficient sediment management and fewer sediment-related problems in sewers. This could reduce the cost of sediment removal from whole catchments by concentrating cleaning efforts at the point where the cost is lowest. Example 16.6 shows how the long-term sediment supply, build-up and washoff model just developed can be used to assess the impact of street sweeping on sediment entry into the piped system.

Example 16.6

A proposal has been made to reduce the road sweeping frequency in an area, from twice a month to once a month. Estimate the impact of this proposal on the sediment load on the road and the sediment washoff rate. Assume a sediment supply rate of 10 g/m² wk and a rainfall removal constant of 0.05 per week.

Solution

Estimate the initial sediment removal constant from equations 16.14 and 16.15:

$$b = b_r + f_s/10 = 0.05 + 0.5/10 = 0.1 \text{ per week}$$

So, initial surface load from 16.13:

$$\chi_u = \frac{\kappa}{b} = \frac{10}{0.1} = 100 \text{ g/m}^2$$

Surface load due to proposed changes, $f_s = 0.25$ ($b_s = 0.025$):

$$\chi_u = \frac{\kappa}{b} = \frac{10}{(0.05 + 0.025)} = 133 \text{ g/m}^2$$

Which is a 33% increase.

Estimate initial average washoff:

$$\zeta = \chi_u b_r = 100 \times 0.05 = 5 \text{ g/m}^2. \text{ wk}$$

Final average washoff:

$$\zeta = 133 \times 0.05 = 6.7 \text{ g/m}^2. \text{ wk}$$

Also a 33% increase.

Street cleaning

The frequency of street cleaning varies from once or more per day in shopping areas to once a year or less on many roads. The efficiency of cleaning in removing street debris is shown in Table 16.9, based on site trials. It is clear that vacuum sweepers are more efficient than manual sweepers and that efficiency drops off with reducing particle size. In everyday practice, it is suspected that efficiencies are lower, especially for smaller particles.

Table 16.9 Efficiency of street cleaning

Particle size range (μm)	Removal efficiency (%)	
	Manual sweeper	Vacuum sweeper
>5600	} 57	90
5600–1000		91
1000–300	46	84
300–63	45	77
<63	25	76
Average	48	84

These smaller particles are most important in terms of water quality control because they contain the majority of the pollutants of concern. Several studies (e.g. Sartor and Boyd, 1972; Pitt, 1979) have shown that very frequent cleaning (several times a week) is required even to deliver modest reductions in solids and metals.

Gully pot cleaning

Gully pot cleaning frequency varies from place to place, but is normally carried out once or twice a year. The efficiency of cleaning varies, but is around 70%. Pots not only trap and retain the sediment for which they are provided, but also other pollutants including oil. Problems arise during dry periods when the retained sediment is degraded anaerobically, allowing NH_4 and COD to build-up in the retained liquor. At the next storm event, the liquor, fine and dissolved solids will be mixed and entrained into the flow, adding significantly to the pollutant load. Thus gullies *can* contribute to a reduction in water quality (Mance and Harman, 1978).

Problems

16.1   Outline the various sources of sediment. Discuss ways in which sediment generation might be reduced.

16.2   Explain how sediment enters, moves through and leaves an urban drainage system.

16.3   What are the main problems caused by sediment deposition?

16.4   How does deposited sediment affect the hydraulic characteristics of a sewer system?

16.5   Define and explain the three phases of sediment movement: entrainment, transport and deposition.

16.6   Sewer sediments have been classified into five types (A–E). Compare and contrast the physical and chemical characteristics of each. What is their significance?

16.7 A 1.5 km long, 1.5 m diameter combined sewer suffers from sediment deposition along its whole length. Measurements reveal an average depth of 300 mm of type A sediment overlain by 20 mm of type C. A storm with a flow of 2.2 m³/s sustained for 30 minutes completely releases the pollutants bound by the sediment bed. Assuming the incoming stormwater has negligible pollution, estimate the average concentration of BOD_5 and COD released.

[567 mg/l, 2993 mg/l]

16.8 Explain the three main modes of sediment transport in sewers. Under what conditions would each mode be most important?

16.9 Explain the main elements of the CIRIA sewer design procedure (for self-cleansing). What are its advantages over conventional approaches? What are its weaknesses?

16.10 Calculate the minimum velocity needed under CIRIA criterion 3 (τ_b = 2 N/m², k_b = 1.2 mm) to cleanse a 1000 mm diameter pipe running half-full. If the pipe has a sediment bed of roughness k_b = 50 mm, what will be the actual shear stress generated by this velocity? [0.88 m/s, 6.9 N/m²]

16.11 Explore the influence and importance of three variables used in the CIRIA equations. What are the implications for sewer design?

16.12 A developer has designed an inner-city development in an area with 750 mm of annual rainfall. Gully pots with 90 l sumps draining an average of 250 m² are proposed. If the local authority sweeps the roads every 2 months, what interval between gully cleaning is required? Assume the following: sediment supply rate = 14 g/m².wk, rainfall removal constant = 0.05 wk⁻¹, bulk density of gully sediments = 1400 kg/m³, pot retention efficiency = 65%, pot cleaning efficiency = 70%. [48 wk]

16.13 What is the maximum flow-rate that can be intercepted by a 450 mm diameter gully pot in order to ensure at least 90% retention of 0.5 mm diameter particles (assume the sediment specific density is 2650 kg/m³ and water kinematic viscosity is 10⁻⁶ m²/s)?

[2.2 l/s]

16.14 A major construction site is about to be established and both the local authority and water authority are concerned at the increased sediment load that will be generated. It is estimated that the initial sediment supply rate of 5 g/m².wk will be quadrupled. What frequency of street sweeping is required to keep the surface load at current levels achieved by sweeping once a month? Assume a rainfall removal constant of 0.06 per week. [2.8 per week]

Key sources

Ackers, J.C., Butler, D. and May, R.W.P. (1996a) *Design of Sewers to Control Sediment Problems*, Report R141, CIRIA, London.

Ashley, R.M., Bertrand-Krajewski, J.-L., Hvitved-Jacobsen, T. and Verbanck, M. (eds) (2003). *Sewer Solids – State of the Art*. Scientific and Technical Report No. 14, IWA Publishers.

Ashley, R.M. and Verbanck, M.A. (1996) Mechanics of sewer sediment erosion and transport. *Journal of Hydraulics Research,* **34**(6), 753–769.

Butler, D. and Clark, P.B. (1995) *Sediment Management in Urban Drainage Catchments,* Report R134, CIRIA, London.

Binnie and Partners and Hydraulics Research (1987) *Sediment Movement in Combined Sewerage and Storm-Water Drainage Systems,* CIRIA Project Report No. 1.

Crabtree, R.W., Ashley, R.M. and Saul, A.J. (1991) *Review of Research into Sediments in Sewers and Ancillary Structures,* Report No. FR0205, Foundation for Water Research.

References

Ackers, P. (1991) Sediment aspects of drainage and outfall design. *Proceedings of the International Conference on Environmental Hydraulics,* Hong Kong, 215–230.

Ackers, P. and White, W.R. (1973) Sediment transport: new approach and analysis. *American Society of Civil Engineers Journal of Hydraulic Engineering,* **99**(HY11), 2041–2060.

Ackers, J.C., Butler D., John, S. and May, R.W.P. (1996b) Self-cleansing sewer design: the CIRIA Procedure. *Proceedings of 7th International Conference on Urban Storm Drainage,* Hannover, September, 875–880.

Arthur, S., Ashley, R., Tait, S. and Nalluri, C. (1999) Sediment transport in sewers – a step towards the design of sewers to control sediment problems. *Proceedings of Institution of Civil Engineers, Water, Maritime and Energy,* **136**, March, 9–19.

Ashley, R.M. and Crabtree, R.W. (1992) Sediment origins, deposition and build-up in combined sewer systems. *Water Science and Technology,* **25**(8), 1–12.

Ashley, R.M., Arthur, S., Coghlan, B.P. and McGregor, I. (1994) Fluid sediment in combined sewers. *Water Science and Technology,* **29**(1–2), 113–123.

Bertrand-Krajewski, J.-L., Briat, P.M. and Scrivener, O. (1993) Sewer sediment production and transport modelling: a literature review. *Journal of Hydraulic Research,* **31**(4), 435–460.

Butler, D. and Karunaratne, S.H.P.G. (1995) The suspended solids trap efficiency of the roadside gully pot. *Water Research,* **29**(2), 719–729.

Butler, D., May, R.W.P. and Ackers, J.C. (1996a) Sediment transport in sewers – Part 1: Background. *Proceedings of Institution of Civil Engineers, Water, Maritime and Energy,* **118**, June, 103–112.

Butler, D., May, R.W.P. and Ackers, J.C. (1996b) Sediment transport in sewers – Part 2: Design. *Proceedings of Institution of Civil Engineers, Water, Maritime and Energy,* **118**, June, 113–120.

Crabtree, R.W. (1989) Sediments in sewers. *Journal of the Institution of Water and Environmental Management,* **3**, December 569–578.

Macke, E. (1982) *About Sedimentation at Low Concentrations in Partly Filled Pipes,* Mitteilungen, Leichtweiss – Institut für Wasserbau der Technischen Universität Braunschweig, Heft 76.

Mance, G. and Harman, M.N.I. (1978) The quality of urban stormwater runoff, in *Urban Stormwater Drainage* (ed., R.P. Helliwell), Pentech Press, 603–618.

May, R.W.P. (1993) *Sediment Transport in Pipes and Sewers with Deposited Beds*, Report SR 320, HR Wallingford.

May, R.W.P., Ackers, J.C., Butler, D. and John, S. (1996) Development of design methodology for self-cleansing sewers. *Water Science and Technology*, **33**(9), 195–205.

Nalluri, C. and Alvarez, E.M. (1992) The influence of cohesion on sediment behaviour. *Water Science and Technology*, **25**(8), 151–164.

Pitt, R. (1979) *Best Practice Management Implementation*. EPA Report No. 600/S2–88/038.

Raudkivi, A.J. (1998) *Loose Boundary Hydraulics*, 4th edn, A.A. Balkema.

Sartor, J.D. and Boyd, G.B. (1972) *Water Pollution Aspects of Street Surface Contaminants*. EPA Report No. R2/72–081.

Verbanck, M.A. (1995) Capturing and releasing settleable solids – the significance of dense undercurrents in combined sewer flows. *Water Science and Technology*, **31**(7), 85–93.

17 Operation, maintenance and performance

17.1 Introduction

In the past, operation and maintenance (O & M) of urban drainage systems has often been inadequate. The temptation has been to assume that if there were no immediate problems, there was no need to spend money. Yet, drainage systems corrode, erode, clog, collapse and ultimately deteriorate to the point of failure and beyond.

Maintenance is needed to maintain the operational function of the system and to extend its working life. Widespread recognition of this fact began to emerge in the 1980s with publication of the first edition of the *Sewerage Rehabilitation Manual* and has continued apace, particularly since privatisation of the water industry in England and Wales (the *SRM* will be considered in more detail in the next chapter). Emphasis has now changed from considering sewer networks as liabilities to recognising them as valuable assets, with O & M needed to maintain a properly performing system.

17.2 Maintenance strategies

There are several reasons for the comprehensive maintenance of a sewer system.

Public health

Maintenance of public health is paramount (see Chapter 1), and the continuing good functioning of the system can help to achieve it. In addition, the system itself should not cause nuisance or a health hazard to either its users or its operators.

Asset management

All systems were costly to construct and would be even more costly to replace. High priority must, therefore, be given to maintaining the physical integrity of the assets.

Maintain hydraulic capacity

A primary function of maintenance is to preserve the as-built hydraulic capacity of the system. This will minimise the possibility of wastewater backing-up into properties or widespread surface flooding. This can be done by cleaning and ensuring, as far as is practicable, that the system is watertight.

Minimise pollution

All combined and storm sewer systems have discharge points to the environment that come into operation periodically. Maintenance has a role in reducing the frequency of operation as far as possible, and in avoiding conditions in the system that cause build-up of pollutants.

Minimise disruption

The privatised water industry is judged by its customers on the efficiency with which it deals with operation and maintenance. Disruption to the general public should be minimised.

Various degrees of sophistication can be built into maintenance strategies, but there are two main categories: reactive and planned (BS EN 752–7: 1998).

17.2.1 Reactive maintenance

In reactive maintenance, problems are dealt with on a corrective basis as they arise (i.e. after failure): the so-called 'fire fighting' approach. This approach will always be required to a certain extent, as problems and emergencies are bound to occur from time-to-time in every urban drainage system. However, reactive maintenance cannot reduce the number of system failures. To achieve this, a planned approach is needed.

17.2.2 Planned maintenance

In planned maintenance, potential problems are dealt with prior to failure. Unlike reactive maintenance, planned maintenance is proactive and has the objective of reducing the frequency or risk of failure. Central to planned maintenance is a comprehensive inspection programme and analysis of existing data.

Planned maintenance is not the same as routine maintenance (operations at standard intervals, regardless of need), but involves identifying elements that require maintenance and then determining the optimum frequency of attention.

17.2.3 Operational functions

The major O & M functions are:

- location and inspection
- cleaning and blockage clearance
- chemical dosing
- fabric rehabilitation – repair, renovation or replacement.

The first three of these functions will be described in this chapter. The last topic of rehabilitation will be considered in Chapter 18.

Inevitably, operational effort needs to be prioritised and a suitable hierarchy of sewer maintenance, such as that described by Read and Vickridge (1997), can be devised (see Table 17.1).

Finally, it should be noted that maintenance of sewer networks holds some specific challenges, even when compared with other industries. These include:

- geographical size of networks (e.g. dispersion and length of pipework)
- physical size of assets (e.g. access, non-man entry)
- aggressive environment (e.g. hazardous gases).

17.2.4 Role of design

Full consideration should be given to the operation and maintenance of systems during the design process. In particular, design should attempt to minimise the degree of maintenance required. That which *is* required should be simplified and allowed for (e.g. access). This process can be facilitated by regular communication between design and operations staff, giving benefits to both parties.

An attempt should be made to balance potential savings in O & M costs against possibly higher construction costs.

Table 17.1 Sewer maintenance hierarchy

Level	Task	Type	Consequences of omission
1	Periodic cleaning	*	Level 2 task needed
2	Blockage removal	**	Failure
3	Repair	*	Level 4 task needed
4	Renovation	**	Level 5 task needed
5	Replacement	**	Failure

* Planned   ** Reactive

17.3 Sewer location and inspection

Sewer location and inspection methods are routine tools of the maintenance engineer. Basic methods for locating sewers and manholes have been available for many years, but these have been improved and newer methods introduced. Inspection, in particular, has been revolutionised by the introduction of remote surveillance equipment. Now, detailed surveys can be carried out in previously-inaccessible locations in a cost-effective way.

17.3.1 Applications

The main applications of sewer location and inspection are:

- periodic inspection to assess the condition of existing sewers (planned maintenance)
- crisis inspection to investigate emergency conditions or the cause of repeated problems along a particular sewer length (reactive maintenance)
- inspection of workmanship and structural condition of new sewers before 'adoption' – see Chapter 7 (quality control).

17.3.2 Frequency

It is impossible to accurately estimate the rate of deterioration of a particular sewer length from a single survey. This is particularly true of insidious problems such as sulphide attack (see Section 17.6). The only reliable way to monitor a sewer's condition is to carry out a series of inspections at given intervals such as those recommended in Table 17.2. The level of inspection chosen will reflect an attempt to balance the risks with the consequences of failure.

Table 17.2 Inspection frequencies for critical sewers (adapted from Sewers and Water Mains Committee (1991))

Condition grade*	Frequency (yr)	
	Category A**	Category B
5	n/a	n/a
4	n/a	5
3	3	15
2	5	20
1	10	20

* The SRM defines condition grades ranging from 'collapse' (5) to 'acceptable structural condition' (1).
** Category A critical sewers are those where collapse repair costs would be highest and amount to about 5% of the total system. Category B sewers, although less critical, would still have substantial collapse costs (see Chapter 18).

17.3.3 Locational survey

The first steps in a maintenance strategy are to check on the accuracy and completeness of existing records of the system, and then initiate a survey on the parts of the system where there is doubt. Before an inspection of any kind can take place, it is necessary to locate the manholes and thereby determine the route of all the sewers in the system.

Manhole location is usually straightforward, although a metal detector may be required if it is suspected that a cover has been buried. The position and level (cover, soffit and invert) of each manhole can be determined using standard land surveying techniques. This procedure can now be substantially speeded up using GPS (global positioning satellite) technology, allowing positional data to be logged on-site in seconds.

Techniques available to determine the route of a sewer range from the simple to the sophisticated. Visual inspection of flow directions in manholes is sometimes sufficient and, if not, dye tracing can be carried out. Electronic tracing is also becoming common (Fig. 17.1). A probe or 'sonde' which emits radio signals is pushed, rodded, jetted or floated along the sewer and its progress tracked from the surface using a hand-held receiver. Using this approach, sewers up to 15 m deep can be traced to a claimed accuracy of ±10%. Interference from signals generated by other buried metallic assets can, however, cause problems.

Ground probing radar is an emerging technique, with potential for sewer location.

17.3.4 Closed-circuit television (CCTV)

CCTV inspection of sewers was first introduced in the early 1970s and has been subject to gradual development and refinement ever since. A small TV camera incorporating a light source is propelled through the sewer and the images are relayed to the surface for viewing and recording (see Fig. 17.2).

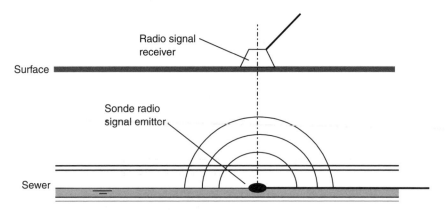

Fig. 17.1 Electronic tracing of sewers

Fig. 17.2 CCTV image of a sewer in poor condition (courtesy of Telespec Ltd, Guildford)

CCTV is a popular choice for sewer inspection, not least because internal investigation of sewer systems can be carried out quickly, and with minimal disruption, avoiding lengthy shutdowns and unnecessary excavation. This method is particularly useful in environments that are too small or hazardous for people to enter. CCTV can also be used to locate and define the cause of a known condition or defect, and to help establish a plan of maintenance. Progress is relatively rapid with typical rates of 400–800 m/day.

The method is commonly used in pipes from 100–1500+ mm in diameter, but is less effective in larger pipes, due to increased lighting requirements and difficulty in achieving adequate resolution of detail with high subject-to-camera distances. However, developments in camera technology are continually increasing the feasibility of monitoring even larger pipes.

Propulsion

Generally, for sewers of <150 mm diameter the camera must be winched between manholes. A line attached to the winching cable is floated or rodded between the manholes and the cable pulled through. The camera is mounted on skids and winched slowly through the sewer. For sewers of diameter greater than 150 mm, it is usual for the camera to be mounted on a self-propelled, remotely-controlled tractor. The tractor speed varies according to the size of the wheels fitted, and will be greater in larger sewers, with typical speeds ranging from 0.1 to 0.2 m/s.

Camera operation

The basic camera technology used in sewer inspection came originally from broadcast TV. Development has centred on gathering the maximum amount of information at reasonable cost in very difficult and dirty conditions. Most cameras are now of the charge coupling device (CCD), solid-state type, housed in strong waterproof cases. These have high sensitivity and can be used for surveying very large sewers (over 3 m diameter).

Lamps attached to the front or sides of the camera head provide lighting. Lenses of several focal-lengths are available, including zooms, and focusing can either be pre-set or remotely-controlled by the operator. The most useful view for a CCTV camera is usually axially along the sewer, but there are specific occasions (e.g. looking up a house connection or at specific problems) when a lateral view is preferable. Several techniques are available for achieving this, including a wide-angle lens, pan and tilt equipment and an electronic 'fish eye'. A typical CCTV camera is illustrated in Fig. 17.3

17.3.5 Manual inspection

Manual inspection is only used in exceptional circumstances and only if inspection cannot be done another way.

Fig. 17.3 A pan and tilt in-sewer CCTV camera (courtesy of Telespec Ltd, Guildford)

If such inspection *is* necessary, a survey begins at one manhole and works along the length of the sewer in increments (typically of 1 m) to the next. Information can be gathered on mortar loss displaced/missing bricks, cracks, sewer shape, connections, silt/debris, etc. Paper-based methods have traditionally been used to record this information, but portable data loggers or hand-held computers with appropriate software are now widely available.

Progress is relatively slow using this procedure (200–400 m/day); it is costly and dangerous. However, high quality information can be obtained.

17.3.6 *Other techniques*

Sonar

Before the advent of in-sewer sonar technology, it was only possible to inspect relatively empty sewers. In many cases, this required the effort and expense of over-pumping. Sonar techniques can acquire an image of the profile of a liquid-filled sewer, without the need for a light source (Winney, 1989). The sonar head is controlled from a surface processor and scans through 360° in discrete increments. Data derived from the acoustic signal can be displayed on a colour monitor and recorded.

Infra-red

Another approach to in-sewer inspection is thermal imagery, in which an infra-red camera is used to gather and focus emitted black-body radiation and convert it into a form visible to the human eye. Again, no external light source is necessary. This technique is rather limited in application, but can be used to inspect for infiltration, based on the assumption that the wastewater and groundwater are at different temperatures.

Sewer profiling

The use of solid-state cameras in conjunction with opto-electronic light-measuring systems enables internal sewer profiles to be accurately mapped. This technique involves a specially configured light head that is attached to the front of the camera, with two light sources that cast a focused circle of light on the internal sewer surface. Any changes in shape can then be detected and accurately quantified using appropriate computer software. This technique is particularly appropriate in monitoring old brick sewers or deformed plastic pipes.

17.3.7 Data storage and management

Conventionally, CCTV images have been stored on video tapes for subsequent playback and study of the images. Unfortunately, tapes have a number of significant drawbacks:

- they do not permit rapid pinpointing of defects along the sewer length
- they require significant space for storage
- there are concerns over the long-term stability of the images.

Recent practice is to transfer existing images or directly store new images in digital format. This allows still pictures or short sequences to be stored on a computer hard disk. Whole sewer lengths can also be stored on CD-ROM disks. This largely overcomes the three major problems mentioned above. It also allows the potential for automated scrutiny of the images.

Sewer surveys generate a great deal of data that requires careful and systematic handling. Software packages are now available to aid in data management, allowing defects, coded in a standardised way (WRc, 1993) at relevant locations in the pipe or manhole, to be stored in an easily-accessible database. Such information can then be used to assess the structural condition of the system systematically. Most packages will produce data files in a format compatible with simulation models. More recently, databases have been upgraded to Geographic Information Systems (GIS) which allow spatial information to be held and graphically displayed. Information on many services can be held on different 'layers'. Also, sewer record databases can now be linked with Computer Aided Drawing (CAD) packages to allow speedy production of drawings.

17.4 Sewer cleaning techniques

17.4.1 Objectives

Sewer cleaning is carried out:

- proactively, to remove sediment in order to restore hydraulic capacity and limit pollutant accumulation
- reactively, to deal with blockages or offensive odours
- to permit sewer inspection
- to aid sewer repair/renovation.

17.4.2 Problems

Blockages

These are defined as full or partial restrictions within the sewer and are most commonly found in smaller diameter pipes. A blockage is normally associated with a system defect (e.g., displaced joint, severe change of direction). The effect of a blockage ranges from partial loss of capacity to complete stoppage.

Sedimentation

Sediment is defined as any settleable particulate material that may, under certain conditions, form bed deposits in sewers and associated hydraulic structures. It is normally associated with large, flat sewers. Sedimentation rarely completely chokes the pipe, but can still have a significant impact on capacity. Chapter 16 discusses the issues associated with sediment deposition in further detail.

Grease/scale

Solidified grease is often associated with non-domestic properties, restaurants being particular culprits. High temperature dishwashers often move the grease from the premises, only for it to cool and solidify further downstream causing loss of hydraulic capacity. Wall scale or encrustation can also cause similar problems.

Tree roots

Sewers are susceptible to intrusion of tree roots, which seek out moist conditions. The roots themselves are a nuisance, both in retarding the flow but also in initiating further blockage with larger solids.

Intruding laterals

Intruding laterals or other connections are common as a result of poor construction practice (see 18.3.5). The intrusion reduces the cross-sectional area causing the same problems as tree roots.

A number of cleaning techniques and methods are in use, depending particularly on location and severity, including rodding, winching, jetting, flushing and hand excavation. A combination of more than one method may well be used in any particular locality.

17.4.3 Rodding or boring

Rodding or boring is primarily a manual procedure, in which short flexible rods are screwed together and then inserted into the blocked sewer. The principal action is the physical contact of the tools, although the compression of air by plungers and dense brushes can contribute. A more recent development is the use of semi-rigid, coiled GRP rods supplied in continuous lengths on a reel. The procedure can be mechanised.

This technique is limited to small diameter pipes (≤225 mm) at shallow depths (≤2.0 m) and must be close to the access point (≤20 m). It is particularly well suited to dislodging blockages and roots.

17.4.4 Winching or dragging

Winching or dragging is a technique involving the use of purpose-shaped buckets that are dragged through the sewer collecting sediment, which is emptied out at a manhole. Although the winch can be manually operated, power driven devices are normally used.

The procedure is capable of removing most materials, even in large pipes, but is most effective in sewers up to 900 mm diameter, up to 50% silted. Care has to be taken that damage is not caused to the sewer fabric.

17.4.5 Jetting

Jetting is a widely used technique that relies on the ability of an applied high-pressure (100–350 bar) stream of water to dislodge material from sewer inverts and walls and transport it down the sewer for subsequent removal. Water under these high pressures is fed through a hose to a nozzle containing a rosette of jets sited in such a way that the majority of flow is ejected in the opposite direction to the flow in the hose. The jets propel the hose through the sewer, eroding the settled deposits in the process. A range of nozzles is available to cope with specific situations. Modern combination units incorporate vacuum or air-displacement lifting equipment to remove the material, as well as to dislodge it without the need for man entry.

Jetting is a versatile and efficient procedure for removing a wide range of materials and is widely used in practice. Concern over the possibility of damaging pipes during the jetting process has resulted in the publication of the *Sewer Jetting Code of Practice* (WRc, 1997).

17.4.6 Flushing

Flushing is a technique in which short duration waves of liquid are introduced or created so as to scour the sediment into suspension and, hence, transport it downstream. Waves may be induced by:

- automatic siphons at the heads of sewers
- dam and release using blow boards/gates
- hydrant and hose
- mobile water tanker.

Of these, the first two are now rarely used. All merely move the dislodged material downstream and do not remove it from the system.

17.4.7 Hand excavation

Historically, large diameter sewers were cleared by manual digging-out of deposited material. Labourers entered the sewer and shovelled sediment into skips that were transported to the surface for emptying. The method is limited in application to larger size pipes (>900 mm) and has significant health and safety implications. It is now used only in exceptional circumstances.

17.4.8 Invert traps

Not widely used, but a potentially useful cleaning option, invert grit traps are intended to intercept sediments travelling as bed-load in combined sewers. Invert traps in existence are typically large rectangular chambers, which although effective at trapping sediments, also collect other near-bed and suspended solids. Performance of these chambers can be improved by the introduction of a cover with an open slot across the width of the invert designed to collect heavier, inorganic sediment only (Buxton *et al.*, 2001).

17.4.9 Comparison of methods

No one method of cleaning is superior to the others on all occasions; each has its advantages and disadvantages. These are summarised in Table 17.3.

17.5 Health and safety

It is important to appreciate the health and safety hazards involved with sewer operation. Many of the practices to be carried out are covered by the Health and Safety at Work Act 1974. In addition, the sewer environment is classified as a 'confined space' under the Confined Spaces Regulations (HSE, 1997). Specific safety advice is available including ICE's *Safety Guide for Men working in Sewers and at Sewage Disposal Works* and *Safe Working in Sewers and at Sewage Works*.

17.5.1 Atmospheric hazards

Atmospheric hazards are probably the most dangerous: explosive or flammable gases may develop at any time. Anaerobic biological

Table 17.3 Relative performance of sewer cleaning techniques (adapted from Lester and Gale, 1979)

Topic	Rodding	Winching	Jetting	Flushing
Sewer size (mm):				
<375	Good	Fair	Good	Good
450–900	Poor	Good	Fair	Fair
Max. cleansing distance (m)	25	100	100	50
No. of manholes required	1	2	1	1
Dislodging materials:				
Invert	Fair	Good	Good	Good
Walls	Fair	Fair	Good	Poor
Joints	Fair	Fair	Good	Poor
Materials encountered:				
Silt	Fair	Fair	Good	Good
Sand/gravel	Poor	Good	Good	Good
Rocks	Poor	Good	Fair	Poor
Grease	Fair	Fair	Fair	Poor
Roots	Good	Good	Fair	Poor
Material removed?	No	Yes	Yes	No
Damage potential	Low	Medium	High	Low
Flooding potential?	No	No	No	Yes

decomposition of wastewater yields methane, which is lighter than air. Petrol vapour, on the other hand, is heavier than air and tends to form pockets at the invert of the sewer. Factories should report the disposal of dangerous chemicals, but they may not report accidental spills or deliberately negligent acts.

The most commonly occurring toxic gas within the sewer is hydrogen sulphide. It is a flammable and very poisonous gas that has a distinctive smell. It is particularly dangerous to workers in sewers because the ability of a person to smell the gas diminishes with exposure and as the concentration increases. The amount of breathable oxygen in a sewer can be reduced or even eliminated by displacement by a heavier gas. If there is no breathable oxygen in a sewer, the life expectancy of a person entering is approximately 3 minutes.

To provide adequate ventilation, the access manhole cover and those upstream and downstream need to be lifted well before an inspection. Those entering the sewer should use a gas detector. Because of the short life expectancy in unfavourable conditions, current best practice is to ensure that gangs can be self-rescuing, rather than relying on Fire Brigade assistance. If a dangerous atmosphere exists within a man-entry sewer the inspection can be carried out using breathing apparatus.

17.5.2 Physical injury

Physical injury is an ever-present hazard in the sewer environment. It can result from falls down manholes or in the sewer, and the dropping, throwing or misuse of equipment. The risk of drowning must not be under-estimated, whether in the residual flow within the sewer or due to a sudden rise in water level after rainfall. This can be avoided by maintaining close contact with the surface at all times.

Acids may find their way into the sewer system and could cause burns unless protective boots and gloves are worn at all times.

17.5.3 Infectious diseases

Infection from tetanus, hepatitis B or leptospirosis is a potential risk that should be planned for by ensuring workers are under medical supervision and inoculated as appropriate. Discarded needles found in sewers should be avoided. Animals, such as rats and insects, can also be a health hazard.

17.5.4 Safety equipment

If the sewer is to be entered for inspection or any other maintenance work, suitable safety equipment must be worn. This includes waterproof clothing, heavy waders, gloves, safety harnesses and hard hats as well as radios, safety lamps and, if necessary, breathing equipment. Portable gas detectors should always be used to test for toxic gases. Two-way radios have been used to keep in contact with members of the team on the surface, but may be ineffective in certain conditions; visual and vocal contact with a surface member is preferable. The surface team must have access to regular weather forecasts.

17.6 Pipe corrosion

Corrosion of concrete, metal and electrical equipment in urban drainage systems can be caused by the generation of hydrogen sulphide. Particularly susceptible locations are points of turbulence following long-retention times e.g. back-drop manholes, wet wells of pumping stations and outlets of wastewater rising mains. Problems are particularly serious in hot, arid climates. In addition, hydrogen sulphide can cause odour nuisance when escaping into the atmosphere, danger to sewer workers (as already described) and acute toxicity to aquatic organisms.

17.6.1 Mechanisms

Wastewater naturally contains sulphur as inorganic sulphate or organic sulphur compounds. The sulphate is usually derived from the mineral content of the municipal water supply or from saline groundwater infiltration. Organic sulphur compounds are present in excreta and household detergents, and in much higher concentrations in some industrial effluents such as from the leather, brewing and paper industries (see Chapter 4).

Bacterial activity quickly depletes any dissolved oxygen that is present and septicity can easily develop. Under anaerobic conditions, complex organic substances are reduced to form volatile fatty acids resulting in a drop in pH. *Desulphovibrio* bacteria in pipe biofilms and sediment reduce organic sulphur compounds and sulphates (SO_4^{2-}) to sulphides (S^{2-}):

$$SO_4^{2-} + C,H,O,N,P,S \rightarrow S^{2-} + H_2O + CO_2 \qquad (17.1)$$

Hydrogen sulphide (H_2S) results from reaction with hydrogen ions in the water and hence is pH dependent, with more being formed in acidic conditions.

$$S^{2-} + 2H^+ \rightarrow H_2S \qquad (17.2)$$

In pipes flowing under gravity, H_2S escaping into the atmosphere from solution in the wastewater tends to rise and accumulate in condensation water on the soffit of the pipe (see Fig. 17.4). There it is oxidised by *thiobacillus* bacteria to form sulphuric acid:

$$H_2S + 2O_2 \rightarrow H_2SO_4 \qquad (17.3)$$

Sulphuric acid can cause serious damage to pipe materials.

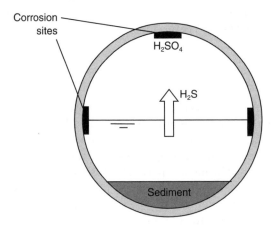

Fig. 17.4 Corrosion in sewers

17.6.2 Favourable conditions

The most favourable conditions for the production of hydrogen sulphide are (Newcombe *et al.*, 1979):

- wastewater with a large proportion of trade wastes with substantial sulphide or organic sulphur contents

- wastewater with a relatively high sulphate concentration
- low pH wastewater – the lower the pH value, the greater is the proportion of molecular hydrogen sulphide
- a high rate of oxygen demand to rapidly consume available dissolved oxygen
- high wastewater temperature which accelerates biological activity
- retention of the wastewater flow under anaerobic conditions. Examples are long sewers with flat gradients, long rising mains, large pump wet wells
- low wastewater velocity that decreases the rate of oxygen re-aeration and increases sedimentation.

Factors tending to increase the emission of hydrogen sulphide are:

- high concentrations of molecular hydrogen sulphide in the wastewater
- high wastewater velocity or turbulence
- high relative velocity and turbulence in the head space *above* the flow
- clean wastewater surfaces with respect to oil films, surfactants, etc.

17.6.3 Sulphide build-up

Pomeroy and Parkhurst (1977) have suggested a formula that produces an index Z that is broadly indicative of the conditions under which sulphide may be formed in gravity sewers under 600 mm in diameter. The 'Z formula' gives:

$$Z = \frac{3(EBOD)}{S_o^{\frac{1}{2}} Q^{\frac{1}{3}}} \frac{P}{B} \tag{17.4}$$

$EBOD$ Effective $BOD_5 = BOD_5 \times 1.07^{(T-20)}$ (mg/l)
T wastewater temperature (°C)
S_o sewer gradient (m/100m)
Q flow-rate (l/s)
P wetted perimeter (m)
B flow width (m)

The formula contains factors representing the main influences on sulphide generation with *EBOD*, accounting for both the influence of temperature and (indirectly) sulphate content of the wastewater. The value of Z and its interpretation is given in Table 17.4 (see also Example 17.1).

17.6.4 Control

A wide range of techniques is available to control the generation or emission of hydrogen sulphide.

Table 17.4 Likelihood of sulphide development

Z	Prevalence of sulphide
<5000	Rarely present.
~7500	Low concentrations likely.
~10 000	May cause odour and corrosion problems.
~15 000	Frequent problems with odour and significant corrosion problems.

Example 17.1

A 500 mm diameter concrete sewer ($n = 0.012$) has been laid at a gradient of 0.1%. At average flow, the pipe runs half full of wastewater. If the wastewater has a BOD_5 of 500 mg/l and the summer temperature is 30 °C, estimate the likelihood that hydrogen sulphide will be generated.

Solution

Geometric properties of flow:

   Area of flow, $A = \pi D^2/8 = 0.098$ m^2

   Wetted perimeter, $P = \pi D/2 = 0.785$ m

   Flow width, $B = D = 0.5$ m

Flow rate, Q (equation 8.23):

$$Q = \frac{A}{n}R^{\frac{2}{3}}S_o^{\frac{1}{2}} = \frac{0.098}{0.012}\left(\frac{0.098}{0.785}\right)^{\frac{2}{3}}0.001^{\frac{1}{2}} = 64.6 \text{ l/s}$$

$$EBOD = BOD_5 \times 1.07^{(T-20)} = 500 \times 1.07^{10} = 984 \text{ mg/l}$$

From equation 17.4:

$$Z = \frac{3(EBOD)}{S_o^{\frac{1}{2}}Q^{\frac{1}{3}}}\frac{P}{B} = \frac{3 \times 984}{0.1^{\frac{1}{2}}64.6^{\frac{1}{3}}}\frac{0.785}{0.5} = 3653$$

This value of Z implies sulphide is unlikely to be present.

Sewerage design

Probably the most effective way of controlling the generation and emission of H$_2$S is to design the sewer system carefully in the first place. In particular, self-cleansing velocities should be designed to occur regularly. Dead spots in manholes and other structures should be avoided.

   Where sulphide generation cannot be avoided, perhaps because of high ambient temperatures, sewer design should avoid excessive turbulence.

Concrete is a material susceptible to H_2SO_4 attack. Boon (1992) suggests that using calcareous aggregates rather than quartz aggregates could considerably extend the life of a concrete sewer. Alternatively, pipes protected by thin epoxy resin coatings have been used, but relatively little data on their longevity is yet available. *In situ* coatings have also been tried.

Vitrified clay or plastic pipes have been shown to be more resistant to corrosion by sulphuric acid. These should be specified at vulnerable locations.

Ventilation

Good ventilation has been mentioned in Chapter 7. It has several benefits including:

- helping to maintain aerobic conditions and hence preventing sulphate reduction
- stripping any H_2S from the atmosphere
- oxidising any H_2S in the atmosphere
- reducing pipe soffit condensation.

The latter point is particularly important – H_2S does not cause corrosion under dry conditions, as bacterial oxidation requires moisture.

Aeration

Dissolved oxygen in the form of air, or directly as molecular oxygen, can be used to oxidise dissolved H_2S. Both methods are widely-used to treat rising mains. The injection point could be at the inlet end of the main with the aim being to maintain aerobic conditions. Alternatively, treatment could be at the discharge point to oxidise already formed sulphides. Both methods will have relatively high capital and running costs.

Disinfection

The addition of chlorine or hypochlorite oxidises any sulphides present and temporarily halts bio-activity, thus preventing further sulphide generation. Experience suggests that such treatment is only moderately effective, probably because wastewater has a high chlorine demand. Continuous addition of high doses of chlorine would be prohibitively expensive.

An alternative is to use hydrogen peroxide, which also oxidises sulphides present and has bactericidal properties. However, its cost is high relative to that of commercial oxygen (Boon, 1992).

Chemical addition

Ferric(III) salts can be dosed to precipitate existing sulphides as insoluble ferrous(II) sulphide. This could lead to problems with additional solids being generated. Additionally, if the salt added is ferric sulphate, the sulphate anion is available for reduction to sulphide and may actually exacerbate the problem. Storage and handling of the corrosive liquid has also proven to be difficult.

Nitrates can be used as an alternative oxidant to molecular oxygen. If sufficient nitrate is available, denitrifiers are encouraged to grow, and under these anoxic conditions will convert nitrates to nitrogen gas. This removes any H_2S present and suppresses further anaerobic degradation. As sufficient nitrate is not normally present in wastewater, this must be added. However, calculation of the required amount is difficult and will depend on the specific (changing) characteristics of the wastewater. A useful by-product can be additional BOD removal. However, care is needed not to add excessive nitrate as this could cause problems at the WTP.

Another possible strategy is the addition of lime to the wastewater. The effect is to increase the pH of the wastewater, thus reducing the proportion of the sulphide present as H_2S. This might be a cost-effective option for small diameter rising mains (Boon, 1992).

17.7 Performance

The need to have an adequately performing urban drainage network would seem to be an obvious requirement, as would be its link with asset maintenance and rehabilitation (discussed in the next chapter). The performance objectives are relatively straightforward and have already been highlighted in this book: to efficiently carry away wastewater from properties, efficiently carry away stormwater from properties and their environs, and safely returning both to the environment. Of course consistently achieving these objectives is less straightforward, as is measuring and demonstrating that it has been done.

Increasingly, the water industry is turning to performance *indicators* as a means of checking whether the system is consistently performing correctly or, put another way, if an adequate service is being delivered to customers. Fenner (2001) lists several reasons for wanting to develop robust performance indicators:

- to represent the effects of complex processes and physical interactions in a simple manner
- to measure progress made towards targets
- to provide benchmarking information to allow comparisons to be made

- to aid the development of investment plans for capital maintenance of assets
- to provide a basis for monitoring and regulating the delivery of minimum levels of service.

In England and Wales, the regulator OFWAT has developed a number of indicators of the standards of service to be achieved. Those relevant to urban drainage are:

- number of sewer collapses
- number of pollution incidents occurring at combined sewer overflows and sewers
- number of properties at risk of flooding due to insufficient sewer capacity.

The last of these (so-called DG5) is one of a suite of seven indicators used as a means of comparing the overall performance of individual companies, with good performers being financially rewarded and poor performers penalised. A more extensive set of indicators has been devised for a wider range of stakeholders (Matos *et al.*, 2003).

Care must be taken in interpreting indicators of this kind. Fenner (2002) expresses the view that performance indicators should be seen as numerical indicators that require interpretation. They can then be used to help answer questions about performance, and ensure consistent levels of service provision between geographic areas.

Problems

17.1 Describe the main operation and maintenance functions in an urban drainage system, and the reasons for carrying them out. What are the particular challenges to be met?

17.2 Explain the differences between reactive and planned maintenance. What are the benefits of combining both approaches?

17.3 Compare and contrast manual sewer inspection with CCTV inspection. What other forms of inspection are available?

17.4 Describe the main types of sewer cleaning equipment. Discuss their main areas of application and effectiveness.

17.5 Explain the dangers to health and safety of sewer workers and detail good working practice.

17.6 'Hydrogen sulphide causes serious corrosion of urban drainage systems.' Discuss the validity of this statement in terms of how, when and where the gas is formed.

17.7 Measurements in an existing sewer reveal the following data: diameter = 300 mm, depth of flow = 240 mm, mean velocity = 0.75 m/s, $BOD = 750$ mg/l and mean temperature = 30°C. Calculate Z to assess the likelihood of H_2S generation. Assume Manning's n is 0.012. [7600]

17.8 List the main methods to control hydrogen sulphide problems and discuss their relative merits.

17.9 How would you measure the performance of an urban drainage system?

Key sources

Ashley, R. and Hopkinson, P. (2002) Sewer systems and performance indicators – into the 21st century. *Urban Water*, **4**, 123–135.

Boon, A. (1992) Septicity in sewers: causes, consequences and containment. *Journal of the Institution of Water and Environmental Management*, **6**, Feb, 79–90.

Rolfe, S. and Butler, D. (1990) A review of current sewer inspection techniques. *Municipal Engineer*, **7**, Aug, l93–207.

Safety Guide for Men working in Sewers and at Sewage Disposal Works, Institution of Civil Engineers.

Sewers and Water Mains Committee (1991) *A Guide to Sewerage Operational Practices*, Water Services Association/Foundation for Water Research.

References

BS EN 752–7: 1998 *Drain and Sewer Systems Outside Buildings. Part 7: Maintenance and Operations.*

Buxton, A., Tait, S., Stovin, V. and Saul, A.J. (2001) The performance of engineered invert traps in the management of sediments in combined sewer systems. *Novatech 2001, Proceedings of the 4th International Conference on Innovative Technologies in Urban Drainage*, Lyon, France, 989–996.

Fenner, R. (2002) Performance is your reality, forget everything else. Editorial. *Urban Water*, **4**, 119–121.

HSE (1997) *Safe Work in Confined Spaces. Approved Code of Practice, Regulations and Guidance*, HSE Books.

Lester, J. and Gale, J.C. (1979) Sewer cleansing techniques. *The Public Health Engineer*, **7**(3), July, 121–127.

Matos, R., Cardoso, A., Ashley, R.M., Molinari, A., Schulz, A. and Duarte, P. (2003) *Performance Indicators for Wastewater Services*. Manual of Best Practice. IWA Publishers.

Newcombe, S., Skellett, C.F. and Boon, A.G. (1979) An appraisal of the use of oxygen to treat sewage in a rising main. *Water Pollution Control*, **78**(4), 474–504.

Pomeroy, R.D. and Parkhurst, J.D. (1977) The forecasting of sulphide buildup rates in sewers. *Progress in Water Technology*, **9**, 621–628.

Read, G.F. and Vickridge, I.G. (eds) (1997) *Sewers. Rehabilitation and New Construction. Repair and Renovation*, Arnold.

Safe Working in Sewers and at Sewage Works, Water Industry Health and Safety Guideline No. 2.

Winney, M. (1989) New sonar survey cracks flooded pipes problem. *Underground Magazine*, March, 17–18.

WRc (1993) *Manual of Sewer Condition Classification*, 3rd edn, WRc, Swindon.

WRc (1997) *Sewer Jetting Code of Practice*, WRc, Swindon.

18 Rehabilitation

18.1 Introduction

18.1.1 *The problem*

If a sewer collapses under a busy street, causing a bus or lorry to fall down a newly-opened hole, and requiring the street to be closed to traffic, something must be done. If sewers regularly flood, causing serious damage to property as well as health risks and public revulsion due to the presence of wastewater, something must again be done. But what?

Clearly the collapsed sewer must be replaced. But should we wait for another disastrous collapse before replacing the rest of that sewer length? Which other sewers may need attention? What part of the system must be improved to prevent the flooding? How do we ensure that only truly necessary work is carried out? Can the work be done without digging up the town centre, leading to public inconvenience and loss of business?

These important and frequently-asked questions are answered by the art, science and practice of sewer rehabilitation.

18.1.2 Sewerage Rehabilitation Manual

There is more work in the UK on improving existing urban drainage systems than on constructing completely new ones. So sewer rehabilitation is not a sideline, it is one of the main areas of activity of drainage engineers. This has been recognised for many years, certainly since 1983 and the publication of the 1st edition of the *Sewerage Rehabilitation Manual*. The current edition is the 4th (Water Research Centre, 2001). It identifies three types of problems that can be solved by sewer rehabilitation: structural, environmental and hydraulic.

The *Sewerage Rehabilitation Manual* (SRM) is a definitive manual of good practice, and this chapter will not attempt to duplicate its contents. We will deal here more with the overall philosophy, and summarise the main techniques that are available. The most important aspect of the philosophy of the *SRM* is the idea that systematic definition of problems, and development of integrated solutions, can lead to savings. This is made possible by:

- the availability of sophisticated tools for problem analysis – particularly computer models of sewer systems
- the availability of technology for in-sewer inspection and monitoring: closed-circuit television inspection systems, devices for monitoring flows, and technology for in-sewer rehabilitation work
- pragmatism in the management of sewer assets: recognition of the value of what is already in the ground even if it has some problems; and systematic methods for establishing priorities in rehabilitation programmes
- the recognition that large-scale sewer reconstruction work is very expensive, and also has a high economic cost resulting from disruption to commerce in town centres or to local residents.

The result is a set of procedures that have become established and improved over time.

An important part of the strategy is identification of priorities. All aspects, from inspection to rehabilitation work, are targeted at selected parts of the system. Yet all work is planned with an awareness of its effect on the drainage area as a whole.

18.1.3 The need for rehabilitation

How does the need for rehabilitation arise? A typical sewer system in the UK consists of an older part in the centre of the town with newer sections added as the town has expanded. Some parts of the older system may now be undersized, and loadings from traffic, for example, may be far greater than anticipated when the pipe was constructed. The material of the pipe itself may be old. Overall, about 15% of the UK sewer system is over 100 years old, but in older cities the proportion is higher. The pipe material and construction may be poor quality: a worrying characteristic of the 25% of the system built between the World Wars. The system may have been poorly maintained in the past; and certainly it is sometimes possible to 'get away with' poor maintenance of sewers because they can still function to some extent in poor structural condition. But this may mean that eventual structural collapse is all the more catastrophic.

Seepage of water into or out of a sewer (infiltration or exfiltration) may increase the risk of sewer collapses and contribute to their seriousness when they eventually occur. Infiltration has been discussed in Section 4.4, and is sometimes associated with leakage from water mains, which may also be in a poor structural condition. Any movement of water in the ground may wash soil particles with it, and over a long period of time this can lead to voids in the ground. A small pipe can cause a large cavity (Fig. 18.1). A cracked sewer may be supported by the surrounding ground

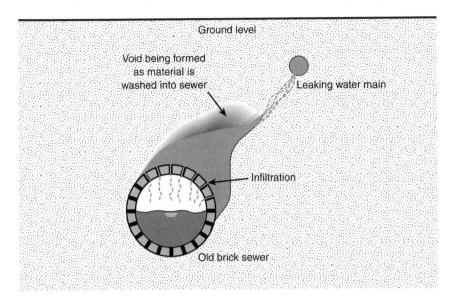

Fig. 18.1 Formation of a void around an old sewer

and not in immediate danger, but if soil is washed in through the cracks by infiltration, then subsidence or collapse will become far more likely.

Symptoms of the need for rehabilitation include:

- collapse
- flooding of roads and properties (of particular seriousness when wastewater is involved)
- increased pollution of watercourses
- blockages
- high levels of infiltration.

In the 1970s, there were increasingly frequent reports of sewer collapses, and in 1977 a national assessment suggested there was a backlog of sewer dereliction to be dealt with (NWC/DoE Standing Technical Committee, 1977). By the 1980s, there were about 5000 collapses and serious blockages per year (Finney, 1990) with the expectation that this rate would increase with time unless something was done to improve the situation. More than fifteen years later, the annual rate had reduced, but was still approximately twenty collapses per 1000 km (OFWAT, 1998).

18.1.4 Repair, renovation and replacement

Sewer rehabilitation can be accomplished by means of sewer *repair, renovation* or *replacement. Repair* is localised work to correct damage to the sewer fabric. *Renovation* is work to improve the performance of a length of sewer, incorporating the original sewer fabric. *Replacement* is construction of a new sewer, incorporating and possibly enhancing the functions of the old.

It is important to point out that there is nothing remarkable about the fact that sewers need rehabilitation. All elements of infrastructure have a certain design life. Within that life, they require maintenance (including repair) and, subsequently, renovation. Ultimately they require replacement. What is perhaps remarkable about sewers is the extent to which these needs have been ignored.

The total length of 'critical sewers' (defined in the next section) that were renovated or replaced in the whole of England and Wales in the 11 years between 1990 and 2001 was about 2300 km, giving an average rate of 210 km per year (Battersby, 2001; based on OFWAT data). At this rate it would take 350 years to rehabilitate the total length of critical sewers (about 74 000 km), implying that this was considered to be the expected life of the asset. In two water company areas the implied asset life was over 1000 years!

The cost of replacing sewers is high, but only about 20% is for the actual pipe, the rest is for excavation and reinstatement. The costs of disruption to other services, people, business and traffic in busy towns can be three times the cost of construction. This gives rise to a popular saying in sewer rehabilitation: 'The greatest asset is the hole in the ground'.

Sewer rehabilitation is concerned with stabilising, improving or strengthening the surrounding of the 'hole' created by the original pipe, sometimes even making the hole itself bigger. Access, where possible, is via the hole.

The most effective rehabilitation scheme will retain as much as possible of the existing system, and will cause minimum disruption to the community. The actual work of inspection and rehabilitation will be selective: it will concentrate on core areas according to an integrated strategy.

The *SRM* gives detailed guidance on implementation of both structural and hydraulic rehabilitation. Read (1997) gives a good overview of the techniques and the companies offering them. A summary of available approaches is given in BS EN 752-5: 1998.

18.2 Preparing for sewer rehabilitation

Both the *SRM* and BS EN 752-5 recommend procedures with the overall sequence:

- initial planning
- diagnostic study
- implementation and monitoring.

Initial planning involves comparing the performance of the system with pre-determined criteria in order to determine the most suitable approach

Table 18.1 Phases in the diagnostic study (adapted from Water Research Centre, 2001)

Phase 1	Information
Phase 2a	Hydraulic investigation, *in parallel with*
Phase 2b	Environmental investigation,
Phase 2c	Structural investigation, *and*
Phase 2d	Operations and maintenance investigation
Phase 3	Develop solutions

for the diagnostic study. The phases of the diagnostic study are given in Table 18.1.

Phase 1 Information

Phase 1 is a review of all existing information. The focus of this is the 'inventory': the dimensions, location and characteristics of the system components, including pipes, manholes and ancillary structures. Existing information on hydraulic, environmental, structural and operational performance is also assessed. Of course the basic inventory information is needed for all aspects of the later investigations, so if sewer records are missing the necessary survey work must be carried out.

Phase 2a Hydraulic investigation

Assessment of hydraulic performance is likely to involve hydraulic modelling of the system. The aim is to help to identify problem areas within the system (giving rise, for example, to surcharging or flooding) and to investigate the possible effects of physical improvements within the system. Flow models are discussed extensively in Chapter 19. As is pointed out there, models are heavily dependent on the accuracy of the data used to specify the physical properties of the system. An important stage (described in Chapter 19) is 'model verification' in which simulations of the model are compared with actual measured performance.

Once the model has been verified, it can be used to assess the hydraulic performance of the system and investigate ways of correcting deficiencies.

Phase 2b Environmental investigation

This is largely the domain of the *Urban Pollution Management Manual* (FWR, 1998), also considered in Chapters 20 and 22. The *SRM* shows how the UPM approach can be coordinated with the *SRM* procedures. The environmental investigation entails collecting data, building water quality and impact models where necessary, and using them to identify the causes of environmental problems.

Phase 2c Structural investigation

Two factors determine whether pre-emptive rehabilitation will be necessary on structural grounds: the likelihood of failure and the consequence of failure. A significant recommendation of the *SRM* has always been that priorities within the system are established by identifying *critical sewers*. These are sewers for which the costs expected to result from failure would be significantly higher than rehabilitation costs. Critical sewers make up 20% of a system on average. They are usually the sewers with larger diameters, predominantly made of brick or concrete (see Table 18.2). Two categories of critical sewer are defined:

- *Category A*, for which the aim is to avoid failures
- *Category B*, for which the aim is to significantly reduce the failure rate.

All critical sewers need periodic inspection. For non-critical (Category C) sewers, maintenance would be reactive: only carried out in response to failure unless there are repeated failures at a particular location.

In general, a critical sewer could have any of the following characteristics:

- above-average depth (3 m or more to invert)
- brick or masonry construction
- large diameter (above 600 mm)
- bad ground

and/or could be in any of the following locations:

- under a busy traffic route
- under buildings, railways, tram routes, canals, rivers, main shopping streets, primary access to industrial sites or motorways
- with difficult access for repair following failure.

Table 18.2 Sewer size and material distribution in the UK (adapted from Moss, 1985)

Diameter or major dimension (mm)	All sewers (%)	Critical sewers (%)
<300	70	10
300–499	13	20
500–900	10	35
>900	7	35
Material		
Clay	75	14
Concrete	15	60
Brick	5	25
Other	5	1

The SRM contains specific guidance on classifying sewers as Category A or B.

It is important to point out that the implication of identifying critical sewers in this way is that the 80% (on average) of a system that is non-critical is generally left to reactive maintenance, with the implication that an increasing number of failures will occur in the future as these sewers continue to age. Planning needs for this part of the system have been considered by Fenner and Sweeting (1999).

Inspection of the internal structural condition of sewers is commonly carried out by closed-circuit television (CCTV). Further details of sewer inspection techniques are given in Chapter 17. There is a standard system in the *SRM* for grading the condition of a sewer.

Phase 2d Operational investigation

There is close linkage between the hydraulic, environmental and structural aspects of the performance of a sewer system. And all these are linked to the operation and maintenance practices being used. All phases of the investigation involve assessment of the sewer system. In phases 2a, 2b and 2c the focus is on the physical performance of the system, whereas in phase 2d the focus is on operational and maintenance procedures.

Phase 3 Develop solutions

The results of the investigations that have been described above are used to produce a rehabilitation plan (or drainage area plan) and an operation and maintenance plan.

The solutions should be integrated where possible (solving different types of problems together), be cost-effective and consider the catchment as a whole.

Rehabilitation will be planned with the aims of overcoming hydraulic and environmental problems in the system, taking account of planned structural rehabilitation and allowing for known plans for future development of the area.

The flow model allows an engineer to study the overall effects of various hydraulic rehabilitation options. The options are considered in Section 18.4. One undersized existing pipe could cause flooding over an extensive part of the system. A proposal for overcoming the problem can be developed, and the model run again to see its benefits. The final solution can be derived from an iterative process of adjusting the proposal and seeing the consequences from the model.

The central *SRM* philosophy – that systematic definition of problems and development of integrated solutions can lead to savings – is strongly evident in the use of flow models to develop plans for hydraulic rehabilitation.

The SRM presents a detailed analysis of the potential advantages and disadvantages of typical options.

Implementation and monitoring

This involves detailed design and construction of rehabilitation works and implementation of the operation and maintenance plan.

The *SRM* provides detailed guidelines for structural design of renovated sewers. The main techniques available for structural rehabilitation are considered in Section 18.3, and for hydraulic rehabilitation in Section 18.4.

Monitoring of hydraulic, environmental and structural performance, and effectiveness of the operation and maintenance plan should continue after rehabilitation has been carried out.

18.3 Methods of structural repair and renovation

This section is intended to provide only an introduction to this fascinating area of developing technology. Technical guidelines on the practice of sewer repair and renovation are available in the *SRM* (Water Research Centre, 2001) and Read (1997). Naturally there are major differences between methods for man-entry sewers and those for non-man-entry sewers.

18.3.1 Man-entry sewers

Repair

The general aims of repair are to correct defects and reduce infiltration in physically-intact sewers.

Pointing is the renewal of mortar in brick man-entry sewers. If the mortar is carried to the point of application by hand, and finished off with a trowel, the process is called 'hand pointing'. This is a labour-intensive and time-consuming process, but produces a good finish. In longer lengths of sewer, it may be more appropriate to use equipment for delivering the mortar under pressure to the point of application ('pressure pointing'). In this process, excess mortar is still normally removed by hand-trowelling.

Other forms of repair in man-entry brickwork sewers include replacing areas of brickwork, and rendering with high-strength mortar.

Renovation

Renovation methods generally involve providing a new lining, inside the old, to a whole length of sewer. The lining may either be constructed *in situ*, or erected from ready-made segments. There is usually a loss of area within the sewer.

Clearly, one possibility for creating a new lining for an old brick sewer is to use *new brickwork*. Where this is done, the new lining must be tied to the old brickwork, and the space between the two filled with grout in order to create one structure. Grout is a general word for a material applied under pressure as a liquid to fill a space and subsequently 'cure' to a hardened state. Grouts used in sewer rehabilitation are generally made from ordinary Portland cement with added PFA (pulverised fuel ash). In some applications a 'chemical grout' is used as an alternative. As well as filling the space between the new lining and the old sewer, grout can also be used to fill any voids outside the original sewer.

If the brickwork is sufficiently intact, the pipe wall can be *rendered* by hand with a lining of mortar (which may include reinforcing fibres). Alternatively, *Ferrocement* may be used consisting of a cement-rich mortar formed on layers of fine reinforcing mesh. Both techniques require relatively thin linings, resulting in limited reductions to the pipe cross-sectional area.

In situ gunite lining is formed by spraying concrete on to the old sewer wall. A reinforcing mesh is placed around the wall first, and is incorporated in the sprayed lining. An *in situ* lining can also be created using *pumped concrete*. Reinforcement and specially-designed steel formwork sections are put in place, and high quality concrete is then pumped into the annular space.

Segmental linings are made up of ready-made segments erected in the sewer. Common materials are glass-reinforced cement, glass-reinforced plastic, precast gunite (sprayed concrete), and polyester resin concrete (effectively a plastic containing an aggregate in the same way as concrete). The space between the new lining and the old pipe wall is, again, filled with grout. Precast segments can be made to fit any cross-sectional sewer shape, not just a straightforward circle. Success with the technique depends on careful location of the many joints.

It should be noted in relation to all work in man-entry sewers that health and safety regulations only allow work in confined space when there is no reasonably practicable alternative.

18.3.2 Non-man-entry sewers

The most innovative developments in sewer rehabilitation have tended to be for application to sewers that are too small for direct access by a person. Some methods involve systems of remote control with high levels of technical sophistication.

Repair

Patch repairs involve remotely placing a patch of fabric containing an appropriate resin material and then curing it in place. In a *resin injection* system, an inflatable packer is used to isolate a pipe defect and to force injected resin into the defect. *Chemical grouting* uses a similar packer device to seal joints and fill any associated voids in the ground. The packer is located at a joint, and collars positioned on either side of the joint inflate to create a seal (Fig. 18.2). Applying air or water pressure can first check the effectiveness of the joint. If pressure loss indicates that the joint is unsatisfactory, chemical grout is released under pressure to fill any void outside the pipe and seal the joint.

An alternative form of chemical grouting treats a whole sewer length at a time. Two chemicals are used, which react together to form a sealing gel. The sewer, laterals and manholes for a particular length are filled with the first chemical. After a suitable period to allow penetration of the ground around the pipe, the chemical is replaced by a second which reacts with the residue of the first to form an impermeable mass around the defects. Any surplus of the second chemical is then pumped out.

Bunting (1997) and the *SRM* provide further information on repair techniques for non-man-entry sewers.

Renovation

Sliplining involves forming a continuous length of plastic (medium or high density polyethylene) lining and pulling it through the existing pipe. One approach is for long lengths of plastic pipe (typically 5 m) to be welded end-to-end on the surface. The resulting continuous length of pipe has some flexibility and is winched along the sewer via a specially-excavated lead-in trench (Fig. 18.3(a)). The winch cable is attached to a nose-cone fixed to the front end of the new lining. As an alternative to welding the lengths on the surface, pipes can be welded in an enlarged trench (Fig. 18.3(b)). The first approach has the disadvantage that it requires space on the surface for assembly of the pipe length; the second requires more excavation (though both approaches require a significant amount). Another

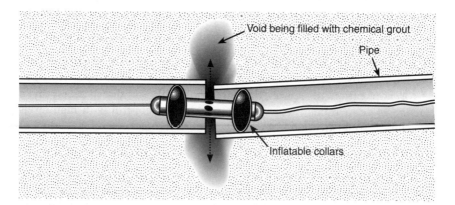

Fig. 18.2 Chemical grouting

alternative is much shorter lengths of pipe (HDPE or polypropylene) with push-fit or screw joints. These are connected within existing manholes (Fig. 18.3(c)). In all cases, any significant space between the new lining and the old pipe is filled with grout.

These processes, as described here, will result in a smaller diameter pipe. If the existing pipe includes imperfections like distorted cross-section or off-set joints, the limitation on the diameter of the new lining may be even greater. It may be appropriate to carry out localised repair of

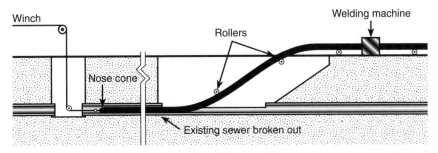

(a) Sliplining – surface welding

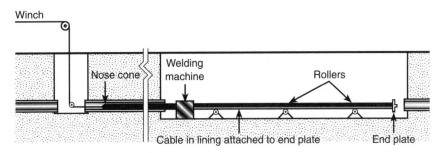

(b) Sliplining – trench welding

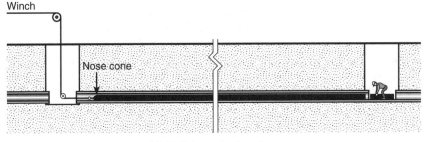

(c) Sliplining – short sections

Fig. 18.3 (a) Sliplining – surface welding;
(b) Sliplining – trench welding;
(c) Sliplining – short sections

imperfections before proceeding with sliplining, in order to increase the practicable diameter of the new lining. The new lining is likely to be smoother than the old, but there is still likely to be a loss of hydraulic capacity, and this is obviously a potential disadvantage.

However, the technique of *pipe-bursting* or *moling* allows the new lining to have a diameter equal to that of the old pipe, or even greater. Pipe-bursting is achieved by a pneumatically or hydraulically operated hammer which breaks the existing pipe from the inside and forces the broken pieces into the ground, to form a void larger than the former pipe interior. The mole can be drawn through the pipe ahead of a new lining. Pipe-bursting with sliplining can therefore produce a new pipe that is bigger and smoother than the old (though at the same gradient) and is a very powerful technique for renovation of non-man-entry sewers. There are circumstances in which pipe-bursting may not be practicable, for example when the old pipe has a concrete surround, or when there is reinforcement in the pipe or surround, or when damage might be caused to neighbouring services, or in certain ground conditions. However the main problem with pipe-bursting in conjunction with sliplining is that reconnection of laterals has to be carried out externally. This creates a need for excavation at each connection and can be a significant disadvantage if the pipe is deep or there are a large number of connections.

Cured-in-place linings are positioned in the sewer in a flexible state, and then cured in place to form a hard lining, usually in contact with the original sewer wall. The most common process involves a fibre/felt liner or 'sock' impregnated with resin which is inserted in the sewer by an inversion process (turning inside-out) under water pressure. When this flexible lining is in place, it is cured by circulating steam or water (Fig. 18.4). These linings do not provide high structural strength, but do avoid the problems of significant reduction in diameter. They are also less likely to require excavation for side connections. It is possible to detect the location of laterals from inside the pipe by observing the shape of the lining, and then to cut through the lining by remote control in order to reform the connection.

Another method using material that is not initially 'pipe-shaped' involves inserting a soft plastic lining (PVC-U) in the sewer in a folded state (so that it has a smaller area). Once in position it is heated and pushed out against the old lining by a rounding device to form its final circular shape.

Lining with spirally-wound pipes is yet another alternative. A long PVC-U strip is fed into a winding machine placed at the bottom of a manhole. The machine winds the strip into a spiral of continuous pipe lining which then travels up the sewer.

18.3.3 *Relative costs of rehabilitation and collapse*

When comparing the cost of rehabilitation *now* with the cost of coping with a sewer collapse in the future, a 'discount rate' must be applied.

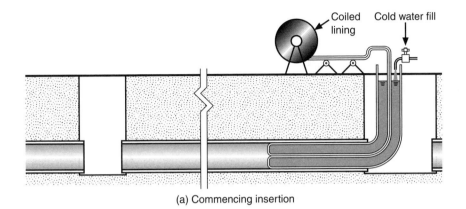

(a) Commencing insertion

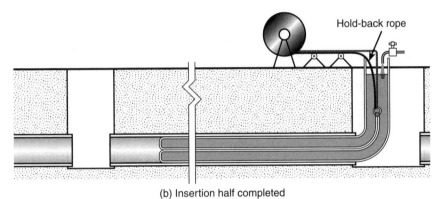

(b) Insertion half completed

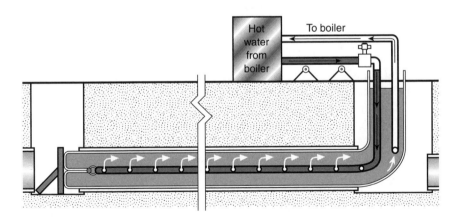

(c) Insertion completed – curing by hot water circulation

Fig. 18.4 Cured-in-place lining

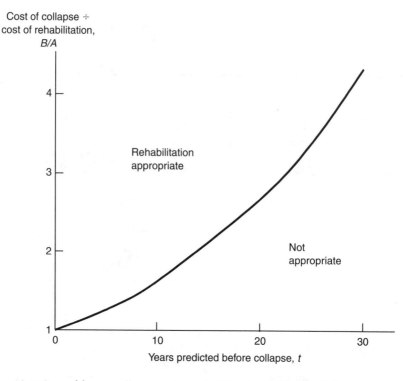

Fig. 18.5 Cost of future collapse against rehabilitation (5% annual discount rate)

In cost terms, it is appropriate to carry out rehabilitation work if

$$A < \frac{B}{\left(1 + \dfrac{r}{100}\right)^t}$$

A estimated cost of rehabilitation
B estimated cost of collapse (disruption + sewer replacement)
t number of years before collapse is predicted to take place
r annual discount rate (%)

Fig. 18.5 represents this comparison for a discount rate of 5%. However, in practice it is difficult to predict:

- the exact time of failure
- the total collapse costs.

So the decision is based more on judgement than precise calculation.

Example 18.1

It is predicted that, if rehabilitation is not carried out, a sewer will last 15 years before collapsing. The cost of rehabilitation is estimated as £300 000. The full cost of collapse is estimated as £800 000. In cost terms, with an annual discount rate of 5%, is rehabilitation appropriate?

Solution

$$\frac{B}{A} = \frac{800\ 000}{300\ 000} = 2.7$$

In Fig. 18.5, the point with co-ordinates (15, 2.7) is above the line, indicating that rehabilitation is appropriate.
Or:

$$\frac{B}{\left(1 + \dfrac{r}{100}\right)^t} = \frac{800\ 000}{\left(1 + \dfrac{5}{100}\right)^{15}} = £385\ 000 \text{ (the 'discounted cost' of collapse)}$$

Cost A (£300 000) is less, so rehabilitation is appropriate.

18.3.4 Choice of method

If the decision is made to go ahead with rehabilitation, the next stage is to choose the most appropriate method. The following points need to be fully considered:

- type of failure
- sewer type and configuration
- change in hydraulic capacity of newly rehabilitated sewer
- direct cost of alternative options
- indirect costs of alternative options
- effective life of newly rehabilitated sewer (see Table 18.3)
- whether to renovate or replace.

Differences in the properties and life expectancy of alternative materials must be considered (Table 18.3).

The *SRM* contains detailed guidance on the design of pipe linings. It also includes a 'Buyer's Guide' which gives information on suppliers and contractors.

Table 18.3 Relative properties and life expectancy of sewer materials (adapted from Moss, 1983)

Pipe material	Corrosion resistance	Chemical resistance	Stiffness	Resistance to brittle failure	Abrasion resistance	Estimated life
Concrete*	xx	x	xxx	x	x	xx
Fibre cement*	xx	xx	xxx	x	x	xx
GRC*	xx	xx	xxx	xx	x	x(x)
PVC-U	xxx	xxx	xx	x(x)	xx(x)	xxx
GRP	xx(x)	xxx	xx	xx(x)	xx	xx
Insituform	xx(x)	xxx	xx	xx(x)	x(x)	xx
HDPE	xxx	xxx	x	xx(x)	xxx	xx
PP	xxx	xxx	xx	xx	xxx	xxx
Resin concrete	xxx	xxx	xxx	x	xx(x)	xx

x = low; xx = medium; xxx = high.
* Some of these properties may be modified by the use of a protective coating or lining.
Brackets indicate that current information is insufficient to differentiate between ratings.

18.3.5 Associated work

Laterals

Household drain connections, or laterals, pose one of the most awkward problems in non-man-entry renovation for two reasons.

* Badly-formed laterals can protrude into the pipe, causing obstruction and potentially preventing insertion of a lining.
* Once lining has taken place, existing connections need to be re-opened.

Both problems are commonly tackled by excavating down to the drain connection. But this is an expensive, time-consuming procedure that negates some of the benefits gained in adopting trenchless methods for renovating the pipe itself.

One device that can be used remotely in the sewer utilises a high-pressure water jet to cut protruding laterals. The device can be winched or driven into the sewer and is monitored by a CCTV camera.

There are two main techniques for re-opening drains following relining: either from the drain or from the sewer itself. The first has the advantage of simplicity, particularly if access to the drain can be made via an inspection chamber.

A difficult problem is identifying which of the laterals are still live. It is important that old disused laterals are sealed.

Services

Even though many sewer rehabilitation techniques are specifically aimed at minimising excavation, the accurate detection of other services is still essential to avoid unnecessary delays and extra costs. There is more information in Section 15.5.

Cleaning

Most rehabilitation operations need to be preceded by sewer cleaning. The available techniques are discussed in Section 17.4.

Overpumping

It is a common requirement that flow in the sewer needs to be diverted via a pump and temporary overground pipe system, especially for work in non-man-entry pipes. Overpumping can be a significant cost element in a sewer rehabilitation scheme. The overpumping system cannot be designed without estimates of flow for wastewater (including industrial) and stormwater, where appropriate, and these should be as accurate as possible. It may be worth using a sewer system model for this purpose. Nedwell and Vickridge (1997) discuss practical aspects of overpumping.

18.4 Hydraulic rehabilitation

To solve problems resulting from hydraulic overloading, and achieve the performance targets set, the *SRM* proposes a range of upgrading options to be considered in sequence.

Reduce hydraulic inputs to sewer system

Some adjustments to a system may be possible that reduce inflow without requiring major works. For example, it may be possible to divert certain inflows from overloaded sewers to points in the system where there is less hydraulic overloading. However, the most significant method of reducing inputs, and one which is gaining importance as a solution to a range of drainage problems, is the use of stormwater management techniques such as pervious pavements and infiltration devices. These are described in Chapter 21.

Maximise capabilities of the existing system

Again there are some simple approaches that can have significant effects. Removal of local constrictions may allow the capacity of significant sections of the system to be increased. Sewer cleaning will also increase capacity, though if the deposition of debris had been caused by the physical nature of the sewer, the deposition will be a continuing problem. There is more information on sewer cleaning in Chapter 17.

Combined sewer overflows (CSOs) are major controls on flow in combined sewer systems. Rehabilitation work in sewer systems often involves improvement, replacement or relocation of CSOs. This may be primarily related to environmental requirements, but often has significant hydraulic effects.

Adjust system to cause attenuation of peak flows

Most sewers flow less than full most of the time, and, by attenuating peak flows, more efficient use can be made of the capacity of the sewer. This can be achieved by providing additional storage within the system in the form of detention tanks (described in Chapter 13), or by installing throttles and controls at key points to make the best use of in-sewer storage (these devices are described in Section 9.1). Also many stormwater management techniques provide attenuation as well as inflow-reduction (Chapter 21).

Increase capacity of the system

When the possibilities above have been exhausted, it is necessary to use what might be considered the most obvious method of overcoming hydraulic overloading – increasing capacity. Many of the structural rehabilitation techniques that have been described in Section 18.3 cause an increase in capacity, either because they reduce roughness and remove imperfections, or, in some cases where pipe-bursting techniques are being used, actually increase the size. This is why, in the planning of sewer rehabilitation schemes, structural and hydraulic effects must be considered together.

When sufficient hydraulic upgrading cannot be achieved by renovation, it may be necessary to replace existing sewers with new sewers of larger capacity. This will, of course, bring with it the high costs and associated disruption discussed in Section 18.1. Existing sewers may also be duplicated by the construction of additional sewers which flow in parallel.

The capacity of a system can also be increased by techniques for overall system management, including real-time control, as considered in Chapter 21.

Problems

18.1 It is predicted that a sewer will last 10 years before collapsing, if rehabilitation is not carried out. The cost of rehabilitation is estimated at £280 000. The full cost of collapse is estimated at £500 000. Using an annual discount rate of 5%, determine whether, in cost terms, rehabilitation is appropriate. Repeat, using an annual discount rate of 7%.   [yes, no]

18.2 Engineers are deciding whether to renovate or completely replace a section of sewer network. Analysis has shown that to replace the failed section with vitrified clay pipes will cost £850 000 but to renovate the same section *in situ* will cost £500 000. The estimated life-span of the clay pipes is 100 years and of the renovated existing pipes, 25 years. Which is the most cost-effective option? (Hint: compare the replacement cost with the renovation cost *plus* the replacement cost discounted over 25 years. $r = 5\%$.)　　　[renovate]

18.3 Why is sewer rehabilitation more common now than it was 50 years ago?

18.4 How can flow models be used in the planning of sewer rehabilitation schemes?

18.5 Describe methods of repair and renovation suited to non-man-entry sewers.

Key sources

Read, G.F. with Vickridge, I.G. (ed.) (1997) *Sewers – rehabilitation and new construction; repair and renovation*, Arnold.

Water Research Centre (2001) *Sewer Rehabilitation Manual*, 4th edition, WRc. (Reduced version available from www.wrcplc.co.uk/srm)

References

Battersby, S. (2001) Unsustainable underinvestment. *Water*, Number 136, 14 November.

BS EN 752–5: 1998 *Drain and sewer systems outside buildings, Part 5 Rehabilitation*.

Bunting N. (1997) Localised repair techniques for non-man-entry sewers, in *Sewers – rehabilitation and new construction; repair and renovation* (eds, G.F. Read with I.G. Vickeridge) (Chapter 11, 233–253), Arnold.

Fenner, R.A. and Sweeting, L. (1999) A decision support model for the rehabilitation of 'non critical' sewers. *Water Science and Technology*, **39**(9), 193–200.

Finney, A. (1990) Refurbishment of sewers. *Chemistry and Industry*, 15 October, 658–662.

FWR (1998) *Urban Pollution Management Manual*, 2nd edition, Foundation for Water Research.

Moss, G.F. (1983) Latest developments in sewer renovation, *The Public Health Engineer*, **11**(2), April, 31–34.

Moss, G.F. (1985) Sewerage rehabilitation; the way forward. *The Public Health Engineer*, **13**(3), July, 157–160.

Nedwell, P. and Vickridge, I.G. (1997) Ancillary works – cleaning and overpumping, in *Sewers – rehabilitation and new construction; repair and renovation* (eds G.F. Read with I.G. Vickeridge) (Chapter 9, 193–203), Arnold.

NWC/DoE Standing Technical Committee (1977) *Sewers and Water Mains – A National Assessment*, Report No 1, National Water Council.

OFWAT (1998) *1997 July returns for the water industry in England and Wales*.

19 Flow models

19.1 Models and urban drainage engineering

There are two chapters in this book on modelling drainage systems. Here, we look mainly at models covering hydrological and hydraulic aspects of modelling: rates of input from rain or wastewater, and the hydraulic response of the sewer system in terms of flow-rate and depth. Some sewer system models also include water quality: the presence and behaviour of pollutants entering and flowing through the system. These types of models are considered in Chapter 20.

The purpose of models in urban drainage engineering is to represent a drainage system and its response to different conditions in order to answer questions about it, usually in the form 'What if ...?'. In a sense, people have been modelling drainage systems all the time they have been using calculations to help them to build systems that would operate successfully. For example, the Rational Method (in Chapter 11) is a simple model of the conversion of rainfall into runoff that can be used to look at the likely effects of different rainfall intensities.

Computer programs for drainage design and analysis emerged in the 1970s, but complex models only became standard tools of drainage engineers when appropriate computing power became available. The model SWMM first appeared in the USA in the early 1970s and has continued to be developed ever since. In the UK, the TRRL Hydrograph method (see Chapter 11) used in the 1970s was computer-based. But there was no standard software package until the early 1980s and the introduction of WASSP, based on the Wallingford Procedure (DoE/NWC, 1981). Its impact on the practice of drainage engineering in the UK has been discussed in Chapter 1: it effectively turned a branch of engineering that had relied heavily on conservative decision-making based on experience, into one in which sophisticated methods of analysis were used to produce better and more informed solutions.

More recent Wallingford packages are HydroWorks and InfoWorks. A popular European package, developed in Denmark, is MOUSE. Many other programs, some concentrating purely on design of new systems, are available. Information on commercial software packages is given in Section 19.5.

There have always been two main uses for sewer system models: design (of new systems) and analysis (of existing systems). In design, the physical details of a proposed drainage system are determined so the system will behave satisfactorily when exposed to specific conditions. In simulation, the physical details of the system already exist, and the model-user is interested in how the system responds to particular conditions (in terms of flow-rate and depth, and the extent of surcharge and surface flooding). The aim is usually to find out if the system needs to be improved, and, if so, how.

The same modelling tool *can* be used for both design and simulation, but specialist software has been developed for each. Design tends to be most concerned with extreme flows, and the question 'is there sufficient capacity?', and this can be accomplished by simpler software than that needed for the job of simulating system responses in detail. Some of the more detailed aspects of modelling covered in this chapter are needed only in simulation models.

As well as hydraulic and hydrological aspects, the main sewer flow modelling packages include versatile data interfaces, GIS compatibility and graphical representation of the system and its response to storms.

19.2 Deterministic models

The models referred to above, SWMM, InfoWorks, MOUSE and so on, are based on accepted mathematical relationships between physical parameters. All involve some element of simplification. (No model covers every raindrop or every variation in the catchment surface.) And they are all *deterministic* – one combination of input data will always give the same output, randomness is not accounted for.

The fact that these models include simplifications and ignore random effects, combined with the uncertainty associated with input data and with field measurements, means that it would be a reckless or naive modeller who stated, 'the results of my model are correct'. That modeller would be far wiser to recommend the results as being *useful*.

There are many reasons for randomness to be significant in urban drainage modelling. Apart from genuine randomness in physical phenomena, we may treat something as random because we do not fully understand it or because the physical relationships are too complex for us to represent in the model. Stochastic models take randomness into account, and give an indication of uncertainty in a simulation. These, and other alternatives to deterministic models, are considered in Chapter 20.

For most sewer simulations that are concerned with hydrological and hydraulic aspects, deterministic models have become the standard tools. Most of the rest of this chapter is devoted to describing their make-up and uses.

19.3 Elements of a flow model

A physically-based deterministic model of sewer flow must represent the *inputs* (rainfall and wastewater flow) and convert them into the information that is needed: flow-rate and depth within the system and at its outlets. The model carries out this conversion by representing (mathematically) the main physical processes that take place. The model must, therefore, be reasonably comprehensive: we could not expect to leave out an important process and still produce accurate results. In order to represent processes in a physically-based mathematical form, a good level of scientific knowledge about the processes is needed. Therefore, sewer flow models are based on a body of research information about runoff, pipe-flow and so on. However, as has already been pointed out, the model is also bound to include some elements of simplification.

At a very general level, therefore, three factors greatly influence the accuracy and usefulness of the simulations by a particular physically-based modelling package: the comprehensiveness of the model, the reliability and completeness of the scientific knowledge on which it is based, and the appropriateness of the simplifications it contains.

The word 'model' tends to be used in a number of ways. We are referring throughout to mathematical, computer-based, models. The mathematical representation of each process can be termed a 'model', as can the combination of all the processes (into a package like InfoWorks or MOUSE). However, these 'models' are simply tools ready to do a job: simulation of a flow in a particular catchment. To do this job, a great deal of data is required about, for example, the surfaces of the catchment and the network of sewers. That specific application of a package to a particular catchment, requiring great effort in checking, calibration and verification, as will be considered later, is also commonly referred to as a 'model'.

We will now consider the main physical processes that must be represented in a package for flow modelling. The main components are shown on Fig. 6.1.

A. Rainfall

The model will be used to find the response of the catchment and the sewer system to particular rainfall patterns. Straightforward examples would be simple constant rainfall, or, more realistically, rainfall with a particular storm profile (variation of rain intensity with time). This would be generated for a specified return period using the types of relationship between intensity, duration and frequency considered in Section 5.3. To model the operation of CSOs, or storage facilities that need to be emptied during dry periods, a typical sequence of wet and dry periods would be studied using time series rainfall (see Section 5.5). Spatial variation of rainfall is also an important consideration in large catchments. During model verification (described more fully in Section 19.6.5), actual raingauge records are used as rainfall input to the model, and the flow simulations by the package are compared with actual flows measured in the system.

B. Rainfall to runoff

The conversion of rainfall into 'runoff', water destined to find its way into the sewer system, is a highly complex process. There are many reasons for rainwater not to become stormwater in the sewer. It may, for example, soak into the ground (even on an 'impervious' surface, via cracks), may form puddles and later evaporate, or may be caught in the leaves of a tree. There is an obvious distinction between water that falls on a roof or a road and that which falls in an undrained garden; but where, for example, there is a grass strip beside a pavement, some water falling on the grass may run off on to the pavement and enter the sewer, whereas some water falling on the pavement may run on to the grass and infiltrate. The methods of representing these processes have been considered in Section 6.2.

C. Overland flow

The main consideration here is not so much the amount of rainwater that will enter the sewer, but how much time it will take to get there. Clearly the extent to which water entering from one area will overlap with that entering from another will have a significant effect on the way the flow in the sewer builds up with time. Again, the physical processes are highly complex, with many ways in which surface irregularities can affect the flow. Overland flow is usually represented in a very generalised form, and methods have been given in Section 6.3.

D. Flow in the sewer system

In a combined system, the stormwater joins the flow of wastewater in the sewer. Realistic simulation of wastewater generation is an important function in a flow-modelling package. Methods have been presented in Sections 4.2 and 4.3.

Packages tend to describe the main body of the sewer system as consisting of 'links' and 'nodes'. The links are generally pipes, in which the model must represent the relationship between the main hydraulic properties: diameter, gradient, roughness, flow-rate and depth. (Links are also used to represent pumps and other features.) The nodes are generally manholes, at which there may be additional head losses and changes of level. In addition to these primary building blocks, the package must also be capable of simulating conditions in more specialised ancillary structures: including tanks and CSOs.

The emphasis throughout is on unsteady conditions: on variations of flow-rate and depth with time. A crucial element in a flow-modelling package is the way in which it simulates these unsteady conditions. Common methods are presented in Section 19.4.

The output of all this modelling effort generally comes in the form of simulated variations of flow-rate and depth with time at chosen points within the sewer system, and at outlets from it. There is usually particular interest in the ability of the sewer system to cope with the simulated flows, and thus on the extent of possible pipe surcharge or surface flooding.

It is possible for the spatial pattern of surface flooding to be simulated by a combination of sewer system modelling and GIS surface data (for example Boonya-aroonnet *et al.*, 2002). There is some background to this in Section 11.2.2. More advanced modelling tools are currently under development (Maksimovic and Prodanovic, 2001).

Table 19.1 summarises the parts of this book that contain more detail on each of the elements of a flow model.

19.4 Modelling unsteady flow

Wastewater flow varies with the time of day, and, during storm conditions, inflow to the sewer system can vary dramatically with time. The representation of unsteady (time varying) flow is an important component in a sewer-flow package.

In unsteady flow in a part-full pipe, there is a far more complex relationship between depth and flow-rate than there is in steady flow (described in Section 8.4). Also, as a storm wave moves through a sewer system it *attenuates* (it spreads out and the peak reduces) and it *translates* (moves along). The relationship between flow-rate (or depth) and time cannot be accurately predicted without taking this effect into account. In addition, accurate simulation of unsteady flow may save on the overdesign that might result from assuming that waves did not change shape. (If you are used to associating the word 'wave' with an effect that lasts for a few seconds, remember that increases in flow in sewer systems resulting from rainfall, i.e. storm waves, may last many hours.)

There are a number of methods available for analysing unsteady conditions in a sewer system. Some are based on approximations, others on attempts to give full theoretical treatment of the physics of the flow. The main methods are for free surface flows. Adjustments for surcharged pipes are considered later.

Table 19.1 Detail on the elements of a flow model

Topic	Chapter/Section
A. Rainfall	5
B. Rainfall to runoff	6.2
C. Overland flow	6.3
D. Flow in the sewer system	8, 19.4

19.4.1 The Saint-Venant equations

For gradually-varied unsteady flow in open channels (including part-full pipes), the full one-dimensional theoretical treatment leads to a pair of equations usually referred to as 'the Saint-Venant equations' after A.J.C. Barré de Saint-Venant, who first published them in the middle of the 19th century. A clear derivation of these equations is available in Chow (1959).

There are two equations: a dynamic equation and a continuity equation.

The dynamic equation can be written:

$$S_f = S_o - \frac{\partial y}{\partial x} - \frac{v}{g} \cdot \frac{\partial v}{\partial x} - \frac{1}{g} \cdot \frac{\partial v}{\partial t} \qquad (19.1)$$

$\longrightarrow|$

1

$\qquad \longrightarrow|$

2

$\qquad\qquad \longrightarrow|$

3

y	flow depth (m)
v	velocity (m/s)
x	distance (m)
t	time (s)
S_o	bed slope (–)
S_f	friction slope (–)

In this form, the components that make up the equation can be identified. The part marked '1' above includes no variations with distance or time and applies to uniform steady conditions. Taking the terms up to '2' includes variations with distance but not with time, and applies to nonuniform steady conditions. The whole equation '3' also includes variation with time and applies to nonuniform unsteady conditions (the ones that interest us here).

Equation 19.1 is commonly presented in terms of flow-rate rather than velocity, and is given below together with the continuity equation.

$$\frac{\partial Q}{\partial t} + \frac{\partial}{\partial x}\left(\frac{Q^2}{A}\right) + gA\frac{\partial y}{\partial x} - gA(S_o - S_f) = 0 \qquad (19.2)$$

$$B\frac{\partial y}{\partial t} + \frac{\partial Q}{\partial x} = 0 \qquad (19.3)$$

Q	flow-rate (m³/s)
A	area of flow cross-section (m²)
B	water surface width (m)

The Saint-Venant equations are valid in situations where the following assumptions are appropriate.

- The pressure distribution is hydrostatic.
- The sewer bed slope is so small that flow depth measured vertically is almost the same as that normal to the bed.
- The velocity distribution at a channel cross-section is uniform.
- The channel is prismatic.
- Friction losses estimated by steady flow equations (see Chapter 8) are valid in unsteady flow.
- Lateral flow is negligible.

19.4.2 Simplifications of the full equations

Equation 19.1, the dynamic equation, includes terms for the bed slope, the friction slope, the variation of water depth and the variation of flow-rate with distance and time. Some of these terms may be more significant than others, giving opportunities for simplifying the equations.

The most drastic simplification is to assume that most of the terms in equation 19.1 can be ignored, and reduce it simply to:

$$S_o = S_f \tag{19.4}$$

This is the equivalent of ignoring all but part 1 of equation 19.1, and implies that the relationship between flow-rate and depth is the same as it would be in steady uniform flow (as in Section 8.4).

Combining equations 19.3 and 19.4 gives:

$$\frac{\partial Q}{\partial t} + c\frac{\partial Q}{\partial t} = 0 \tag{19.5}$$

The wave, called a 'kinematic wave', does not attenuate, but translates at the wave speed, c.

A less drastic simplification involves ignoring just the variation of flow-rate with time (the 'unsteady' effects), that is using equation 19.1 up to '2'. The equivalent of equation 19.5 is now:

$$\frac{\partial Q}{\partial t} + c\frac{\partial Q}{\partial t} = D\frac{\partial^2 Q}{\partial x^2} \tag{19.6}$$

This 'diffusion wave' travels at the same speed, c, as the kinematic wave but, as a result of the term on the right-hand-side of the equation, is subject to diffusion. The diffusion coefficient, D, regulates the attenuation of the wave as it propagates downstream. For simplicity, c and D are usually regarded as constant although they do vary slightly.

Table 19.2 Hydraulic conditions accounted for by simplifications of wave equations

Accounts for	Kinematic wave (1)	Diffusion wave (2)	Dynamic wave (3)
Wave translation	✓	✓	✓
Backwater	✗	✓	✓
Wave attenuation	✗	✓	✓
Flow acceleration	✗	✗	✓

Table 19.2 summarises the various simplified equations and their application. Ponce *et al.* (1978) quantify the range of application of the simplifications as follows:

Kinematic: **Diffusion:**

$$TS_o \frac{v_o}{d_o} > 171 \qquad\qquad TS_o(gd_o)^{0.5} > 30$$

T duration of flood wave (s)
v_o initial velocity (m/s)
d_o initial depth (m)

Table 19.2 gives the range of hydraulic conditions accounted for in the methods.

19.4.3 Numerical methods of solution

The main equations derived in the last two sections are partial differential equations since Q (or v) and y are functions of both distance (x) and time (t). The most common method of solution is using finite differences, involving dividing distance and time into small, discrete steps. This can be represented on a schematic two-dimensional grid showing x and t together, as on Fig. 19.1.

On Fig. 19.1, the distance step is shown as Δx and the time step as Δt. The points where the lines intersect are 'calculation nodes'. The nodes marked with a circle are the distance nodes for time = 0 (say, the start of the calculation). Flow conditions for these are likely to be known: for example, baseflow all along the pipe before the beginning of storm flow. The nodes marked with a square are the time nodes for distance = 0 (say, the upstream end of the pipe). Flow conditions for these may also be known: for example, inflow of stormwater varying with time.

To calculate the conditions at the node marked with the arrow, we can use the known values at neighbouring nodes. Once this calculation is complete, it can be repeated for successive distance nodes (the nodes to the right of the arrow on the same horizontal line). When we have calculated conditions at all distance nodes for that time step, we can proceed

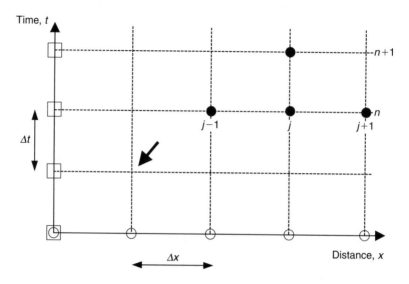

Fig. 19.1 Numerical solution of flow equations: x-t grid

to the next time step (the horizontal line above the one we have dealt with).

We will have to take into account the *boundary conditions* – the hydraulic conditions at the limits of our system, for example the relationship between flow-rate and depth at the inlet and outlet of the pipe system.

There are many ways in which finite difference calculations can be carried out. In Fig. 19.1 the *j*th distance node and the *n*th time node are marked.

Suppose we are solving equation 19.5:

$$\frac{\partial Q}{\partial t} + c\frac{\partial Q}{\partial x} = 0$$

We will write the value of Q at the *j*th distance node and the *n*th time node as Q_j^n.

Suppose we have reached the stage where we wish to calculate Q_j^{n+1}. If we assume that Q varies linearly within each distance and time step, we can write:

$$\frac{\partial Q}{\partial t} \text{ as } \frac{Q_j^{n+1} - Q_j^n}{\Delta t},$$

$$\text{and } \frac{\partial Q}{\partial x} \text{ as } \frac{Q_{j+1}^n - Q_j^n}{\Delta x} \text{ (forward difference),}$$

or $\dfrac{Q_j^n - Q_{j-1}^n}{\Delta x}$ (backward difference),

or $\dfrac{Q_{j+1}^n - Q_{j-1}^n}{2\Delta x}$ (central difference).

If we substitute these types of expression into equation 19.5, it can be rearranged with the unknown, Q_j^{n+1}, by itself on the left-hand side, and solved directly. In solutions where this is the case, the method is described as '*explicit*'.

There are many different explicit finite difference methods, some incorporating 'half-steps', for example the intermediate calculation of $Q_{j+\frac{1}{2}}^{n+\frac{1}{2}}$, to improve accuracy and stability.

More complex formulations, in which the unknown value appears on both sides of the finite difference equation, are described as '*implicit*'. Even though each set of equations is more difficult to solve, implicit methods can be used with longer time steps than explicit methods, and therefore often have the advantage of being more computationally efficient. More information on these methods can be found in the texts recommended at the end of this section.

These approaches to numerical solution can suffer from various forms of inaccuracy, especially when the input data contains rapid changes. Two common problems are illustrated in Fig. 19.2. Fig. 19.2(a) shows 'numerical diffusion', where values are smoothed out in what should be a zone of rapid change. Fig. 19.2(b) shows 'numerical oscillation': small fluctuations

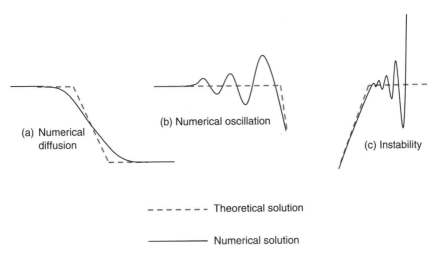

(a) Numerical diffusion

(b) Numerical oscillation

(c) Instability

- - - - - - - Theoretical solution

———————— Numerical solution

Fig. 19.2 (a) Numerical diffusion;
(b) Numerical oscillation;
(c) Instability

at points of change. These problems arise because the method requires variations that are actually continuous to be treated as a series of linear steps. The problems are overcome by selection of appropriate solution methods and suitable time and distance steps. Explicit schemes usually have to satisfy the 'Courant condition' to maintain stability:

$$\frac{\Delta x}{\Delta t} \geq c \tag{19.7}$$

Fig. 19.2(c) shows instability, in which errors introduced by the finite difference method are amplified as the calculation proceeds. This may cause the solution to go completely out of control. A common cause is the use of a time step that is too long.

More information on numerical methods of solving partial differential equations can be found in a number of books, some highly specialised. Accessible introductions to the subject are given by Chadwick and Morfett (1998), Koutitas (1983), Vreugdenhil (1989) and Yen (1986).

19.4.4 Surcharge

It is common for pipes in drainage systems to experience surcharge – to run as full pipes under pressure rather than open channels with a free surface (see Section 8.4.5). It is important that this condition is modelled appropriately because pipe surcharge is an important warning signal to engineers when they are using models to investigate alternative proposals for rehabilitation schemes.

The methods that have been described so far in this section are for free-surface flows. In a surcharged pipe, equation 19.3, for example, would present problems; the terms B (water surface width) and y (flow depth) would be meaningless: B is equal to zero, and y is always equal to pipe diameter regardless of flow-rate.

A concept that allows equations 19.2 and 19.3 to continue to be applied in surcharge conditions is the *Preissman slot* (Yen, 1986). An imaginary slot (Fig. 19.3) is introduced above the pipe which allows y to exceed pipe diameter and give the effect of pressurised flow.

The width of the slot b is calculated precisely to suit the conditions, and must not be so wide that it has a significant effect on continuity. For a circular pipe of diameter, D, this is given by:

$$b = \frac{\pi D^2 g}{4c^2} \tag{19.8}$$

where c is the speed of a wave in a pressure pipe (m/s).

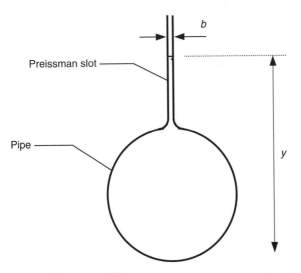

Fig. 19.3 Preissman slot

19.5 Computer packages

Some engineers and many researchers use tailor-made models, developed for very specific applications. However, standard packages are used most, with a few well-established names particularly prominent in different parts of the world.

The following sections give just an outline of some of the standard flow packages that have been most widely used. Many aspects of specific packages quickly go out of date, and are updated or replaced by new versions. It is inappropriate to give too much detail here; the intention is simply to provide a 'snapshot'. Guidance on use of a package, and the details of its theoretical basis, can be sought from the supplier's website (see 'Useful websites', at the back of this book). An alternative source is the user documentation. A useful general source of information on UK modelling practice is the series of user notes produced by the Wastewater Planning User Group (WaPUG), available via their website (also given in 'Useful websites').

19.5.1 SWMM

The US Environmental Protection Agency Storm Water Management Model, SWMM (pronounced 'swim'), is the most widely-known large-scale sewer system model in the United States. It covers both hydraulic and quality aspects. The core code for the model is in the public domain, and this makes it different from the best-known European models which have been developed within particular organisations and then marketed as packages. Software companies market packages that include SWMM together with user-friendly interfaces and other attractions but they do not have 'ownership' of the core code.

The main computational blocks in SWMM are:

- *RUNOFF* which generates surface runoff and pollutants resulting from rainfall
- *TRANSPORT* which routes flows and pollutants through the sewer system; the basis of the modelling of unsteady flow is the kinematic wave approximation
- *EXTRAN* which alternatively routes flows (but not pollutants) using the full Saint-Venant equations.
- *STORAGE/TREATMENT* which routes flow and pollutants through a storage or treatment facility.

SWMM first appeared in the early 1970s and after 30 years of piecemeal development the core computational engine is being rewritten in a major project coordinated by the EPA.

19.5.2 *Wallingford packages*

Before the first Wallingford package, a computer-based method was in common use in the UK: the TRRL Hydrograph Method. However, the first piece of 'modelling software' in the UK, WASSP, was released in the early 1980s. The documentation included an explanation of the calculation methods (the Wallingford Procedure (NWC/DoE, 1981), Volume 1 – still a very useful document).

Subsequent developments were WALLRUS and SPIDA. HydroWorks was released in the mid-1990s, still based in part on elements of the Wallingford Procedure. HydroWorks was designed to be a comprehensive sewer system model, including facilities for quality modelling.

InfoWorks

In the late 1990s, Wallingford Software released InfoWorks. This was a development of HydroWorks with in-built GIS (geographical information system) facilities. It was designed to integrate system modelling with asset and business planning. The key development in InfoWorks was its incorporated data interface, which greatly simplified the task of inputting pipe and manhole data for the sewer system. InfoWorks also includes a quality module and a real-time control module.

19.5.3 MOUSE

The MOUSE package (<u>Mo</u>delling of <u>U</u>rban <u>Se</u>wers) has been developed by the Danish Hydraulic Institute. It is popular all over Europe and in many other parts of the world. Flow modelling is based on a number of modules, covering surface runoff, hydrodynamic network modelling, advanced hydrological modelling for continuous simulation, real time control and long-term hydraulic simulation with statistical analysis. There is also a GIS interface module. The related quality model is MOUSE TRAP.

19.5.4 SOBEK

SOBEK is software produced by the research institute WL/Delft Hydraulics, based in the Netherlands. It is named after an ancient hydraulic modeller, the Egyptian crocodile river god Sobek. Crocodiles were supposed to be able to predict the precise extent of flooding from the Nile each year in order to lay their eggs safely. The SOBEK range covers various hydraulic modelling applications. SOBEK-Urban is a modelling tool for urban drainage systems, including modules covering water flow (hydraulic modelling in pipes and channels), rainfall run-off and real-time control.

19.5.5 Models for system design

There are several packages aimed primarily at system designers. Most drainage design is for specific domestic, commercial and industrial developments. The areas tend to be relatively small, and the Rational Method is often appropriate. Some design packages also allow for simulation of system performance for use in drainage area planning, but it is more common for model-users to carry out this function using the other established simulation packages.

The best design packages are geared to the needs of design engineers, and are highly automated and responsive. They allow the designer to optimise the system quickly and to review the result on a graphical display.

Popular software is produced by the UK company Micro Drainage, including a very user-friendly and 'intelligent' Rational Method package, 'WinDes', allowing automatic adjustment and rationalisation of the sewer system while it is being designed. WinDes has facilities for including SUDS devices, an AutoCAD interface, and a cost and quantities package. WinDap is a linked package for drainage area planning (with the emphasis on simulation rather than design).

19.6 Setting up and using a system model

19.6.1 *Types of system flow model*

Sewer system flow models are used in a variety of applications: general planning purposes, preparing for and designing rehabilitation schemes (as considered in Section 18.2), design of new systems, and forming the basis of quality models (to be covered in Chapter 20).

Setting up models is expensive, and drainage authorities have to be careful about defining the aims of each modelling exercise so that the work is carried out at the right level of detail (and expense). Three general types of flow model are common for particular applications (WaPUG, 2002).

1. *Overall planning*

The model would be highly simplified, used to allow overall assessment of a catchment, or as the initial stage of a more detailed study, or to provide approximate inputs or background information for the study of one component of the system.

2. *Development of a drainage area plan*

The model would be used for overall assessment of a catchment at a greater level of detail, and for identification of parts of the system in need of particular attention.

3. *Detailed design*

The model would be used for detailed investigations and detailed design of new elements of a system.

For application 3 it may be necessary for every pipe in the system to be represented, with few simplifications. For 1 and 2, a simplification of the network will usually suffice. In the case of flow models for sewer rehabilitation schemes, sewer lengths identified as critical (see Section 18.2) will have priority for inclusion in the model.

Software packages are available for storing sewer survey data and displaying it graphically or preparing it for input to a sewer system model (see Section 17.3.6).

19.6.2 *Input data*

A flow model requires physical definition of the sewer system (at the level of detail defined as above), information about the catchment for runoff calculations, and about specific inflows. In a conventional system in which pipes run

Table 19.3 Data requirements for flow models

Element	Data type
Manholes	reference number
	location (map reference)
	ground level
	storage volume
	head-loss parameter
Pipes	reference number
	connectivity information
	length
	shape
	size
	roughness
	invert level
Catchment	total area contributing
	pervious/impervious areas
	slope of ground
	soil data
	details of gullies
	flooded areas
Inflows	dry weather flow
	infiltration
	industrial inflows
	input hydrographs for areas not modelled
Ancillary structures	data to define hydraulic performance
CSOs	geometry
	inflow/outflow arrangements
pumping stations	geometry
	trigger levels
	pump characteristics
outfalls	hydraulic characteristics

between manholes, this generally requires the type of data listed in Table 19.3. Both manholes and pipes may have reference codes (see Chapter 7).

If a package is used for design, the input data above is for the proposed system, and the model simulates the response of that system to specified rainfall conditions.

If a package is used to model an existing sewer system, the input data is that which exists for the system. If the system has poor records, it may be necessary for extensive sewer survey work to be carried out. This is expensive, but good sewer records are of great value, not just for modelling (see Section 17.3). Significant effort may be needed to define the catchment area data, especially where there is incomplete information on the connection of particular properties or sub-areas to the sewer system. Increasing use is being made of geographical information systems (GIS) for storing and processing this type of data.

The model will simulate the response of the system (represented by the data above) to specified rainfall patterns. The rainfall data may take a number of forms:

- for verification, rain gauge records
- for design, selected design storms (see Section 5.4)
- for simulation, sets of synthetic storms (Section 5.4) or time–series rainfall (Section 5.5)

19.6.3 Model testing

Initial checks on a catchment model are needed to make sure that the model is behaving satisfactorily in mathematical terms, and to eliminate obvious mistakes in the input data. It is necessary to check against instability for a variety of extreme conditions. In addition, the overall volume entering the system, or part of the system, should be compared with the overall volume leaving.

19.6.4 Flow surveys

To create a successful working model of an existing system, we also need data on how the catchment actually responds in particular rainfall conditions. Flow surveys give records of the hydraulic performance of the system, and the conditions (mainly rainfall) that produced that performance. If that rainfall is used as input to the model, a comparison between the resulting simulation and the flow survey data should tell us how much confidence we may have in our model.

Appropriate rainfall measurement is a very important part of a sewer flow survey, and is covered in Section 5.2.

In-sewer measurement is carried out at selected sites, and appropriate selection of sites is essential. There are a number of considerations. The number of sites must suit the level of detail of the model; and their location must allow comparisons with the simulation at locations of particular interest, but must also be appropriate from a hydraulic and a practical point of view: with good access, and without excessive disruption to the flow, or deposited sediment.

A typical monitoring point is in a manhole, with the data-logging equipment located near the top of the manhole, and a sensor in the sewer that monitors depth and velocity. A type of sensor that is common in the UK lies on the invert and is wedge-shaped to avoid causing build-up of solids. It measures depth using a pressure transducer, and velocity using a 'Doppler shift' system. (An ultrasonic signal is transmitted against the oncoming flow and is reflected back by solid particles or air bubbles. The reflected signal changes in frequency and the magnitude of the change is proportional to the particle velocity, thus giving water velocity.) These give reasonable results and are widely used. Accuracy is in the order of $\pm 10\%$ on flow-rate under ideal conditions.

Velocity may also be measured using an electromagnetic meter (where a current is induced in a conductor moving across a magnetic field that is

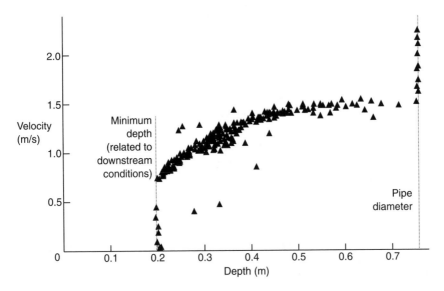

Fig. 19.4 Velocity against depth 'scattergraph'

directly proportional to the speed of movement of the conductor). Meters can be hand-held or mounted on the sewer wall. An alternative method of detecting depth is a top-down ultrasonic meter, positioned above the water surface.

Fig. 19.4 gives a plot of velocity against depth for a large number of readings at a monitoring point in a sewer system. This can be used to check the data for consistency. A similar 'scattergraph' is formed by plotting log flow-rate against log depth (Water Research Centre, 1987).

Records for a number of significant storms are needed for verification of a flow model. More information on flow surveys is available from Water Research Centre (1987) and Saul (1997).

19.6.5 Model verification

It is useful at this point to discuss some of the terminology used during the model development and building process, particularly the distinction between the terms 'calibration' and 'verification'. During calibration of a model, the most appropriate values of the various model parameters are sought. In verification, the model parameters and their values have now been established and input is run through the model to produce results that are compared with known conditions (e.g., flow survey data). In this way, the agreement between computed and observed values can be verified. The verification is, strictly speaking, only valid for that particular location and only over the range of the available data.

Now, in theory, if a model is deterministic, it does not need calibration. If all the input parameter values are accurate and the physics of the processes are simulated sufficiently well, accurate results should be produced without calibration. In practice, however, many of the input parameters are not or cannot be accurately ascertained and the physics is only approximated, thereby making it necessary to resort to default values which may not be representative of the site in question.

Price and Osborne (1986) define verification as the art of demonstrating that a model, which incorporates previously calibrated sub-models, correctly represents the reality of the particular system being studied. They see calibration as occurring only during model development, whereby the physical phenomena represented by the various sub-models are tested, under varied conditions, on many catchments to ensure approximation to observed data with sufficient accuracy. Verification is then carried out by the model user (the drainage engineer), primarily to demonstrate the physical details of the sewer system are correctly incorporated in the model.

Fig. 19.5 is a plot of actual and simulated hydrographs produced during verification of a flow model.

19.6.6 Documentation

Setting up and developing a sewer system model is an investment in the gathering of information. Some of the benefits may be physical, in the form of improved system performance as a result of successful rehabilita-

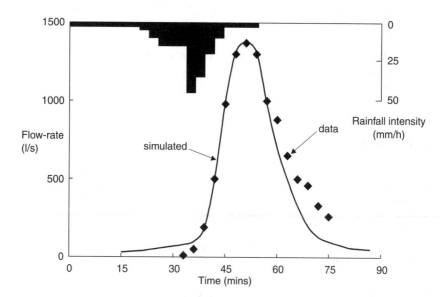

Fig. 19.5 Model verification: hydrograph

tion work. But many of the benefits will remain as information: data on the properties of the system, understanding of its response to particular conditions, and the potential to predict its performance under new conditions. The value of this information will be related to the quality of the documentation of all stages of model development. The conventions used in the documentation will depend on the type of model and on the standard practice within the organisation carrying out the modelling.

19.7 Flow models in context

19.7.1 The modeller

Although sewer system modelling involves application of mathematical methods, most commentators stress the fact that success is heavily reliant upon the skill, experience, judgement – even intuition – of the human beings who set up and run the models.

Osborne *et al.* (1996), comparing SWMM and Wallingford software in a large-scale application, comment that 'the two approaches turn out to be similar and a good engineer can get sensible results using either approach. However an inexperienced engineer will probably get bad results using either approach.'

A German comparison of sewer flow and quality models (Russ, 1999) concluded that 'the reliability and achievable accuracy depend more on the user's qualification, experience and care than on the performance of the model'.

19.7.2 Uncertainty

Sources of uncertainty in a sewer system flow model include the mathematical representations in the model, numerical problems, the level of detail, the assumed values of physical parameters, the input data, and the definition of the initial state of the system. Studies have been made of the impact of uncertainties on model output simulations (for example, Lei and Schilling, 1993) and these put some notions of accuracy in their proper context. One study (Willems and Berlamont, 1999) has found that, in a particular application, replacing the full Saint Venant equations with an extremely simple pipe model made a difference of less than 5% to the total uncertainty.

19.7.3 Hydroinformatics

Hydrological and hydraulic modelling can be seen merely as components in the overall management of information and knowledge in urban drainage. This bigger picture is the subject of 'Hydroinformatics', a growing discipline which aims to harness the benefits of all forms of information technology in water management. The forms of information technology of particular relevance to urban drainage are

- geographical information systems (GIS)
- other forms of data management and graphical presentation
- artificial intelligence, expert systems
- general information sources, the Internet, etc.
- document management.

Sewer system model packages (for example, Wallingford Software's InfoWorks) are developing components and interfaces to give potential benefits in these areas.

Problems

19.1 What is meant by 'a physically-based deterministic model of flow in a sewer system'? What should such a model cover? To what uses might it be put?

19.2 What has been the impact of computer models on urban drainage engineering since they were first introduced?

19.3 Substitute in equation 19.5 using finite difference forms proposed in Section 19.4.3 (with forward differences for both terms). Rearrange to give Q_j^{n+1} on the left-hand side.

19.4 Explain why 'calibration' and 'verification' mean different things (in the context of sewer-flow models). How is verification carried out in practice?

19.5 'Models are always wrong.' So what's the point of using them (in urban drainage)?

Key sources

Chadwick, A.J. and Morfett, J.C. (1998) *Hydraulics in Civil and Environmental Engineering*, 3rd edn, E & FN Spon.

Saul, A.J. (1997) Chapter 8: Hydraulic Assessment, in *Sewers – Rehabilitation and New Construction, Repair and Renovation* (eds G.F. Read with I.G. Vickridge), Arnold.

References

Boonya-aroonnet, S., Weesakul, S. and Mark, O. (2002) Modeling of urban flooding in Bangkok. *Global Solutions for Urban Drainage: Proceedings of the 9th International Conference on Urban Drainage*, Portland, Oregon, September, on CD-ROM.

Chow, V.T. (1959) *Open-channel hydraulics*, McGraw-Hill.

DoE/NWC (1981) *Design and analysis of urban storm drainage – The Wallingford Procedure*. Standing Technical Committee Report No 28.

Koutitas, C.G. (1983) *Elements of Computational Hydraulics*, Pentech.

Lei, J. and Schilling, W. (1993) Propagation of model uncertainty. *Proceedings of the 6th International Conference on Urban Storm Drainage*, 1, Niagara Falls, Canada, September, 465–470.

Maksimovic, C. and Prodanovic, D. (2001) Modelling of urban flooding – breakthrough or recycling of outdated concepts? *Urban Drainage Modeling: Proceedings of the Speciality Symposium of the EWRI/ASCE World Water and Environmental Resources Congress*, Orlando, Florida, May.

Osborne, M., Dumont, J. and Martin, N. (1996) Beckton and Crossness catchment modelling: hydrologic modelling – two routes to the same answer. *Proceedings of the 7th International Conference on Urban Storm Drainage,* 1, Hannover, September, 443–448.

Ponce, V.M., Li, R.M. and Simons, D.B. (1978) Applicability of kinematic and diffusion wave models. *American Society of Civil Engineers, Journal of the Hydraulics Division*, 104, HY3, 353–360.

Price, R.K. and Osborne, M. (1986) Verification of sewer simulation models, in *Urban Drainage Modelling* (eds C. Maksimovic and M. Radojkovic), Pergamon Press, Oxford.

Russ, H.-J. (1999) Reliability of sewer flow quality models – results of a North Rhine-Westphalian comparison. *Water Science and Technology*, 39(9), 73–80.

Vreugdenhil, C.B. (1989) *Computation Hydraulics: an introduction,* Springer-Verlag, Berlin.

Water Research Centre (1987) *A guide to short term flow surveys of sewer systems*, WRc/WAA.

Willems, P. and Berlamont, J. (1999) Probabilistic modelling of sewer system overflow emissions. *Water Science and Technology*, 39(9), 47–54.

Yen, B.C. (1986) Hydraulics of sewers, in *Advances in Hydroscience*, Vol 14 (ed. B.C. Yen), Academic Press.

WaPUG (2002) *Code of practice for the hydraulic modelling of sewer systems*, 3rd edn. www.wapug.org.uk

20 Quality models

20.1 Development of quality models

Sewer systems can be direct contributors of pollution to watercourses. Stormwater outfalls can discharge pollution, as discussed in Section 6.4; and serious short-term pollution can arise from combined sewer overflows, as considered in Chapters 2 and 12. The effects can be counteracted by measures such as the inclusion of additional storage within the system and by rehabilitation or replacement of CSO structures. In order to design these measures to achieve high standards, and to evaluate them fully, it is necessary to have information on the rate at which pollutants flow into the structures from the sewer system, and how the pollutants are distributed in the storm flow. This information is provided by models of *quality* (distribution of pollutants) in sewer flow.

Simplified models of quality have been linked to flow models of sewer systems for some time. For example the *Sewerage Rehabilitation Manual* in 1986 gave factors which could be multiplied by pollutant concentrations in dry weather flow to give estimated average concentrations in storm flow. These could then be used, in conjunction with a flow model, to give total pollutant loads to watercourses resulting from CSO discharge. However, time-dependent effects like a first flush cannot be predicted using this method.

By the mid-1980s, deterministic *flow* models had become so popular and widespread in their use, such a natural tool for the drainage engineer, that it seemed an appropriate step to develop deterministic models of sewer-flow *quality* to a comparable level of detail. In the same way that flow models gave output in the form of hydrographs (flow or depth) at specified points, quality models would give pollutographs, the variation of concentration of pollutants with time. In the same way that flow models are used to assess alternative proposals for hydraulic rehabilitation of sewer systems, quality models would be used to assess proposals for reducing pollution, particularly from CSOs.

In the UK, these plans became part of the Urban Pollution Management programme – a series of linked research projects in the 1980s and 1990s which, as well as development of a sewer-quality model, also included extensive in-sewer monitoring, and development of models covering rainfall, wastewater treatment and river quality. The outcome was the *Urban Pollution Management (UPM) Manual* (FWR, 1st edition: 1994, 2nd edition: 1998) which presented a comprehensive set of procedures for controlling pollutant discharges from sewer systems, covered in more detail in Chapter 22.

Physically-based deterministic quality models have become available for general use, though significant problems limit their widespread acceptance in drainage engineering. There is still research effort devoted to investigating and developing deterministic quality models, and some engineers and companies are committing considerable resources to using the models in practice. Yet, some experts feel that truly comprehensive, deterministic quality modelling is unachievable.

The problem is that the physical processes in the sewer are so complex that:

- they are not fully understood
- it may simply be over-ambitious or inappropriate to attempt to represent them in a physically-based deterministic model, and
- data input and monitoring/verification requirements may be too time-consuming and expensive to be worthwhile.

Ashley *et al.* (1999) list the principal problems relating to the development of sewer process models as follows:

- the difficulty of actually measuring the processes in the field, for example measuring *in situ* yield strength of sediment deposits
- the limited amount of observations economically or logistically possible even when measurement methods are effective
- the extreme temporal and spatial variability of all aspects of the phenomena related to sewer processes, for example the heterogeneous nature of sewer sediments, resulting in the fact that 'global' generalisations can be misleading.

There are obvious benefits from quality models, and even simple approximate models may be useful. Detailed quality models may become established, but it seems likely that they will not always be the full-scale, physically-based deterministic models that were once envisaged.

However, full-scale quality modelling *is* carried out in appropriate circumstances, and most of the major sewer modelling packages include quality aspects as indicated in Chapter 19. This chapter considers how these things are achieved, and discusses some of the requirements that are common to all methods of quality modelling. Alternative approaches to modelling quality are considered at the end of the chapter.

20.2 The processes to be modelled

The main aim of a sewer-quality model is to simulate the variation of concentration of pollutants with time at chosen points in a sewer system. These simulations will be used to improve the performance of the system, for example by aiding the design of CSOs to optimise the retention of pollutants in the sewer system.

The main quality parameters, modelled separately, will be the standard determinands described in Chapter 3, including suspended solids, oxygen demand (BOD or COD), ammonia, and others depending on the model.

Pollutants find their way into combined sewers from two main sources: wastewater and the catchment surface. Once in the system, the material may move unchanged with the flow in the sewers, be transformed or become deposited. Deposited pollutants may subsequently be re-eroded, usually in response to an increase in flow.

The main elements of the system that influence the quality of the sewer flow are indicated schematically in Fig. 20.1 (for a combined sewer).

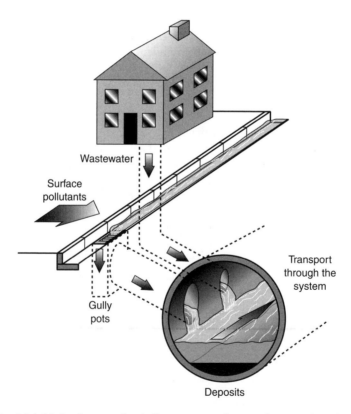

Fig. 20.1 Main elements that influence sewer-flow quality (combined sewer)

Wastewater inflow

The flow-rate and the concentration of pollutants vary with time in dry weather flow in a fairly repeatable daily (diurnal) pattern (see Chapters 4 and 10). The variation is related to patterns of human behaviour, and there tend to be significant differences between weekdays and weekends. Industrial and commercial flows may be present, in addition to domestic contributions. Infiltration can be a significant fraction of dry weather flow (Chapters 4 and 10), and tends to dilute the wastewater.

Catchment surface

In periods of dry weather, there is a build-up of pollutants on roads, roofs etc., and these are washed into the drainage system by the next rainfall (Chapter 6). In general terms, the amount entering the drainage system depends on the quantity of material accumulated on the catchment surface, on the intensity of the rain, and on the nature of the overland flow.

Gully pots

As with catchment surfaces, pollutants tend to build up in gully pots in dry weather. Material remaining in the liquor is regularly washed into the drainage system, even by minor storms. Only high return period storms are thought to disturb previously-deposited, heavier solids.

Transport through the system

Once they have been carried into the pipe system, and provided they are not attached to solids that are deposited, pollutants are transported by the moving liquid. It is generally assumed that the pollutants move at the mean liquid velocity, though there are cases when this might be inappropriate, as will be considered later.

Pipe and tank deposits

At low flows in sewers, especially during the night, solids (and pollutants attached to them) may settle and form deposits. At higher dry weather flows or during storms they may be re-eroded, releasing suspended solids and dissolved pollutants into the flow. The resulting increase in pollutant concentration depends on the characteristics of the system, of the catchment, of the dry weather and storm flows, and the antecedent dry period. As is clear from Chapter 16, the deposition and erosion of sediments is a complex subject, with many different types of sediment and associated pollutants. Flow patterns in tanks also lead to deposition, and the subsequent erosion can have a significant effect on quality.

The mechanisms by which pollutants are introduced to the flow, and are subsequently transported with it, are heavily dependent on hydraulic conditions. Therefore physically-based deterministic quality models need to be based on hydraulic models of the sewer system. The main deterministic quality models in current use are based on established flow models. The accuracy of quality simulations is strongly affected by the accuracy of the hydraulic simulation on which they were based.

In some systems, a significant feature in the variation of sewer-flow quality with time is the first flush in early storm flows, which may contain particularly high pollutant loads. This effect has been considered in Section 12.3.2. It is important that it is represented in a model of sewer-flow quality.

Table 20.1 lists the chapters of this book that deal with the processes that should be represented in a sewer quality model.

20.3 Modelling pollutant transport

20.3.1 Advection/dispersion

Transport of pollutants with the flow is normally represented by one of two alternative forms of equation:

$$\frac{\partial c}{\partial t} + v\frac{\partial c}{\partial x} = 0 \tag{20.1}$$

or

$$\frac{\partial c}{\partial t} + v\frac{\partial c}{\partial x} = \frac{\partial}{\partial x}\left[D\frac{\partial c}{\partial x}\right] \tag{20.2}$$

x distance (m)
t time (s)
c concentration of pollutant (kg/m³)
v mean velocity of flow (m/s)
D longitudinal dispersion coefficient (–)

Equation 20.1 represents *advection*, movement of pollutants at the mean velocity of flow. Equation 20.2 additionally includes *dispersion*, spreading

Table 20.1 Chapters relevant to quality modelling

Topic	Chapter
Wastewater inflow	4, 10
Catchment surface	6, 11
Sediment	16
Flow modelling	19
First foul flush	12

out of pollutants relative to mean velocity. These two mechanisms are illustrated in Fig. 20.2.

Fig. 20.2(a) shows a slug of pollutant in a pipe (equally distributed horizontally and vertically) which is moved along at the mean flow velocity without spreading out: *advection only*. Fig. 20.2(b) shows a similar slug of pollutant which is moved along the pipe at the mean flow velocity and at the same time spreads out: *advection* and *dispersion*.

Nearly all practical examples of sewer flow are strongly dominated by advection. For this reason not all sewer quality models include dispersion.

Numerical methods for solving these equations are similar to those outlined in Section 19.4.3.

20.3.2 Completely mixed 'tank'

An alternative to equations 20.1 and 20.2 is to treat each pipe length as a conceptual tank in which pollutants are fully mixed with the flow. The governing equation is:

$$\frac{d(Sc)}{dt} = Q_I c_I - Q_O c_O \tag{20.3}$$

c, c_I, c_O concentration in pipe length, at inlet to pipe, at outlet (kg/m³)
Q_I, Q_O flow at inlet to pipe, at outlet (m³/s)
S volume of liquid in pipe length (m³)

This equation can be solved to give progressive estimates of concentration in each pipe as a whole. It does not include a distance or a velocity

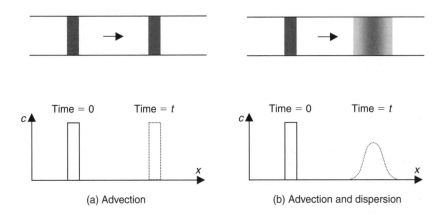

(a) Advection   (b) Advection and dispersion

Fig. 20.2 (a) Advection;
(b) Advection and dispersion

term, and is therefore not capable of explicitly modelling the progress of pollutants at mean velocity.

20.3.3 Sediment transport

Pollutants transported in the body of the flow by advection and dispersion are either dissolved or suspended. Pollutants in solution will remain in that form whatever the flow regime, although they may be transformed by bio-chemical processes (discussed in more detail in the next section). Pollutants in suspension may, however, be affected by the flow regime. At low flows they may concentrate in a near-bed layer or deposit to form a sediment bed, and at high flows they may re-erode. Larger, heavier solids may never (or rarely) achieve suspension, yet be transported as bed-load. The complex hydraulics of sediment movement has already been discussed in Chapter 16.

All sewer quality models have at least some representation of the movement of solid-associated pollutants through the system in terms of:

- mechanics: entrainment, transport and deposition
- sediment bed
- solid attachment.

Mechanics

A relatively straightforward way to model entrainment, transport and deposition of pollutants is by using one of the sediment transport equations which predict the volumetric sediment carrying capacity of the flow c_v (e.g., Macke's equation for suspended solids transport, equation 16.4). Thus, at each time step, for an incoming pollutant concentration c, if:

- $c < c_v$: all the incoming pollutant is transported, and if deposited sediment is available this may be eroded up to the carrying capacity c_v
- $c > c_v$: only c_v/c of the incoming pollutant is transported. The remainder is subject to deposition.

The most straightforward approach is to consider just one type of solid. In principle, many solid size fractions could be represented, although considerable data would be needed for calibration/verification. Representing just two fractions (coarse and fine) can improve model performance.

We have seen in Chapter 16 that sediment can travel in suspension and as bed-load. Some models aim for a more realistic representation by using equations for total sediment transport. However, the mechanics of both types of transport process can be represented separately.

Sediment bed

In the appropriate hydraulic conditions, pollutants will move out of suspension and become deposited on the pipe invert. Most models allow for the development of a sediment bed. At its simplest, this can be considered to have no impact on the hydraulics of the flow. A more realistic representation will include loss of flow area and increased hydraulic roughness that can be communicated to the flow model. A further elaboration could be to include the hydraulic effects of the sediment bed forms (Chapter 16).

The simplest structure of the bed is a simple, single layer of sediment that is supplied by deposition and removed by erosion. The settling velocity of the solid may be used to fix the rate of deposition. Erosion may be controlled by the sediment transport capacity of the flow (as mentioned above) or by a defined critical shear stress. A more refined approach is to represent two layers; one containing stored (type A) sediment, and the other, erodible type C deposits (see Section 16.5).

Time-dependent effects such as consolidation and cohesion have also been introduced in a simple form in some models. The near-bed solids layer is not represented in any commercially available model.

Solid attachment

Pollutants may be modelled in two forms: dissolved and solid-attached, for the reason mentioned at the beginning of this section. Potency factors f are often specified (Huber, 1986) which simply relate concentration of solid-attached pollutant (c_s) to solid concentration c, as $c_s = fc$.

Pollutants that do not move at mean velocity

The assumption that pollutants move at the mean velocity of flow applies to a pollutant that is fully dispersed over the cross-section. This is unlikely to be the case for bed-load, or for some gross solids, or (in dry weather flow) for the near-bed layer. None of these special cases is covered by the current sewer-quality packages, even though these components in the flow have the potential to make significant contributions to storm pollution and aesthetic pollution. Approaches to including gross solids in quality models are considered below.

20.3.4 Gross solids

A basis for predicting the behaviour of gross solids has been described in Section 10.5. This information has been used by Butler *et al.* (2003) to develop a model of gross solid transport. The model is primarily a research tool, but could form the basis of practical software for engineers with potential applications in all aspects of the engineered improvement of systems, including the design of storage facilities and the development of efficient screening systems.

The model represents two aspects of gross solid transport: advection (movement with the flow, but not necessarily at the mean water velocity) and deposition/erosion. The fact that gross solids in actual sewers possess an almost infinite variety of properties, sizes and states of physical degradation means that simplification is needed. The model represents this range by a finite number of distinct solid *types*. For each solid type, the model requires advection and deposition properties: the values of α and β in equation 10.8, and the critical value of velocity and depth for deposition as described in Section 10.5.1. Hydraulic conditions in a sewer system are modelled using standard software.

The computational basis of the gross solids model is to 'track' the progress of individual solids or groups of solids through the system. At any point in distance and time, the mean flow velocity is known from the hydraulic model. This can be converted to solid velocity using the relationship specified for the particular solid type. The velocity of a solid in any instantaneous position is thus known, and this can be used to progressively track its movement through the system. If depth or velocity over any section decreases to below the value specified as causing deposition, the progress of solids in the section is halted until the value is again exceeded.

The model has been combined with a simplified model of solids transport in the smaller pipes in the upstream parts of the catchment, and another of solids behaviour in CSO structures, to create a comprehensive model of gross solids loadings throughout a combined sewer system (Digman *et al.*, 2002). The work has also included collection of an extensive set of field data for comparison with model simulations.

20.4 Modelling pollutant transformation

In a gravity sewer, the main transformation processes act within or between the atmosphere, the wastewater itself, the biofilm attached to the pipe wall, and the sediment bed (see Fig. 20.3). In a pressure sewer, there is no atmospheric phase and the biofilm is distributed around the pipe perimeter.

Of particular importance are those processes associated with the biodegradation of organic material. These are caused by micro-organisms occurring either on the pipe wall as a biofilm or in suspension in the wastewater itself. The biofilm will be more influential in smaller pipes and suspended biomass in larger sewers. These are aerobic processes, requiring the presence of adequate dissolved oxygen (DO) levels. Thus, parameters such as BOD or COD to represent the organic material and DO to represent the toxic state of the wastewater need to be modelled. Anaerobic processes are not normally modelled in detail (but see Section 17.6).

All sewer quality models have at least some representation of pollutant

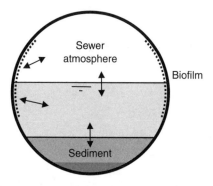

Fig. 20.3 In-sewer pollutant interactions

transformation through the system. This can range from simplistic to sophisticated as follows:

1 Conservative pollutants
2 Simple decay expressions
3 River modelling approach
4 WTP modelling approach.

20.4.1 Conservative pollutants

Conservative pollutants are those that are not affected by any chemical or biochemical transformation processes. Pollutant concentration may still vary due to the processes of advection and dispersion.

Some sewer quality models omit representations of pollutant degradation or biochemical interactions. The justification for this is not that all pollutants *are* conservative, but that these processes are relatively insignificant. This may be a reasonable assumption for short-retention systems, but will be inaccurate in systems with, for example, long, well-aerated outfalls (see Chapter 22).

20.4.2 Simple decay expressions

A second simplified approach is to model the transformation of individual pollutants using a simplified, summary model of the reactions. A first-order decay model is a common example of this, where:

$$\frac{dc}{dt} = -kc \qquad (20.4)$$

| c | pollutant concentration (g/m^3) |
| k | rate constant (h^{-1}) |

Thus the pollutant concentration X decreases with time and temperature, such that:

$$k_T = k_{20}\theta^{T-20} \tag{20.5}$$

k_T	rate constant at $T°C$ (h^{-1})
k_{20}	rate constant at 20 °C (h^{-1})
θ	Arrhenius temperature correction factor (–)

This approach ignores any interactions that may occur between the various substances e.g. DO and BOD.

20.4.3 River modelling approach

An obvious approach to modelling transformations in sewers is to turn to the extensive body of knowledge in river quality modelling (e.g. Rauch *et al.*, 1998; Chapra, 1997). River models vary greatly in complexity, but seek to represent similar processes to those encountered in sewers: advection, dispersion, sedimentation, aeration and conversion.

Oxygen balance

An example of this approach is to consider the oxygen balance in the sewer (Almeida, 1999). DO in the flow results as a balance between oxygen supplied by aeration from the atmosphere and that consumed by the micro-organisms in the wastewater and biofilm. Processes in the sediment bed may also exert an additional oxygen demand. This can be represented as an *oxygen balance* as follows:

$$\frac{dc_0}{dt} = K_{LA}(c_{0,S} - c_0) - (r_w + r_b + r_s) \tag{20.6}$$

$c_{0,S}$	saturation dissolved oxygen concentration (g/m^3)
c_0	actual dissolved oxygen concentration (g/m^3)
K_{LA}	volumetric reaeration coefficient (h^{-1})
r_w	oxygen consumption rate in the bulk water (g/m^3.h)
r_b	oxygen consumption rate in the biofilm (g/m^3.h)
r_s	oxygen consumption rate in the sediment (g/m^3.h)

Reaeration

Reaeration is a naturally occurring process of diffusion. Oxygen in the atmosphere is dissolved into the liquid up to saturation levels that depend mainly on temperature. Aeration may be increased by turbulence caused by manhole backdrops or pumps. Pomeroy and Parkhurst (1973) have derived an empirical relationship for reaeration in sewers based on a number of hydraulic parameters:

$$K_{LA} = 0.96\left(1 + 0.17\left(\frac{v^2}{gd_m}\right)\right)\gamma(S_f v)^{3/8}\frac{1}{d_m} \qquad (20.7)$$

v mean velocity (m/s)
d_m hydraulic mean depth (m)
γ temperature correction factor (1.00 at 20 °C)
S_f hydraulic gradient (–)

Oxygen consumption in the bulk flow

The oxygen consumption of wastewater (also known as the oxygen uptake rate) varies with the age and temperature of the wastewater but (provided conditions are aerobic) is independent of oxygen concentration. Typical values are 1 to 4 mg/l.h.

Oxygen consumption in the biofilm

Oxygen consumption by biofilms is a complex phenomenon affected by substrate and oxygen availability, among other factors. Pomeroy and Parkhurst (1973) expressed consumption empirically as follows:

$$r_b = 5.3(S_f v)^{\frac{1}{2}}\frac{c_0}{R} \qquad (20.8)$$

where R is the hydraulic radius (m).

Oxygen consumption in the sediment

Anaerobic processes in the sediment bed will produce oxygen-demanding by-products resulting in a *sediment oxygen demand* (SOD). This will be exerted when the sediment bed is eroded.

   The most important point about using a river modelling approach is that, although many processes are common in rivers and sewers, the details and objectives are often very different. This means, at the very least, that river models need to be recalibrated using appropriate data.

20.4.4 WTP modelling approach

Recent research seeks to represent pollutant transformations in the drainage system using methods developed for WTPs (e.g. Frontreau *et al.*, 1997). Henze *et al.* (1986) give a detailed explanation of WTP modelling based on Monod kinetics. A major difference in this approach is that COD fractions (see Chapter 3) are used, rather than BOD commonly found in river models. In addition, the governing equations (similar in many cases to those used in rivers) are succinctly represented in matrix notation.

One of the potential benefits of this approach, should it prove to be feasible, is to unify the model parameters between the various components of the system. This will facilitate integrated modelling (considered in Chapter 22).

20.5 Use of quality models

20.5.1 Levels of detail

The *UPM Manual* (FWR, 1998) recommends the following three alternative levels of detail in a quality model.

1   A full *quality* model like InfoWorks or MOUSE TRAP
2   A full *flow* model, combined with simplified treatment of quality using average pollutant concentrations (as described in Section 12.4 and Table 12.2; further detail in the *UPM Manual*).
3   A simplified conceptual quality model. The *UPM Manual* contains information on SIMPOL (see Chapter 22).

20.5.2 Input data and model calibration/verification

A deterministic quality model of the type being described here must be linked to a flow model. Data requirements for a flow model have already been considered in Section 19.6.2.

The concept of verifying a pre-calibrated quality model is a less realistic proposition than for a flow model (as presented in Chapter 19). Variations and uncertainties are much larger, making it harder to transfer experience or default values from one catchment to another, even though apparently similar. Hence, a major issue is that these models require a great deal of data, and the acquisition of this data is highly resource demanding.

A typical deterministic quality model (type 1 in Section 20.5.1) requires the type of input data specified in Table 20.2.

Table 20.2 Input data for a quality model

Wastewater	For a typical day in the week and at the weekend: • variation of flow with time • variation of pollutant concentrations with time
Surface pollutants	Land use characteristics Particulate characteristics
Deposits	Initial sediment depth Characteristics of sediment

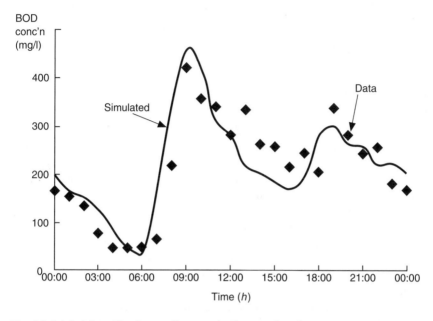

Fig. 20.4 Model verification: pollutograph (dry weather flow)

In spite of the problems of using default values, the extreme expense of gathering full catchment-specific data means that quality modellers usually have to rely heavily on default data. One feature of the calibration/verification of quality models is the replacement of default values by measured catchment values. This is done to the extent that is necessary to gain the desired accuracy, and calls for considerable judgement by the modeller.

It is common to consider dry weather flow first in this process. The simulations are compared with measured data, and default values replaced by measured catchment values, until the model is deemed to be verified. Fig. 20.4 shows actual and simulated dry weather flow quality data used in the verification of a model. The process is then applied to storm flow.

20.5.3 *Data collection*

Taking samples from the flow at particular locations in the system and analysing them for pollutant concentrations is an important operation in the development of quality models for specific systems. Samples from the dry weather flow may be needed to provide catchment-specific input data, and samples from both dry weather and storm flow are needed for calibration/verification.

Other catchment-specific data collection will normally be needed. This may include taking samples of particulate material on the catchment surface, and measuring the depth and properties of the sediment layer in pipes.

There is guidance on the collection of field data to support quality models in the *UPM Manual* (FWR, 1998).

20.6 Alternative approaches to modelling

Most of this chapter has concentrated on physically-based deterministic models. As suggested in Section 20.1, this type of quality model has not gained the full acceptance achieved by physically-based deterministic flow models. It may be that there will never be one dominant model type for sewer quality, but that a wide range of model types, each suiting a different type of application, will become established. One problem is in generalising the processes that affect quality; and it is likely that many quality modelling tools will remain catchment-specific.

This section summarises the alternative approaches to modelling. They can be used for modelling flow as well as quality, though it is in the field of quality modelling that the search for successful methods is most intense.

Every urban drainage model receives input data and produces output data. For any system, a modeller needs access to data on the actual behaviour: conditions recorded for historical events, and the consequences observed to result from those conditions. To reproduce the relationship between conditions and consequences in the model, the conditions become the input and the consequences the output.

The model converts the input into the output using a set of mathematical procedures sometimes called a *transfer function* (therefore representing the relationship between conditions and consequences). The differences between alternative approaches to modelling is the extent to which the model actually attempts to represent the system itself as opposed to simply representing the relationship between input and output.

As we have seen, a physically-based model represents the system by creating a mathematical equivalent of each major physical process in the system – rainfall to runoff, transport of pollutants, etc. Other types of model concentrate purely on the input and output by finding statistical or other relationships between them. In these cases, the transfer function has no physical basis – we cannot point to an equation and say 'That is where rainfall is converted to runoff'. The formulation and potential success of the model has a mathematical explanation, but not a physical one. The model is a '*black box*'. It follows that one disadvantage of a black box model is that it cannot easily be used to look at options to upgrade a sewer system, since the physical properties of the system are not contained in the model.

General alternatives to physically-based deterministic models are summarised below.

Empirical models

An empirical model is based on observation rather than theory. It usually represents the real system by simple relationships that rely for their accuracy on parameters that are calibrated using observed data.

Conceptual models

In a conceptual model, the physical system is represented by highly simplified 'concepts', for example representation of the physics of pipe flow by a simple tank system. Detailed treatment of individual processes is replaced by overall global representation.

Grey box models

This is a model in which some physical relationships are defined, but which relies mostly on non-physical ('black box') relationships derived from observed data.

Stochastic models

A stochastic model includes randomness. Unlike a deterministic model, it does not necessarily give the same output for the same input.

Simple empirical models may be designed to be stochastic, and this is appropriate since the measured data on which the empirical model is based is certain to contain random elements. A complex physically-based model can also have stochastic characteristics by introducing random influences on some elements of input and showing the effect on the output.

The output from a stochastic model will not be a single answer, but a range of answers, possibly represented by a mean and standard deviation. Since it is naive to suppose that any model of a sewer system can give a single correct answer, it can be claimed that all models should include stochastic elements.

Artificial neural networks

Artificial neural networks are a product of developments in Artificial Intelligence. Computer signals are passed between artificial 'neurons'. Each neuron receives signals from a large number of other neurons, applies a weighting to each input, then applies a transfer function before outputting signals to more neurons. The network trains itself to reproduce the relationships in example data (input and output). Given enough training on good data, the network can make useful predictions for new cases, without any need for the artefacts of *human* intelligence such as physical parameters and equations. The approach has been used successfully in sewer modelling (Loke *et al.*, 1996).

Problems

20.1 Explain why quality modelling of a sewer system is more difficult than flow modelling.

20.2 Describe the physical processes that should be covered in a deterministic sewer quality model.

20.3 Classify and describe the various approaches to pollutant transformation modelling. What are their relative merits?

20.4 'Attempting deterministic modelling of sewer-flow quality is a waste of time.' Many experts seriously hold this view. What do *you* think?

20.5 Describe alternatives to physically-based deterministic models for sewer-flow quality. What are their advantages and disadvantages?

Key sources

Ashley, R.A., Hvitved-Jacobsen, T. and Bertrand-Krajewski J.-L. (1999) Quo vadis sewer process modelling. *Water Science and Technology*, 39(9), 9–22.

FWR (1998) *Urban Pollution Management Manual,* 2nd Edition, Report FR/CL0009, Foundation for Water Research.

References

Almeida, M. do C. (1999) *Pollutant transformation processes in sewers under aerobic dry weather flow conditions*, Unpublished PhD thesis, Imperial College, University of London.

Butler, D., Davies, J.W., Jefferies, C. and Schütze, M. (2003) Gross solid transport in sewers. *Proceedings of the Institution of Civil Engineers, Water and Maritime Engineering*, 156 (WM2), 175–183.

Chapra, S. (1997) *Surface water-quality modeling*, McGraw-Hill.

Digman, C.J., Littlewood, K., Butler, D., Spence, K., Balmforth, D.J., Davies, J.W. and Schütze, M. (2002) A model to predict the temporal distribution of gross solids loadings in combined sewerage systems. *Global Solutions for Urban Drainage: Proceedings of the 9th International Conference on Urban Drainage*, Portland, Oregon, on CD-ROM.

Frontreau, C., Bauwens, W. and Vanrolleghem, P. (1997) Integrated modelling: comparison of state variables, processes and parameters in sewer and wastewater treatment models. *Water Science and Technology*, 36(5), 373–380.

Henze, M., Grady Jr., C.P.L., Gujer, W., Marais, G.V.R. and Matsuo, T. (1986) *Activated sludge model No 1*, Report, IAWPRC, London.

Huber, W.C. (1986) Deterministic modeling of urban runoff quality, in *Urban Runoff Pollution* (eds H.C. Torno, J. Marsalek and M. Desbordes), NATO ASI Series G: Ecological Sciences – Vol 10, Springer-Verlag.

Loke, E., Warnaars, E.A., Jacobsen, P., Nelen, F. and Almeida, M. (1996) Problems in urban storm drainage addressed by artificial neural networks. *Proceedings of the 7th International Conference on Urban Storm Drainage*, 3, Hannover, September, 1581–1586.

Pomeroy, R.D. and Parkhurst, J.D. (1973) Self purification in sewers. *Advances in*

Water Pollution Research. Proceedings of the 6th International Conference, Jerusalem, Pergamon Press, 291–308.

Rauch, W., Henze, M., Koncsos, L., Reichart, P., Shananan, P., Somlyody, L. and Vanrolleghem, P. (1998) River quality modelling. I State of the art. *Water Science and Technology*, 38(11), 237–244.

21 Stormwater management

21.1 Introduction

21.1.1 The fundamental issue

The fundamental issue relating to stormwater management has already been referred to a number of times in this book. Fig. 1.3 illustrates the combined effect of covering areas with impervious surfaces like roofs and roads, and carrying rainwater runoff away in a piped system. The result is an increase in the risk of flooding and pollution of the natural watercourse to which the runoff is discharged. In Section 2.1 we introduced a way of reversing this trend: to drain developed areas in a more natural way, using the infiltration and storage capacities of semi-natural devices such as infiltration trenches, swales and ponds – SUDS as they are known collectively in the UK (standing for Sustainable Urban Drainage Systems, or SUstainable Drainage Systems). Earlier chapters, especially Chapter 11, have dealt with piped systems for stormwater, so in this chapter we consider the alternatives to pipes.

This fundamental issue of stormwater management, and of urban drainage as a whole, is effectively the same as the fundamental issue for much of civil engineering and the environment-related professions. In developed countries, we have adapted our environment in order to pursue goals such as better transport links, or more protection from disease or flooding, but have achieved these advances largely by *imposing on* nature. As the negative effects have become more apparent, we have reviewed our approaches with the aim of working more in harmony with nature. As well as in urban drainage, this change of approach is also apparent in areas such as river engineering, coastal defence and alternative energy. The Institution of Civil Engineers, whose definition of civil engineering used to be 'to divert the great sources of power in nature for the use and convenience of man' now describes the profession as 'the practice of improving and maintaining the natural and built environment to enhance the quality of life for present and future generations'.

21.1.2 SUDS

We have concentrated so far on just one of the benefits of this move away from 'hard engineering solutions' (i.e. conventional piped sewer systems): that of reversing the trend in Fig. 1.3 by reducing the speed and peakedness of urban runoff so that flooding and erosion of the natural watercourse to which the stormwater is discharged are less likely. In some cases the result of using SUDS devices will be that all the drainage will be by natural means and there will be no need for storm sewers. In other cases there will still be some discharge to a sewer system but the reduced runoff load will mean we need fewer or smaller storm sewers, or that an existing storm sewer is less likely to surcharge. If the existing system is a combined one, it will mean less storm flow entering the system and therefore fewer CSO spills. SUDS can also help to counteract the effects of climate change (Section 5.6).

Another benefit is in the area of water quality. The reduction in erosion will improve quality of the water in the natural watercourse, and SUDS devices themselves may improve the quality of runoff through filtration and biological action.

Further benefits are that SUDS preserve or enhance natural vegetation and wildlife habitats in urban areas; they may recharge soil moisture and groundwater; and they may be used to provide stored water for reuse.

This can be summarised by identifying three main areas of benefit (CIRIA, 2001) as *quantity*, *quality* and *amenity*. There is more to SUDS than just drainage.

21.1.3 Development

Countries such as Australia, the USA and Sweden have been using these approaches for many years. Concerted developments in the UK started in the late 1980s, and in 1992 a series of guides with the title *Scope for Control of Urban Runoff* was produced (by CIRIA) which gave guidance on a range of options including those described in Section 18.4 for hydraulic rehabilitation of the sewer system, together with those in Chapter 13 for storage, and those described in this chapter. During the 1990s the acceptance of SUDS advanced more rapidly in Scotland than in England and Wales, and when a major set of guidance documents was published in 2000, two separate design manuals were released (CIRIA, 2000): one for Scotland and Northern Ireland and one for England and Wales. Shortly after that, a best-practice manual was published giving more general guidance for a wider audience (CIRIA, 2001).

A range of important issues still surround SUDS (to be considered in Section 21.6). Depending on one's point of view these can be seen either as disadvantages of SUDS or as barriers to wider acceptance that need to be removed.

21.2 Devices

The common types of device are now described separately. Section 21.3 considers how they can be linked together, 21.4 gives guidance on detailed aspects of design and 21.5 provides an outline of water quality issues.

The ultimate aim is to find the right tools for the job, used in the best combination. The result may be a system of drainage based entirely on SUDS and involving no conventional piped drainage. Alternatively it may contain elements of source control in combination with oversized sewers or underground storage tanks, discharging heavily attenuated storm flows to a conventional drainage system. Or, in a densely developed area, there may be no scope for SUDS devices at all.

Approaches to overall decision-making in relation to SUDS are well set out in the design manuals by CIRIA (2000).

21.2.1 Inlet control

Stormwater can be controlled at source by detaining it at the point at which it runs off the catchment, essentially within the curtilage of the individual property. This is achieved by throttling (restricting) the inflow to the drainage system. Systems in use include rooftop ponding, downpipes and paved area ponding.

On the roof

Stormwater can be retained on flat roofs, thus exploiting their storage potential by using flow restrictors on the roof drains. Obviously, this will induce an additional live load to be taken into account in the structural design, and increase the need for watertightness of the roofing materials. Unfortunately, in practice, flow restrictors can become blocked, leading either to overtopping or prolonged ponding.

Maskell and Sherriff (1992) report that the attenuation of runoff using roof storage can reduce peak sewer flows by 30–40%. Roof storage has little or no direct positive effect in reducing pollutant concentrations.

A variation on the theme of dedicated roof rainwater storage is a 'green roof'. This consists of a planted area that has significant storage potential, encourages evapo-transpiration and provides the added benefit of water quality improvement as stormwater travels through the soil (Fig. 21.1).

Downpipes

An alternative to detaining water on the roof structure itself is to store it at the foot of the downpipe in localised storage, either above or below ground. Small volumes (a water butt will have a capacity of about 350 l) used in large numbers, can have effects comparable with rooftop ponding.

Fig. 21.1 Green roof, Portland, Oregon

An additional advantage of such devices is that the retained water can be put to good use, such as garden watering or WC flushing.

An alternative to providing storage at the base of the downpipe is to discharge runoff away from the building and over stable pervious areas (such as lawns, swales, porous pavements) rather than directly to the pipe system. Therefore, the surface runoff is delayed, infiltration is increased and pollutants are removed to a certain extent. This can be carried out in existing urban areas as well as new-build.

Paved area ponding

In principle, similar benefits to rooftop ponding can be gained by exploiting the storage available on paved areas. Potential sites include car parks, paved storage yards and other large impervious surfaces. The advantage over roof storage is that much larger surfaces are available and ponding depth can be greater. Disadvantages relate particularly to the nuisance value of ponded water and, in extreme cases, possible damage and safety issues. In addition, unless the system is properly maintained, it will not function. Methods to mobilise ground level storage usually involve restricting flow into the sewer system via gullies with orifices or vortex regulators (see Chapter 9).

21.2.2 Infiltration devices

The two most common infiltration devices are soakaways and infiltration trenches. A soakaway is an underground structure which can be stone filled, formed with plastic mesh boxes, dry wall lined, or built with precast concrete ring units (see Fig. 21.2). It is recommended (Beale, 1992) that any filling has a void ratio, *e* (defined as the ratio of interstitial volume to soil volume), of at least 30%. An infiltration trench is a linear excavation lined with a filter fabric, backfilled with stone and possibly covered with grass. Runoff is diverted to the soakaway or trench and either infiltrates into the soil or evaporates (see Fig. 21.3). The device provides storage and enhances the ability of the soil to accept water by creating a surface area of contact. Soakaways and infiltration trenches should not be located within 5 m of the foundations of buildings, or under roads. As a result of their shape characteristics, trenches are usually more efficient than soakaways at controlling runoff.

Soakaways and trenches can be used in any area that has pervious subsoils such as gravel, sand, chalk and fissured rock. However, trenches installed on land gradients greater than about 4% need 'flow checks' at

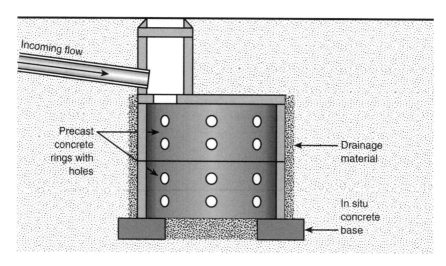

Fig. 21.2 Soakaway

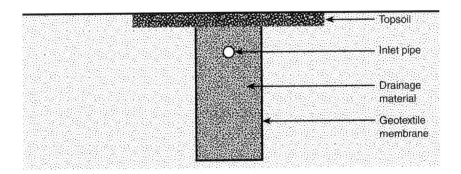

Fig. 21.3 Infiltration trench

regular intervals. These systems are only suitable in areas where the water table is low enough to allow a free flow of the stormwater into the subsoil at all times of the year. The base of the soakaway or trench should therefore be at least 1 m above the groundwater level. Areas with no natural watercourses usually have suitable subsoils.

As well as disposing of the stormwater, soakaways and trenches can reduce the concentration of some of the pollutants it contains, by processes of physical filtration, absorption and biochemical activity. Mean annual removal efficiencies for suspended solids, metals, PAH, oil and COD of 60–85% have been recorded for infiltration trenches draining highway runoff (Colwill *et al.*, 1985).

There are some restrictions on infiltration where there is a risk of pollution to groundwater resources (see Section 21.6).

21.2.3 Vegetated surfaces

The main types of vegetated surfaces used in stormwater management are filter strips and grassed swales (see Fig. 21.4). *Swales* are grass-lined channels used for the conveyance, storage, infiltration and treatment of stormwater. Runoff enters directly from adjoining buildings or other impermeable surfaces. The runoff is stored either until infiltration takes place, or until the filtered runoff is conveyed elsewhere, to the sewer system, for example. *Filter strips*, also known as 'vegetative buffer strips', are gently sloping areas of ground designed to promote sheet flow of stormwater runoff.

To function well, swales require shallow slopes (<5%) and soils that drain well. Typically, they have side slopes of no greater than 1 in 3, allowing them to be easily maintained by grass cutting machinery. The bottom width is usually about 1 m, they are 0.25–2 m deep, and can be readily incorporated into landscape features. Filter strips should allow a

Fig. 21.4 Swale (courtesy of Prof. Chris Jefferies)

minimum flow distance of about 6 m. Swales and strips delay stormwater runoff peaks and provide a reduction in runoff volume due to infiltration and evapo-transpiration. Typical velocities should be below 0.3 m/s to encourage settlement.

They are often used as a pre-treatment in combination with other control measures. Pollutants are removed by sedimentation, filtration through grass and adsorption onto it and infiltration into the soil. Runoff quality can be considerably improved and Ellis (1992) found that a swale of length 30–60 m could retain 60–70% of solids and 30–40% of metals, hydrocarbons and bacteria. Controlled experiments on grass swales in Australia (Fletcher *et al.*, 2002) have demonstrated reductions in total suspended solids *concentrations* of between 73% and 94% and in total suspended solids *loads* of between 57% and 88%, together with significant reductions in total nitrogen and total phosphorus.

For guidance on design, see CIRIA (2000).

21.2.4 Pervious pavements

Pervious pavements are used mostly for car parks (Fig. 21.5), and can also be used for other surfaces where there is no traffic or very light traffic. A typical arrangement for the pavement structure is illustrated in vertical section in Fig. 21.6. There are a number of alternatives for the

Fig. 21.5 Car park with pervious pavement (courtesy of Formpave Limited)

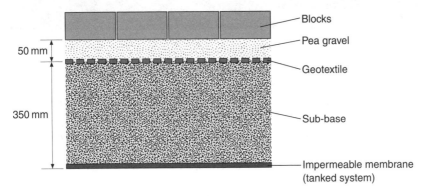

Fig. 21.6 Typical pervious pavement structure, vertical section

surface layer. It could consist of one of a variety of types of block, or could be a layer of porous asphalt. Blocks may be **porous,** allowing water to seep through them via pores in the material itself, or **permeable,** where the material is not porous but the blocks are laid in such a way that water can pass *between* them. Permeable blocks may fit tightly, with water passing through narrow slots between blocks (Fig. 21.7), or may be laid with a pattern of larger voids which are filled with soil and grass, or gravel.

Below the surface layer of blocks is a bedding layer of sand or small-size gravel, separated from the sub-base below by a layer of geotextile. (The bedding layer is not necessary with porous asphalt.) The sub-base consists

Fig. 21.7 Permeable surface blocks (courtesy of Formpave Limited)

of crushed rock or other sufficiently hard material, or a plastic mesh structure.

At the lower surface of the sub-base is a pervious geotextile, if infiltration to the ground is intended, or an impervious geomembrane if it is not. If there is no infiltration, the pavement structure is a 'tanked system', providing considerable attenuation to the storm flow but still requiring arrangements for outflow. While water is stored below the pervious surface, up to 30% can be lost to evaporation (CIRIA, 2000).

The storage potential of a pervious car park structure can also be exploited by receiving further runoff from roof surfaces or other impermeable surfaces. It is recommended that this additional inflow is evenly distributed on to the pervious pavement, and that the water is introduced to the pavement structure either by being released so that it flows through the surface layer, or via a silt trap, to prevent clogging of the sub-surface layers. There is also the potential for rainwater stored in the sub-base to be used for applications such as toilet flushing and garden/landscape watering. An example is described in Chapter 24, Box 24.2. When a pervious pavement structure is constructed at a slope, there is a potential loss of storage volume which may need to be counteracted by including ridges across the sub-base.

Infiltration rates through permeable pavement surfaces, especially new ones, are generally high. For new surfaces, rates are commonly in excess of 1000 mm/h, and sometimes considerably higher, depending on the type of surface. Studies from a number of countries reported by Pratt *et al.* (2002) indicate that the infiltration rate of a pavement surface may reduce over its life to 10% of its original (new) value. To limit blockage at the surface, it is recommended (CIRIA, 2000) that the surface should be cleaned by vacuum sweeping twice a year. However Rommel *et al.* (2001) showed that other parts of the system, especially the geotextile, were more likely than the surface to limit infiltration through the structure as a whole.

Monitored sites have consistently demonstrated substantial reduction and delay in storm peaks and reduction in overall volumes. The sites have included tanked systems (Abbott *et al.*, 2003; Schlüter and Jefferies, 2001) and systems with infiltration to the ground (Macdonald and Jefferies, 2001).

Pervious pavements may also have a positive effect on water quality by providing mechanisms that encourage filtration, sedimentation, adsorption and chemical/biological treatment, as well as storage (Pratt *et al.*, 2002). Laboratory and field tests have demonstrated excellent performance in retention and degradation of oils (Pratt *et al.*, 1999).

Because of their nature and uses, there are also important structural design considerations. The best source of guidance on this and other aspects of pervious pavement performance is the CIRIA report by Pratt *et al.* (2002).

21.2.5 Filter drains

Filter drains are linear devices consisting of a perforated or porous pipe in a trench of filter material. They have traditionally been constructed beside roads to intercept and convey runoff, but they can be used simply as conveyance devices. They may or may not allow infiltration to the ground, in the same way as pervious pavements.

21.2.6 Ponds

The general function of ponds has been introduced in Chapter 13, and the principles for sizing storage in that chapter can be applied to ponds. Ponds are appropriate for use with reasonably uncontaminated flows. They are classified as wet or dry depending on whether a permanent pool of water is maintained.

Wet ponds

Wet ponds have a permanent volume of water incorporated into the design (see Fig. 21.8). This type of pond can have aesthetic, recreational value (e.g. sailing, fishing) and environmental benefits such as returning wildlife

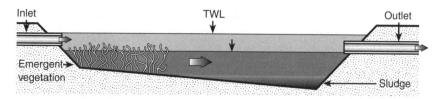

Fig. 21.8 Simplified wet detention pond

habitats into urban areas, in addition to their flood control function. Most wet ponds are on-line.

Wet ponds can play a significant role in pollution control since sedimentation and biological processes may enhance the water quality of the outflow. This is because many pollutants are attached to suspended solids which are themselves captured by sedimentation.

The depth of the pond is usually limited to 1.5–3.0 m to avoid thermal stratification (Lawrence *et al.*, 1996). Shallow side-slopes and dense marginal vegetation help ensure safety.

Dry ponds

Dry ponds do not have a permanent pool of water stored between storm events. They consist of excavated, berm-encased or dished areas, lined with grass or porous paving. Naturally-formed versions are called *water meadows*. They may operate either as retention basins (with no fixed outlet with drainage by infiltration alone) or detention basins with some form of outflow arrangement back into the drainage system (e.g. fixed or mechanical hydraulic controls). Most dry ponds are off-line. In many cases, dry ponds are hardly noticed by the public since they are often multi-functional, also operating as recreation areas and only filling during exceptional storms. Dry ponds have relatively low pollutant removal efficiencies because of the re-erosion of previously deposited solids during filling. They are smaller than wet ponds.

Wet/dry ponds

These act as a combination of the two previous pond types. Part of the storage area contains water at all times, and part only fills at times of high flow. 'Extended' detention basins, for example, often have a permanent pool incorporated for aesthetic reasons.

21.3 SUDS applications

Management train

The CIRIA design manual (CIRIA, 2000) provides guidance on how SUDS devices should be used in combination, and how they should be selected for a particular application. The use of SUDS devices in combination is an important theme in the guidance, and the result is termed a 'surface water management train', also referred to as a 'treatment train'. The recommended sequence of possibilities, with devices appropriate for each stage, is given in Fig. 21.9. It is preferable to find a drainage solution as close to the top of this diagram as possible, but if all drainage needs cannot be achieved at a particular stage, the designer must move further down the list. A good example of the management train principle in practice is described by Bray (2001a), where runoff from different sections of a motorway services site passes through separate trains of linked devices including trenches, swales, ponds and wetlands.

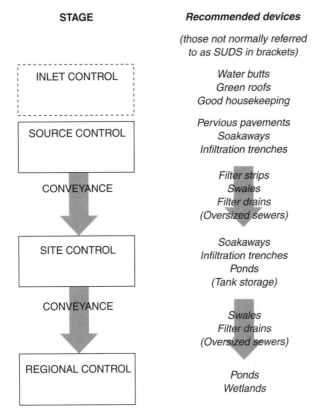

Fig. 21.9 Surface water management train

Retrofit

In existing sewered catchments there may also be benefits in *retrofitting* SUDS devices. The potential benefits include reduction of flooding, removing the need for increasing the capacity of a system, or a reduction in the frequency of CSO spills (Stovin and Swan, 2003).

21.4 Elements of design

We consider here elements in the design of SUDS devices. The detailed design of a whole SUDS system is beyond the scope of this book, but sources of further guidance are given throughout this chapter.

21.4.1 Rainfall

Clearly all SUDS design work involves some prediction of rainfall. Appropriate methods have been presented in Section 5.3. Predicting the impact of climate change (Section 5.6) is as relevant to SUDS as it is to other methods of drainage.

21.4.2 Storage volume related to inflow and outflow

A simple method for determining storage volume by considering storms of different durations is given in Section 13.3. More detailed methods of considering the interaction between inflow, outflow and storage are demonstrated in Sections 13.4 and 13.5.

21.4.3 Infiltration from a pervious pavement sub-base

CIRIA Report R156 (Bettess, 1996) gives the following method for 'plane infiltration systems'. The formula below gives the maximum depth of water in the sub-base.

$$h_{max} = \frac{D}{n}(Ri - f)$$

h_{max}	maximum depth (m)
D	duration of rainfall (h)
n	porosity of sub-base (volume of voids ÷ total volume)
R	ratio of drained area to infiltration area
i	rainfall intensity (m/h)
f	infiltration rate (m/h)

Accurate estimation of soil infiltration rate is difficult because rates depend on numerous factors such as:

- soil particle size and grading
- the presence of organic material
- plant, animal and construction activity
- soil history.

In particular, considerable differences in value may occur at different times of the year and with different antecedent conditions. The best, but still not ideal, information can be obtained from undertaking on-site trials. General soil data, for use in preliminary calculations, is given in Table 21.1. Infiltration systems are not suitable in soils with infiltration rates less than 0.001 mm/h.

Table 21.1 Typical soil infiltration rates (after Bettess, 1996)

Soil type	Rate (mm/h)
Gravel	10–1000
Sand	0.1–100
Loam	0.01–1
Chalk	0.001–100
Clay	<0.0001

The recommendation in CIRIA 156 is that the infiltration rate is determined from on-site tests, and that a factor of safety is then applied in calculations to account for progressive siltation (Table 21.2).

If there is no infiltration from the bottom of the sub-base (a tanked system), and outflow by other means can be ignored, the expression is even simpler.

$$h_{max} = \frac{DRi}{n}$$

Table 21.2 Factor of safety applied to measured infiltration rate

Area drained (m²)	Factor of safety, related to consequence of failure		
	No damage or inconvenience	Minor inconvenience	Major consequences
<100	1.5	2	10
100–1000	1.5	3	10
>1000	1.5	5	10

Example 21.1

Determine the maximum depth of water in the sub-base of a pervious car park for the following storm event: rain intensity 50 mm/h, duration 1 h. Infiltration rate to soil has been measured as 15 mm/h. Area drained is 500 m², area of pavement is 250 m². Surface flooding would cause minor inconvenience. Porosity of sub-base material is 0.4.
   Repeat for a tanked system.

Solution

Safety factor is 3. Therefore $f = 15/3 = 5$ mm/h.

$$h_{max} = \frac{1}{0.4}(2 \times 50 \times 10^{-3} - 5 \times 10^{-3}) = 0.238 \text{ m}$$

For tanked system:

$$h_{max} = \frac{1 \times 2 \times 50 \times 10^{-3}}{0.4} = 0.250 \text{ m}$$

21.4.4 Infiltration from a soakaway or infiltration trench

The main design methods available for design of non-plane infiltration systems are BRE Digest 365 (BRE, 1991), CIRIA Report R124 (Leonard and Sherriff, 1992) and CIRIA Report R156 (Bettess, 1996). BRE Digest 365 and CIRIA 124 are based on the same procedure, but there are some differences between this and the recommendations in CIRIA 156. These differences include the procedures for the use of test pits to measure infiltration rates. Also, as described above, CIRIA 156 recommends factors of safety to be applied to the infiltration rate in calculations, whereas BRE Digest 365 and CIRIA 124 assume that infiltration takes place through the sides but not through the base (which gives an implicit factor of safety by assuming that the base will become clogged with fine particles). BRE Digest 365 and CIRIA 124 assume that infiltration from the system into the soil is constant and corresponds to that when the system is half full of water, whereas CIRIA 156 includes infiltration through the base and uses a more complex representation of infiltration overall. Since the sample calculations in the CIRIA design manual (2000) are based on the method in BRE Digest 365, it is that method (which leads to a more straightforward calculation than CIRIA 156) which is presented in the most detail here.

BRE Digest 365 method

Storage for runoff from the critical storm, at a given return period (10 years is suggested), is given by:

$$\text{Storage} = \text{runoff volume} - \text{infiltration during storm} \tag{21.1}$$

So, if the critical storm has duration, D (h), storage, S (m³), is given by:

$$S = iA_iD - fa_{50}D \tag{21.2}$$

i	rainfall intensity (m/h)
A_i	impervious area (m²)
f	soil infiltration rate (m/h)
a_{50}	effective surface area for infiltration (m²)

This can be applied to an infiltration trench as follows. A trench has a width of B_d, an effective depth (i.e. depth below invert of the incoming pipe) y and a length L. Assuming discharge *from* the trench *to* the soil is through the sides (length and ends) only, and that the average water level is half the effective trench depth, the effective infiltration surface area is:

$$a_{50} = y(B_d + L) \tag{21.3}$$

Active storage volume S in the trench is:

$$S = yB_dLn \tag{21.4}$$

n	porosity of trench material (free volume ÷ total volume)

After the rainfall event, it is recommended that the infiltration rate should be sufficient to empty at least half the stored volume within 24 h.
   The suggested design procedure is:

1 determine the soil infiltration rate
2 adopt a device cross-section
3 determine the required storage volume by considering a range of durations of 10-year design storms
4 review the suitability of the design and check that the device will half empty within 24 hours.

A calculation using this method is presented as Example 21.2.

Example 21.2

Design an infiltration trench to serve an individual property draining an impermeable area of 100 m² if it is filled with an aggregate which gives a free volume that is 0.3 of total volume. The soil infiltration rate is estimated from an on-site trial pit at 10.0 mm/h. Rainfall statistics for the 10-year storm are as derived in Example 5.2.

Solution

1   $f = 10.0$ mm/h (no factor of safety is used, but it is assumed that there is no infiltration through the base).
2   Say: Width $(B_d) = 1$ m
    Effective depth $= 1$ m
    ... determine L
3   Equation 21.2:

$$S = iA_iD - fa_{50}D$$

Substitute using equations 20.3 and 20.4

$$yB_dLn = iA_iD - fy(B_d + L)D$$

Sample calculation for $D = 1$ h: $i = 24.8$ mm/h

$$1 \times 1 \times L \times 0.3 = 24.8 \times 10^{-3} \times 100 \times 1 - 10 \times 10^{-3} \times 1 \times (1 + L) \times 1$$

giving $L = 7.97$ m

Considering the range of durations of 10-year design storms from Example 5.2:

Storm duration, D (h)	(min)	Intensity, i (mm/h)	Length, L (m)
	5	112.8	3.13
	10	80.4	4.44
	15	62	5.12
	30	38.2	6.25
1		24.8	7.97
2		14.9	9.23
4		8.6	9.96
6		6.1	9.93
10		4	9.64
24		2	8.27

The critical case is at 4 h: length 9.96 m – say 10 m
4   Time for emptying is evaluated assuming half the stored volume discharges through the effective surface area for infiltration:

$$t = \frac{0.5B_dLn}{y(B_d + L)f} = \frac{0.5 \times 1 \times 10 \times 0.3}{1 \times (1 + 10) \times 10 \times 10^{-3}} = 13.6\,\text{h}$$

which is acceptable (<24 h).

CIRIA 156 method

For a particular rainfall event discharging to an infiltration device, maximum depth of water in the device is given by:

$$h_{max} = a(\exp(-bD)-1)$$

in which

$$a = \frac{A_b}{P} - \frac{iA_D}{Pf}$$

and

$$b = \frac{Pf}{nA_b}$$

D storm duration (h)
A_b area of base (m^2)
P perimeter of infiltration device (m)
i rainfall intensity (m/h)
A_D impermeable area from which runoff is received (m^2)
f infiltration rate (m/h) = measured rate ÷ factor of safety

Time for half emptying (t in h) is given by:

$$t = \frac{nA_b}{fP}\ln\left[\frac{h_{max} + \dfrac{A_b}{P}}{\dfrac{h_{max}}{2} + \dfrac{A_b}{P}}\right]$$

21.4.5 Modelling

The main applications for models of drainage systems are in *design* and *simulation*, as described in Chapter 19 (in the context of conventional sewered catchments). This is equally true for drainage systems based on SUDS, and also to sewered systems that include SUDS.

SUDS devices are modelled in a simplified form in the Source Control module of the design package WinDes by Micro Drainage (Section 19.5.5). This allows designers to incorporate SUDS in proposals that also include conventional pipe elements, and to investigate SUDS as alternatives.

Kellagher *et al.* (2003) report using InfoWorks (Section 19.5.2) to model pervious pavements by calibrating the infiltration runoff module using observed data recorded in the field. The difference in model parameters established for two sites suggests that this approach needs some site-specific calibration to give accurate results. They also point out that whereas existing drainage models are one-dimensional, it may be necessary to have a two- or three-dimensional model of a pervious pavement system.

The effects of SUDS on existing sewered catchments can be modelled using conventional sewer system software. Swan *et al.* (2001) report studies to test scenarios for retrofitting SUDS devices in existing catchments using HydroWorks software. And Davies *et al.* (2001) describe studies to determine the impact of existing SUDS devices on flows in a small catchment using Micro Drainage software.

21.5 Water quality

Mass balance

The basis of water quality design for local disposal techniques is essentially a matter of mass balance of pollutants, in many ways similar to hydraulic design with an additional term:

$$\text{accumulated mass} = \text{mass inflow} - \text{mass outflow} \pm \text{mass change}$$

$$(21.5)$$

The mass inflow for a given time period is the product of the flow volume and the mean pollutant concentration. The mass change is a function of the chemical, biological or physical changes that take place within the device. Change can be both positive and negative and it is possible that, for a given storm event, there is a net *export* of pollutants due to release of pollutants in some way bound within the device (e.g. in sediments or biomass).

The difficulties in the design become apparent when the complexities of urban rainfall and runoff are compounded with the great number of pollutant types and concentrations, and with the many reactions that can take place. In addition, the pollutant removal rates of devices vary considerably.

For practical purposes, it can be assumed that the percent of pollutant mass captured is proportional to the percent of runoff flow volume captured (i.e. diverted away from the piped system). Capture of about 10–15 mm of effective rainfall (i.e. runoff) should capture in the range of 80–90% of the annual effective rainfall and thus that percentage of the pollution. The reason is that the majority of the storms in any given year only produce a small amount of runoff. So, unlike stormwater quantity management, where infrequent major storms are of most concern, in water quality control it is total *volumes* that are of most significance.

Treatment volume

A feature of storage is that it encourages the settlement of some of the suspended sediment and associated pollutants contained in the incoming stormwater. In practice, removal efficiencies vary considerably, but well-designed ponds have the capacity to make reductions in many pollutants.

For significant treatment to take place, the captured runoff must be retained for at least 24 hours. Martin *et al.* (1999) suggest the following expression to estimate required treatment volume, based on data obtained from the Wallingford Procedure:

$$V_t = 9(M5 - 60 \, min)(SOIL/2 + PIMP(1 - SOIL/2)/100)$$

V_t	basic treatment volume (m³/ha)
SOIL	a soil index for the UK based on the WRAP classification
M5−60 min	5-year return period, 60-minute duration (standard) rainfall depth
PIMP	percentage imperviousness of the catchment served

Dry detention ponds are designed to fully contain the treatment volume V_t, and to drain down within 24 h. Wet retention ponds require a volume of $4V_t$ to allow time for biological treatment in addition to physical processes (Martin *et al.*, 1999).

Other important factors for optimum pollutant reduction (Ellis, 1992) are that:

- the ratio of pond volume to mean storm runoff volume is between 4 and 6
- the storage volume exceeds 100 m³ per hectare of effective contributing area
- the ratio of pond surface area to effective contributing area is 3–5%.

21.6 Issues

We consider here some of the wider issues that are involved when SUDS are considered as drainage options. Some of these issues are described as 'barriers' in the UK, and the growth in the use of SUDS has been associated with their progressive removal. In 2003 an initiative to resolve many of the remaining problems was promoted by the document *Framework for Sustainable Drainage Systems (SUDS) in England and Wales* (National SUDS Working Group, 2003). Certainly in some other countries there have been fewer barriers, and in the UK it is notable that progress has been more rapid in Scotland than in England and Wales.

Engineering practice

The widespread use of SUDS in place of conventional piped drainage represents a change in engineering practice. Most engineers are not afraid of change per se but they must be confident that their designs will do their job properly. Design guidance (for example by CIRIA), together with experience of successful operation of existing schemes, plays a vital role in giving engineers confidence in these approaches. (Examples of case studies of successful implementation include those presented by CIRIA (2000), and, specifically related to pervious pavements, Pratt *et al.* (2002).) The use of SUDS is promoted by the UK Building Regulations (DTLR, 2001a), which state that 'surface water drainage should discharge to a soakaway or other infiltration system where practicable'.

Adoption

As described in Section 7.4.2, if a developer constructs a piped drainage system, it is normal for the system to be 'adopted' by the sewerage undertaker who will then include the system as a revenue-earning asset and accept responsibility for operation and maintenance.

It is not clear that the same procedure is suitable for all SUDS devices because many could be considered as landscape features (rather than drainage features) and therefore more appropriately the responsibility of the local authority. A trial agreement in Scotland has been that the local authority should maintain above-ground features (like swales) and the sewerage undertaker should maintain below-ground features (like infiltration trenches).

Maintenance

Accepting responsibility to maintain SUDS devices in perpetuity is seen by some as a risk because of a lack of data on long-term performance. Possible deterioration in performance, perhaps due to blockage by silt, makes others question the appropriateness of the word 'sustainable' for such devices. (For more thoughts on this controversial word, see Chapter 24.) However all infrastructure requires maintenance. The maintenance needs of piped systems have been described in Chapter 17, and the obvious fact that SUDS also need to be maintained does not necessarily put them at a disadvantage. As Bray (2001b) argues, 'the management of SUDS schemes has been seen as a major barrier to their use in the UK although there is little evidence to support this concern. ... Where SUDS design anticipates maintenance of the system then current landscape care techniques are available to look after any development site.'

Planning

UK planning guidance in *Development and flood risk*, Planning Policy Guidance Note 25 (PPG25), DTLR (2001b), is strongly supportive of the use of SUDS in new developments. The result is that local authority planning departments expect to see SUDS as part of the plans unless there are reasons why this is not practicable, and this will affect whether planning permission is granted.

However, planning policy guidance also encourages high-density developments (PPG3: DETR, 2000), and this could be a discouragement to the inclusion of SUDS in some locations because of lack of space.

Groundwater pollution

The risk of groundwater pollution from infiltrated runoff is rightly a potential concern. The groundwater protection policy of the Environment Agency (1998a) gives detailed guidance in terms of the required aquifer protection. There is generally no objection to infiltration of roof drainage, but, depending on groundwater vulnerability, an oil interceptor may be required for other impermeable areas, or infiltration simply may not be allowed.

21.7 Other stormwater management measures

Oil traps

Oil interceptors are underground chambers with compartments designed to separate oil and water, and retain the oil fraction. They are usually provided for small catchments, particularly where heavy oil or petrol spills are expected, e.g. petrol stations. Outflow is routed to the drainage system.

In most types of interceptor, the flow rate of the effluent is reduced, and the lighter oil fraction separated by gravitational means. Interceptors suitable for treating stormwater should also contain a storage section for silt and suspended matter. The captured oil and sediment must be regularly removed as, otherwise, a heavy storm event may cause oil accumulated from previous storms to be discharged to the receiving water. Current designs include more efficient tilting plate separators and coalescing filters.

Details on the sizing of interceptors, related to level of pollution, volume of effluent, size of catchment and rainfall intensity/duration is available in the literature (e.g. Environment Agency, 1998b).

Constructed wetlands

Constructed or *artificial wetlands* (including reed beds, reed marshes and vegetative systems) are shallow areas of excavated land filled with earth, rock or gravel, saturated with water or covered by shallow flowing water at some time during the growing season, and planted with selected aquatic plants. The key roles of the plants are to transmit oxygen from the atmosphere to the root system (thus the soil) and to encourage microbial growth. Wetlands require relatively large areas of flat to gently sloping (less than about 5%) land.

It is generally agreed that wetlands are simple and inexpensive to build, but there is some disagreement on their ease of operation. In order to remain effective, wetlands need long-term programmes of maintenance, involving the planting and extraction of the aquatic plants. It has been estimated that a wetland has a life of about 15–20 years.

In addition to substantially attenuating and reducing runoff flows, wetlands can significantly improve water quality by removing large quantities of particulate and soluble contaminants (including SS, metals, excess nutrients and bacteria) through biological action and sedimentation. Wetlands also trap silt, and promote recovery of DO in the effluent. Results show that wetlands can, on an annual basis, remove almost all bacteria and suspended solids, and just under half the total phosphorus and nitrogen load (Ellis, 1992). They also have valuable secondary uses, in providing wildlife habitats and recreational/educational areas. However, wetlands are sensitive and must be carefully managed to avoid vegetation die-off.

Due to the complexity of the processes present, wetlands are not fully understood. However, the basic wetland permanent pool will typically be sized at $3V_t$. Depth of water is typically 0.5–0.75 m. A sediment removal forebay is normally included. Further details are given in Nuttall (1997).

Non-structural measures

Stormwater quality can be affected not only by the use of engineered control devices, but also by management practices. This can include altering current maintenance programmes (e.g. street sweeping, gully pot cleaning – see Section 16.7) or discouraging or preventing inappropriate practices (e.g. in pesticide management, chemical storage).

Public education

SUDS are far more obviously part of the urban landscape than underground pipe systems, and therefore, to be successful, they must be welcomed by the public. Amenity benefits are only genuine if they are perceived as such by the public. A survey in Scotland of public attitudes to SUDS (Apostolaki *et al.*, 2001) found that public opinion varied according to the type of device, with more positive attitudes towards ponds than swales, for example. The only real concerns over ponds were over safety aspects. A low understanding of the purpose of swales was thought to contribute to the negative attitudes.

Public education has the potential to increase the popularity and acceptance of SUDS. It can also help protect the stormwater system and the environment by discouraging irresponsible disposal of contaminants such as car oil, anti-freeze and domestic chemicals.

Problems

21.1 What are SUDS, and why is their use being promoted?

21.2 Outline the range of SUDS devices, and describe how the main types of device operate.

21.3 What is meant by the term 'surface water management train'?

21.4 A 0.9 m wide, 1.0 m effective depth infiltration trench is required to drain an impermeable area of 120 m^2. The trench fill material is known to have a porosity of 0.33, but the soil infiltration rate can only be estimated as 25–50 mm/h. Calculate the required length of trench. Rainfall statistics are the same as those used in Example 21.1.
[10.1 m]

21.5 Explain the function and operation of constructed wetlands in stormwater management.

21.6 What non-structural methods are available to improve the quality of stormwater from residential developments? What are their benefits compared with structural measures?

Key sources

CIRIA (2000) *Sustainable Urban Drainage Systems – Design Manual for Scotland and Northern Ireland*, CIRIA C521 and *Sustainable Urban Drainage Systems – Design Manual for England and Wales*, CIRIA C522.

EA/SEPA (1998) *Nature's Way. A Guide to Surface Water Best Management Practices*, Environment Agency/Scottish Environmental Protection Agency.

Ellis, J.B. (1989) The management and control of urban runoff quality. *Journal of the Institution of Water and Environmental Management*, 3, April, 116–124.

Leonard, O.J. and Sherriff, J.D.F. (eds) (1992) *Scope for Control of Urban Runoff. Volume 3: Guidelines*, Report R124, CIRIA, London.

Maskell, A.D. and Sherriff, J.D.F. (eds) (1992) *Scope for Control of Urban Runoff. Volume 2: A Review of Present Methods and Practice*, Report R124, CIRIA, London.

Pratt, C.J. (1995) A review of source control of urban stormwater runoff. *Journal of the Chartered Institution of Water and Environmental Management*, 9, April, 132–138.

Urbonas, B. and Strahe, P. (1993) *Stormwater Best Management Practice and Detention for Water Quality, Drainage and CSO Management*, Prentice Hall.

References

Abbott, C.L., Weisgerber, A. and Woods Ballard, B. (2003) Observed hydraulic benefits of two UK permeable pavement systems. *Proceedings of the 2nd National Conference on Sustainable Drainage*, Coventry University, June, 101–111.

Apostolaki, S., Jefferies, C. and Souter, N. (2001) Assessing the public perception of SUDS in two locations in Eastern Scotland. *Proceedings of the 1st National Conference on Sustainable Drainage*, Coventry University, June, 28–37.

Beale, D.C. (1992) Recent developments in the control of urban runoff. *Journal of the Institution of Water and Environmental Management*, **6**, April, 141–150.

Bettess, R. (1996) *Infiltration Drainage – Manual of Good Practice*, Report R156, CIRIA, London.

Bray, R. (2001a) Environmental monitoring of sustainable drainage at Hopwood Park motorway service area M42 junction 2. *Proceedings of the 1st National Conference on Sustainable Drainage*, Coventry University, June, 58–70.

Bray, R. (2001b) Maintenance of sustainable drainage – experience on two EA demonstration sites in England. *Proceedings of the 1st National Conference on Sustainable Drainage*, Coventry University, June, 93–104.

Building Research Establishment (1991) *Soakaway Design*, Digest 365.

CIRIA (2001) *Sustainable Urban Drainage Systems – Best Practice Manual for England, Scotland, Wales and Northern Ireland*, CIRIA C523.

Colwill, D.M., Peters, C.J. and Perry, R. (1984) *Water quality in motorway run-off*, TRRL Supplementary Report 823, Transport and Road Research Laboratory.

Davies, J.W., Forster, P., Millerick, A., Pratt, C.J., Higgins, P. and Scott, M.A. (2001) The potential for retrofitting of source control technologies to limit the impact of discharges through traditional urban pipe drainage systems. *Novatech 2001, Proceedings of the 4th International Conference on Innovative Technologies in Urban Drainage*, Lyon, France, June, 673–680.

Davis, L.D. (1963) The hydraulic design of balancing tanks and river storage ponds. *Journal of Institution of Municipal Engineers*, **90**(1), 1–7.

DETR (2000) Planning Policy Guidance Note 3: Housing.

DTLR (2001a) The Building Regulations 2000, Approved Document H, Drainage and waste disposal. The Stationery Office.

DTLR (2001b) Planning Policy Guidance Note 25: Development and flood risk.

Ellis, J.B. (1991) The design and operation of vegetation systems for urban source runoff quality control. *Proceedings of Standing Conference on Stormwater Source Control: Quantity and Quality*, Vol III, Coventry.

Ellis, J.B. (1992) Quality issues of source control. *Proceedings of CONFLO 92: Integrated Catchment Planning and Source Control*, Oxford.

Environment Agency (1998a) *Policy and Practice for the Protection of Groundwater*. The Stationery Office.

Environment Agency (1998b) *Use and Design of Oil Separators in Surface Water Drainage Systems*, Pollution Prevention Guidance, PPG 3.

Fletcher, T.D., Peljo, L., Fielding, J., Wong, H.F. and Weber, T. (2002) The performance of vegetated swales for urban stormwater pollution control. *Global*

Solutions for Urban Drainage: Proceedings of the 9th International Conference on Urban Drainage, Portland, Oregon, September, on CD-ROM.

Kellagher, R., Woods Ballard, B. and Weisgerber, A. (2003) Modelling of pervious pavements. *Proceedings of the 2nd National Conference on Sustainable Drainage*, Coventry University, June, 91–99.

Lawrence, A.I., Marsalek, J., Ellis, J.B. and Urbonas, B. (1996) Stormwater detention and BMPs. *Journal of Hydraulic Research*, **34**(6), 799–814.

Macdonald, K. and and Jefferies, C. (2001) Performance comparison of porous paved and traditional car parks. *Proceedings of the 1st National Conference on Sustainable Drainage*, Coventry University, June, 170–181.

Mance, G. and Harman, M.N.I. (1978) The quality of urban stormwater runoff, in *Urban Stormwater Drainage* (ed. R.P. Helliwell), Pentech Press, 603–618.

Martin, P., Turner, B., Waddington, K., Pratt, C., Campbell, N., Payne, J. and Reed, B. (1999) *Sustainable Urban Drainage Systems – Design Manual for Scotland and Northern Ireland*, Report C521, CIRIA, London.

National SUDS Working Group (2003) *Framework for Sustainable Drainage Systems (SUDS) in England and Wales.* May.

NRA (1992) *Policy and Practice for the Protection of Groundwater*, National Rivers Authority, Bristol.

Nuttall, P.M. (1997) *Review of the Design and Management of Constructed Wetlands*, Report 180, CIRIA, London.

Pratt, C.J., Newman, A.P. and Bond, P.C. (1999) Mineral oil bio-degradation within a permeable pavement: long term observations. *Water Science and Technology*, **39**(2), 103–109.

Pratt, C.J., Wilson, S. and Cooper, P. (2002) *Source Control Using Constructed Pervious Surfaces – Hydraulic, Structural and Water Quality Performance Issues.* CIRIA C582.

Rommel, M., Rus, M., Argue, J., Johnston, L. and Pezzaniti, D. (2001) Carpark with '1 to 1' (impervious/permeable) paving: performance of 'Formpave' blocks. *Novatech 2001, Proceedings of the 4th International Conference on Innovative Technologies in Urban Drainage*, Lyon, France, June, 807–814.

Schlüter, W. and Jefferies, C. (2001) Monitoring the outflow from a porous car park. *Proceedings of the 1st National Conference on Sustainable Drainage*, Coventry University, June, 182–191.

Startin, J. and Lansdown, R.V. (1994) Drainage from highways and other paved areas: methods of collection, disposal and treatment. *Journal of the Chartered Institution of Water and Environmental Management*, **8**, Oct, 518–526.

Stovin, V.R. and Swan A.D. (2003) Application of a retrofit SUDS decision-support framework to a UK catchment. *Proceedings of the 2nd National Conference on Sustainable Drainage*, Coventry University, June 171–180.

Swan, A., Stovin, V., Saul, A. and Walker, N. (2001) Modelling SUDS with deterministic urban drainage models. *Proceedings of the 1st National Conference on Sustainable Drainage*, Coventry University, June, 202–213.

22 Integrated management and control

22.1 Introduction

This chapter is concerned with the likely future direction of the management of urban drainage systems. Modern approaches not only consider the system as a whole – for example, the effect of upstream inflows or CSO settings on downstream capacity; they also consider the processes that interact with the drainage system – rainfall or water use at one end, and wastewater treatment and river quality at the other, and how they may be controlled.

A practical outworking of this philosophy is the UK Urban Pollution Management procedure. This represents the best in current and future practice. However, research must move ahead of practice, and new developments are continually being proposed and explored. This chapter considers the real-time control of urban drainage systems – well understood, but not yet widely practised in the UK. Technological and methodological developments now make it possible to model and optimise the effect of control on the urban wastewater system as a whole, and this is also described.

End-of-pipe wastewater treatment is not included in this book, but this chapter does cover state-of-art thinking on in-sewer wastewater treatment. In accord with the rest of the chapter, an integrated approach is advocated.

22.2 Urban Pollution Management

Consideration of the system as a whole is the basis of a practical approach for dealing with wet weather discharges from sewer systems, named (in the UK) Urban Pollution Management.

Pollution is identified in the *Sewer Rehabilitation Manual* (WRc, 2001) as one of the three main types of problem that can be solved by rehabilitation of existing sewer systems (Chapter 18). The *SRM* contains thorough coverage of methods of solving the other two types of problem: structural and hydraulic, but less on pollution problems.

A major programme of research was sponsored jointly by the water companies and the water regulator (NRA/EA) in the late 1980s and early 1990s. It was aimed at providing guidance on the solution of wet weather pollution problems, particularly by creating modelling tools, and the outcome was the Urban Pollution Management (UPM) procedure. This is set out in the *UPM Manual*, first released in 1994, and subsequently updated in a second edition (Foundation for Water Research, 1998).

The *UPM Manual* sets out a methodical procedure for dealing with pollution problems caused by sewer systems in four stages:

- initial planning
- data collection and modelling
- development of solutions and testing for compliance
- obtaining approval and detailed design.

There are three recurring themes in the guidance. The first is that analysis should be *holistic*, covering all elements in the system that determine the pollution impact of a sewer system: rainfall, the sewer system itself, the wastewater treatment works and the receiving river. The second is that the level of detail of any study, and in particular of the models used, should be appropriate and that, in the right circumstances, a holistic approach may also be *simple*. Third, the approach should be underpinned by relevant *environmental standards* with models able to demonstrate compliance with those standards.

For a detailed study, the UPM procedure recommends the creation, verification and application of physically-based deterministic models of each of the main elements of the urban wastewater system. Much of the original development work in the UPM programme was aimed at ensuring that modelling packages were in place to cover these elements. The second edition provides generic advice on model requirements and capabilities.

Such a modelling undertaking can involve a large commitment of resources, as should be apparent from the discussion of just one element – sewer-flow quality modelling – in Chapter 20. The UPM procedure is pragmatic, however, and there is guidance on the decision-making involved in choosing the appropriate level of detail.

The model *SIMPOL*, which is fully defined and supported within the Manual, offers a simplified approach intended specifically for integrated wastewater planning (Crabtree *et al.*, 2003). *SIMPOL* is a spreadsheet-based model covering the complete system in a highly simplified form, as a series of storage volumes and throttles. It also provides a simplified representation of river quality impacts. It is recommended that *SIMPOL* (or any other simplified model) should be used for analysis of simple problems, or, for more complex cases, should be calibrated against a small number of runs by detailed models, and then used by itself to look at a wide range of further conditions at low cost (Dempsey *et al.*, 1997).

The *UPM Manual* gives guidance on all stages of the procedure, starting with assessment of CSO performance and effect on river quality. It provides appropriate standards for intermittent discharges (covered in Chapter 3 of this book). It gives detailed advice on the four elements of modelling: rainfall, sewer quality, wastewater treatment and river impact. It gives guidance on applications of the procedure for cases of discharges to rivers and to bathing water, and gives a number of examples for each case. The Manual also contains information on engineering tools for achieving solutions, for example ensuring effective operation of CSOs, including the use of screens and storage (covered in Chapter 12).

22.3 Real-time control

Most urban drainage systems are managed in an essentially passive mode. That is, fixed elements are provided to operate in one way only without the opportunity for intervention of any kind. Systems are conventionally designed (see Chapter 11) to handle flows from storms of high return period (low frequency) and, therefore, contain large amounts of storage volume (even without dedicated stormwater detention). For most storm events, this storage is not fully utilised. So there should be significant scope to improve the performance of the system by actively managing this storage in response to changing input (e.g. rainfall) and output (e.g. flooding, overflows, interaction with treatment plant).

22.3.1 Definition

Schilling *et al.* (1989) consider an urban drainage system to be operated under real-time control (RTC) when 'process data, which is currently monitored in the system, is used to operate flow regulators during the actual process'. Thus, system information (e.g. rainfall, water level) is continuously collected and processed and is used to make decisions about operation of devices (e.g. pumps, weirs) in the system in real-time to limit the occurrence of adverse effects.

22.3.2 Equipment

The main hardware elements to be found in RTC systems are:

- *sensors* that monitor the ongoing process
- *regulators* that manipulate the process
- *controllers* that activate the regulators
- *data transmission systems* that carry measured data from sensors to controllers and signals from controllers to regulators.

Together, these four elements form the *control loop* that is common to all RTC systems (considered in Section 22.3.3).

Sensors

Although there is potentially a large number of sensor types, only a few are suitable for RTC of drainage systems. These include rain gauges, water level gauges, flow gauges and limit switches. Requirements include the suitability for continuous recording, remote data transmission, robustness and reliability.

Level measuring devices are the most commonly used sensors. They are indispensable for determining the state of storage devices or converting depths to flow rates. Use of water quality sensors, though showing considerable potential, is still in its infancy. Further development of these sensors is needed.

Regulators

Sewer flow regulators include a variety of pumps (constant or variable speed), moveable weirs and penstocks (see Chapters 9 and 14). Some of the more specialised devices are described by Schilling *et al.* (1989).

Controllers

A controller is required for every regulator. It accepts an input signal and adjusts the regulator accordingly. Controllers can be broadly classified into two categories: continuous and discrete. The most common method of discrete control is the two-point controller that has only two positions: on/off or open/closed. An example of two-point control is shown in Fig. 22.1 for a simple pumping station. Here, the pump switches off at a low level and on at a high level as shown.

The disadvantage of two-point control is that the frequency of switching can become excessive. Three-point controllers have been developed to overcome this problem and they are typically used for regulators such as

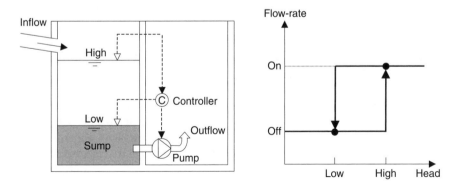

Fig. 22.1 Two-point control at a pumping station

automatic penstocks and moveable weirs. The most commonly used controller for continuously variable regulator settings is the *PID* (proportional-integral-derivative) *controller*. Simplified versions (P, PI and PD) are also available.

Data transmission systems

Some form of data transmission (telemetry) system is required for RTC systems, either transmission by wire (e.g. public telephone) or wireless transmission. Digital data transmission is replacing analogue systems.

22.3.3 Control

Classification

RTC systems can be broadly classified as local, global or integrated. Under local control, regulators use process measurements taken directly at the regulator site (e.g. by a float) but are not remotely manipulated from a control centre, even if operational data is centrally acquired. An example is an automatic penstock being operated in relation to water level monitored by an ultrasonic level monitor. In global systems, regulators are operated in a co-ordinated way by a central computer with respect to process measurements throughout the entire system. For example, mobilisation of upstream and downstream storage volumes is linked in order to avoid emptying the upper tank into the already-full lower tank. The newest class (and least proven in practice) is that of integrated control where control can be influenced by process measurements derived from systems outside the sewer system (e.g. WTP, river).

Control loop

The control loop is a basic element of any RTC system. In this loop, measured values of the controlled variable are compared with the value of the *set point*. The outcome of this comparison then determines how the variable will be adjusted. Two main types of control loop can be distinguished (see Fig. 22.2): feedback and feed forward.

- Feedback control: control commands are actuated depending on the measured deviation of the controlled process from the set point. Unless there is a deviation, the feedback controller is not actuated.
- Feed forward control: this anticipates the immediate future values of these deviations using a model of the process controlled, and activates the controls ahead of time to compensate. Its accuracy therefore depends on the effectiveness of the model.

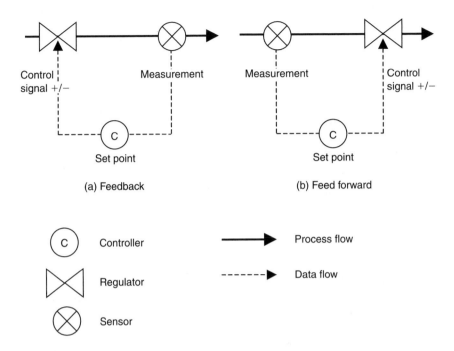

Fig. 22.2 Control loop

Control strategy

As we have seen, controllers adjust regulators to achieve minimum deviations of the regulated process variable (e.g. flow, level) from the set points.
A 'control strategy' can be defined as either the time sequence of all regulator set points, or the set of control rules in a RTC system (Schilling *et al.*,
1989). Strategies can either be defined as *off-line* where fixed rules are
devised or *on-line* where rules are continuously updated depending on fast
computer forecasts of the system state. Clearly, the simplest strategy is to
keep the set points constant, but time-varying set points will probably give
better performance, allowing the system to react to the non-regular transient storm events to which it is subjected.

   Information collection and interpretation is a vital part in the processing of any control strategy. Historical data is very useful, but forecast
information can be even more valuable to allow the system to become
ready for the expected load. Thus, possible sources of information are
(Van de Ven *et al.*, 1992):

• Flow, level and quality measurements in upstream sewers: system reaction must be within the time of flow
• rainfall measurements and results from rainfall/runoff models: avail-

able reaction time is extended to the time of concentration of the catchment

- rainfall forecasts: these gain additional time depending on the forecast time horizon.

In the utilisation of such data, care should be taken since measurements will include errors.

A control strategy can be developed in various ways. A useful first approach is to incorporate the experience of operations personnel. Alternatively or additionally, a trial-and-error (heuristic) approach can be taken. By specifying an initial control strategy (e.g. the default, fixed set point strategy) and then performing multiple simulation runs, the initial strategy can usually be improved.

General strategies are:

- preferential upstream storage: wastewater/stormwater is retained first in the upper reaches of the network to reduce downstream flooding impacts
- preferential downstream storage: wastewater/stormwater is retained first in the lower reaches of the network to minimise upstream CSO impacts
- balanced storage: the various storage elements are evenly filled throughout the catchment.

Although experienced operators can achieve near optimum results, the process of gaining experience is a lengthy one. The experience gained can be stored and formalised in a computer-based expert system.

Alternatively, decision matrices or control scenarios may be formulated. *Decision matrices* are tables containing control actions for all possible combinations of the systems inflow and state variables. *Control scenarios* are similar in that they consist of a set of instructions that may be presented in the form of 'if ... then ... else' rules (e.g. Almeida and Schilling, 1993). Relatively simple rules can be amended heuristically to improve system performance further.

More rigorous approaches can also be taken to strategy development that rely on mathematical optimisation techniques. In these methods, operational objectives are translated into the minimisation of an 'objective function' subject to (physical) constraints. For example, a simple RTC optimisation procedure would be to minimise the sum of all CSO discharge volumes V_i over a time horizon t_i to t_f:

$$\sum_{i=t_i}^{t_f} V_i \rightarrow \min$$

The aim is to identify the optimum solution and, therefore, the 'best' performance of the system, given all the constraints and assumptions.

22.3.4 *Comparison*

RTC options can be considered as part of the drainage area studies referred to in Chapters 18 and 19. Most urban drainage simulation programs allow incorporation of control strategies, but with varying degrees of sophistication, allowing comparison of 'conventional' static solutions and various RTC options. The resulting model should be subject to the usual calibration and verification exercise.

The various alternatives can be characterised by their specific performance, and a cost–benefit analysis carried out. The decision can then be taken as to whether or not to implement an RTC system. Whichever alternative is chosen, the risk of failure should not exceed the risk level in the existing system.

22.3.5 *Applicability*

Perhaps the greatest single indicator of the benefit to be derived from RTC of a particular catchment is the magnitude of the useful storage. In under-designed systems, with low storage volumes, control will be of little benefit. On the other hand, in an over-designed system with large storage volumes, flooding and CSO discharges will be infrequent anyway and little additional benefit will accrue. Properly designed systems with sufficient but not excessive storage that can be 'activated' should produce the best results. Appropriate distribution of storage around the catchment is also important.

A second major factor is the hydraulic load to which the network is subjected. Assuming storage volume is fixed, RTC will provide no benefit for minor storm events that are handled effectively by the passive system. For large storm events, the benefits are more limited because, once all the storage capacity is fully utilised, the only remaining option is to allow discharges to the receiving water. However, even for large events, RTC offers some potential to more evenly distributed CSO discharges and to better control the first foul flush (see Chapter 12). RTC is, however, *most* effective at reducing the frequency of CSO spills from smaller but still significant storms. These minor spills are common occurrences in conventional passive systems where controls are pre-set at the design stage (often many years earlier).

Other network characteristics that favour RTC application are:

- spatially distributed inputs (rainfall)
- spatially distributed storage
- larger, flatter, more looped sewer networks
- many controllable elements (e.g. storage tanks, pumps, overflows).

22.3.6 Benefits/drawbacks

RTC has the following major benefits (WRc, 1998; Schilling, 1996):

- reduction in the risk of flooding by utilising the full storage of the system
- reduction in pollution spills by detaining more wastewater within the system
- reduction in capital costs by minimising the storage and flow-carrying capacity requirements of the system
- reduction in operating costs by optimising pumping and maintenance costs
- enhancement of WTP performance by balancing inflow loads and allowing the plant to operate closer to its design capacity.

Indirect benefits (Kellagher, 1996) include flexibility to respond to changes in the catchment or local failure in the system, and better understanding of the performance of the network (see Section 17.7).

Typical RTC benefits include large reductions in CSO spills at the most sensitive locations, reduced frequencies of overflow operation (by about 50%) and reduced annual CSO volumes (by 10–20%). Secondary benefits include lower energy costs (from less pumping), improved wastewater treatment, control of sewer sediments, and better system supervision, understanding and record keeping (Nelen, 1993). However, operational experience and verified benefits have not been widely reported.

Drawbacks are relatively few, and reluctance to use RTC more widely seems to rest mainly on lack of operational experience. However, there are also legitimate concerns regarding the increased maintenance commitment needed and issues surrounding the risk of failure. Currently, in the UK, local control is very common but globally-controlled systems are rare.

22.4 Integrated modelling

Conventional practice has been to operate and, therefore, model the various components of the engineered urban wastewater cycle (urban drainage, WTP and receiving water body) in isolation. Each component has been engineered to meet the needs of its users and the environment, but with little feedback or cross-reference to other components. Although at a research level the importance of the interactions has been realised for some time, the necessary means to quantify them has not been available until recently. These means are powerful and widely accessible computer hardware coupled with deterministic models of each of the elements of the integrated system.

22.4.1 *Why do we need integrated models?*

The need for an integrated model of the system is similar to the need for a model of any of the individual elements. In addition, representing and understanding the urban system as a whole potentially allows better, more cost-effective solutions to be engineered.

The urban wastewater system as a whole contains numerous elements that can be utilised to prevent water pollution. For example, there is little point in 'wasting' storage volume in tanks and pipes, or treatment capacity, on weak wastewater. If the available capacity could primarily be used to capture the most polluted flows, the potential for pollutant discharge reduction can be increased. Storage other than in the urban drainage system may also be exploited, for example, the time lag of processes in the WTP and the receiving water. If the WTP and CSOs discharge into the same receiving water, carefully-timed release of effluents should minimise overall pollutant discharges.

22.4.2 *Integrated control*

With the advent of modelling tools that can represent the dynamic hydraulic and water quality processes within the entire system, opportunities are created for controlling the performance of the system as a whole. Therefore 'integrated control', with the objective of minimising detrimental impacts on the receiving water, becomes possible and is characterised by two aspects (Schütze *et al.*, 1999).

- *Integration of objectives:* control objectives within one subsystem may be based on criteria measured in other subsystems (e.g. operation of pumps in the drainage system directed at minimising oxygen depletion in the receiving water body).
- *Integration of information:* control decisions taken in one subsystem may be based on information about the state of other subsystems (e.g. operation of pumps in the drainage system based on WTP effluent data).

Real-time control of the components of the integrated system, therefore, takes into account the state of the whole system when utilising the control devices available in order to reach operational objectives defined for any location in the system. Table 22.1 summarises some of the measures and objectives of control, as well as methods applied to determine control strategies.

Fig. 22.3 shows diagramatically how individual sub-systems can be placed in an integrated framework to include control and optimisation. Here, the integrated model has been applied to a catchment and various degrees of control simulated. These include a base case with local control

Table 22.1 Components of control of the urban wastewater system (after Schütze
et al., 1999)

Subsystem	Devices	Objectives	Decision-finding methods
Sewer system	• Pumps • Weirs • Gates	• Prevention of flooding • CSO reduction (frequency, volumes, loads)	• Heuristics, intuition • Self-learning expert system • Off-line optimisation
Treatment plant	• Weirs, gates • Return sludge rate • Waste sludge rate • Aeration	• Equalisation of flows • Maintenance of effluent standards • Process maintenance	• On-line optimisation • Model-based control • Application of control theory
Receiving river	• Weirs • Gates	• Improved water quality • Flood protection	

only, an optimised variation of this with fixed set points, and an example
of simple integrated control. The performance of each of the control
scenarios is optimised to allow fair comparison.

Fig. 22.4 illustrates the effects of these scenarios in terms of:

- the duration (% of run time) of the oxygen concentration below a
  4 mg/l threshold value at any location in the receiving water (a river in
  this case)
- the minimum DO concentration in the river during the simulated time
  period.

It can be seen that control with optimised fixed set points leads to
improved performance over the base case. Further improvement is
achieved for the integrated control scenario.

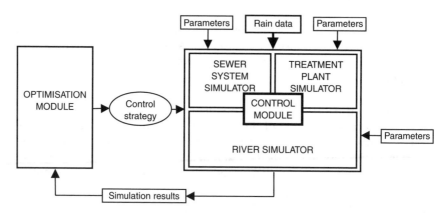

Fig. 22.3 Overview of the 'Integrated Simulation and Optimisation Tool'
(reproduced from Schütze *et al.* [1999] with permission of publishers
Pergamon Press and copyright holders IAWQ)

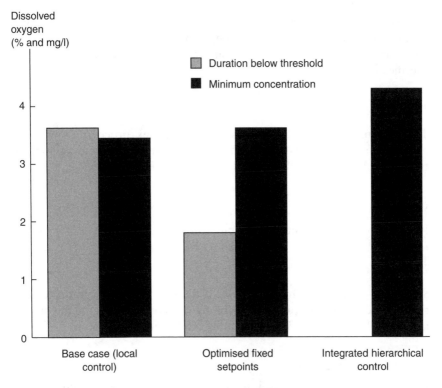

Fig. 22.4 Effect of control scenario on river DO (adapted from Schütze *et al.* [1999] with permission of publishers Pergamon Press and copyright holders IAWQ)

22.5 In-sewer treatment

Historically, the sewer network has been conceived solely as a transportation device, and designed in accordance with the principles of hydraulics. However, it is clear that the prevailing conditions in many systems are suited to provide at least partial treatment of the flow. This leads to the conclusion that systems might be deliberately designed not only to transport, but also to treat the wastewater they convey.

In-sewer treatment should, therefore, be viewed either as an alternative or a complement to traditional end-of-pipe treatment systems or the newer source control methods described in Chapter 21. If correctly designed, substantial benefits should accrue: not only in terms of reduced costs but also in environmental improvement.

22.5.1 Transformations

A sewer acts as a plug flow reactor with a system retention time that may equal or exceed that at the WTP, depending on the configuration of the network. A number of transformation processes occur, even without being specifically engineered.

Physical

The movement of particulate matter is a complex process involving (in addition to sediment transport) particle degradation, mixing, agglomeration, flocculation and turbulent buffering (Crabtree *et al.*, 1991).

Chemical

Processes include dissolution, precipitation and hydrolysis.

Biochemical

Suspended biomass and wall biofilm (slime) act on the degradable fractions of the wastewater, both hydrolysing the slowly biodegradable material and biologically oxidising the readily biodegradable material.

22.5.2 Self-purification

Early work by Pomeroy and Parkhurst (1973) found that BOD_5 was reduced by 26% from 192 to 141 mg/l in a naturally well-aerated gravity trunk sewer with a detention time of 4 h. Thomas *et al.* (1985) noted a 20% COD reduction in a well-ventilated and aerated system. Studies on a long sewer in Portugal, which has many drops and bends, indicated average dissolved COD removal of 19% (3% per kilometre) with slightly lower total COD removal (Almeida and Butler, 1999). Balmer and Tagizadeh-Nasser (1995) estimate that a substantial fraction of the organic matter in the wastewater will be oxidised before reaching the WTP, provided there is 3–10 m length of foul sewer *per capita*.

These findings are corroborated by laboratory reactor studies (Malik, 1998) and laboratory channel experiments (Cao and Alaerts, 1995; Manandhar and Schroder, 1995), where even greater removal efficiencies are obtained.

The potential of the piped system to act as a treatment process is clear, but, by actively promoting certain conditions, its potential can be enhanced.

22.5.3 Treatment methods

The idea of utilising the sewer as a treatment system or part of the overall treatment of wastewater has been discussed and, to a limited extent, practised for several decades (e.g. Pomeroy, 1959; Boon *et al.*, 1977; Raunkjaer *et al.*, 1995). The most feasible process option appears to be aerobic biological treatment and, in an attempt to increase its efficiency, three strategies have been adopted: air/oxygen addition, enhancement of the attached biomass and seeding with fresh wastewater or activated sludge.

Oxygen addition

Oxygen is normally transferred from the atmosphere to the wastewater in gravity sewers. In small sewers, this natural aeration should be sufficient to maintain aerobic conditions. However, in large trunk sewers with slack gradients, the wastewater micro-organisms are likely to become oxygen limited. Intensifying turbulence can enhance surface aeration and, hence, biological activity. This can be achieved either by increasing the mean flow velocity (requiring higher pipe gradients) or by introducing backdrops or other flow discontinuities (Almeida *et al.*, 1999).

As described in Chapter 16, air or oxygen injection is a common way of reducing septicity problems associated with pressure mains. It has the positive by-product of enhancing the oxidation of biodegradable material. For example, as early as 1959, Pomeroy found that air injected in a rising main for sulphide control reduced the BOD_5 of the wastewater by 44%. Reductions of 30–55% and 30–75% for total and soluble BOD_5 respectively were noted by Tanaka and Takeneka (1995) in a 6–7 h detention time main. Even higher removal rates have been noted for locations where pure oxygen addition is practised (e.g. Boon *et al.*, 1977).

Biofilm enhancement

Cao and Alaerts (1995) have shown that the relative importance of suspended and attached biomass is dependent on the ratio of biofilm area to wastewater volume; biofilm being more important in small pipes. In gravity sewers, additional surfaces could be introduced into the flow to provide sites for biofilm growth. The main drawback with this is the additional headlosses that would be generated.

In rising mains, using several small pipes in place of one large one will increase biofilm area but, again, more head loss will be generated. The rate of biodegradation in pressure mains is lower due to the anaerobic conditions, but the biofilm plays a relatively bigger role than in gravity sewers and this can increase the overall conversion rates (Nielson *et al.*, 1992).

Biomass seeding

In this approach, a small activated sludge plant could be built at the head of a long outfall sewer. Surplus sludge could then be added to the sewer to increase the biomass concentration in the flow, and hence the process intensity in the system. Alternatively, the waste sludge from a downstream WTP could be recycled to the head of the sewer, or the sewer itself could be used as a step-feed activated sludge reactor as suggested for Tel Aviv by Green *et al.* (1985). Laboratory studies to model this latter approach showed dissolved COD removal efficiencies could be as high as 80–90%.

Problems

22.1 Outline the basic philosophy and main elements of the procedure presented in the *UPM Manual*.
22.2 Define real-time control, and describe the main hardware components needed to implement it in an urban drainage system.
22.3 What are the advantages and disadvantages of urban drainage RTC? In what situations is RTC likely to be beneficial?
22.4 Explain how you would go about developing an RTC strategy and give an example of a possible practical approach. What sources of data would you need?
22.5 List the main parts of the urban wastewater system and consider ways in which they interact with each other. What are the potential benefits of integrated system control?
22.6 Indicate the main in-sewer transformation processes and their importance. Discuss how these could be engineered to enhance wastewater treatment.
22.7 What are the benefits of in-sewer wastewater treatment?

Key sources

FWR (1998) *Urban Pollution Management Manual*, 2nd edn, Foundation for Water Research, FR/CL0009.
Inman, J.B. (1979) Sewage and its pretreatment in sewers. Chapter 7 in *Developments in Sewerage – 1* (ed. R.E. Bartlett), Applied Science.
Marsalek, J. and Schilling, W. (1998) Operation of sewer systems, in *Hydrodynamic Tools for Planning, Design, Operation and Rehabilitation of Sewer Systems* (eds J. Marsalek, C. Maksimovic, E. Zeman and R. Price), NATO ASI Series 2. Environment, Vol 44, Kluwer Academic, 393–414.
Nielson, P.H., Raunkjaer, K., Norsker, N.H., Jensen, N.A. and Hvitved-Jacobsen, T. (1992) Transformation of wastewater in sewers – a review. *Water Science and Technology*, 25(6), 17–31.
Schilling, W. (ed.) (1989) *Real-Time Control of Urban Drainage Systems. The State-of-the-Art.* IAWPRC Scientific and Technical Reports No 2. Pergamon Press.

Schütze, M.R., Butler, D. and Beck, M.B. (2002) *Modelling, Simulation and Control of Urban Wastewater Systems*, Springer-Verlag.

References

Almeida, M. do C. and Butler, D. (1999) Aerobic biodegradation in a prototype sewer, in *EWPCA Symposium on Sewage Systems – Cost and Sustainable Effective Solutions*, Munich, Germany.

Almeida, M. do C. and Schilling, W. (1993) Derivation of If … Then … Else rules from optimized strategies for sewer systems under real-time control, in *Proceedings of 6th International Conference on Urban Storm Drainage*, Niagara Falls, Canada, 1525–1530.

Almeida, M. do C., Butler, D. and Matos, J.S. (1999) Reaeration by sewer drops, in *Proceedings of 8th International Conference on Urban Storm Drainage*, Sydney, Australia 738–745.

Balmer, P. and Tagizadeh-Nasser, M. (1995) Oxygen transfer in gravity flow sewers. *Water Science and Technology*, 31(7), 127–135.

Boon, A.G., Skellett, C.F., Newcombe, S., Jones, J.G. and Foster, C.F. (1977) The use of oxygen to treat sewage in a rising main. *Water Pollution Control*, 76(1), 98–112.

Cao, Y.S. and Alaerts, G.J. (1995) Aerobic biodegradation and microbial population of a synthetic wastewater in a channel with suspended and attached biomass. *Water Science and Technology*, 31(7), 181–189.

Crabtree, R.W., Ashley, R.M. and Saul, A.J. (1991) *Review of Research into Sediments in Sewers and Ancillary Structures*, Foundation for Water Research, Report No FR0205.

Crabtree, R.W., Dempsey, P., Clifforde, I.T., Quinn, S., Henderson, B. and Wilson, A. (2003) Integrated modelling of urban watercourses. *Proceedings of the Institution of Civil Engineers, Water and Maritime Engineering*, 156 (WM3), 265–274.

Dempsey, P., Eadon, A. and Morris, G. (1997) SIMPOL: A simplified urban pollution modelling tool. *Water Science and Technology*, 36(8–9), 83–88.

Green, M., Shelef, G. and Messing, A. (1985) Using the sewerage system main conduits for biological treatment. *Water Research*, 19(8), 1023–1028.

Kellagher, R. (1996) *Evaluating Real Time Control for Wastewater Systems. A Managerial and Technical Guide*. Report SR479, HR Wallingford.

Malik, M. (1998) Substrate removal by in-sewer oxygenation. *Journal of the Chartered Institution of Water and Environmental Management*, 12, February, 6–12.

Manandhar, U.K. and Schroder, H. (1995) Sewage circulating reactor – an approach to recirculating wastewater in sewers. *Environmental Technology*, 16(3), 201–212.

Nelen, F. (1993) On the potential of real-time control of urban drainage systems. *Water Science and Technology*, 27(5–6), 111–122.

Pomeroy, R.D. (1959) Generation and control of sulphides in filled pipes. *Sewage and Industrial Wastes*, 31(9), 1082–1095.

Pomeroy, R.D. and Parkhurst, J.D. (1973) Self purification in sewers, in *Advances in Water Pollution Research*, Pergamon Press, 291–308.

Raunkjaer, K., Hvitved-Jacobsen, T. and Nielsen, P.H. (1995) Transformation of organic matter in a gravity sewer. *Water and Environment Research*, **67**(2), 181–188.

Schilling. W. (1996) Potential and Limitations of Real Time Control, in *Proceedings of 7th International Conference on Urban Storm Drainage*, Hannover, Germany, 803–808.

Schütze, M., Butler, D. and Beck, M.B. (1999) Optimisation of control strategies for the urban wastewater system – an integrated approach. *Water Science and Technology*, **39**(9), 209–216.

Tanaka, N. and Takeneka, K. (1995) Control of hydrogen sulphide and degradation of organic matter by air injection into wastewater forced main. *Water Science and Technology*, **31**(7), 273–282.

Thomas, O., Blake, G., Wittenburg, M., Mazas, N. and Oudari, I. (1985) A new role for sewers: purification *in situ*. *Tribune du Cebedeau*, **38**(499/500), 13–19.

Van de Ven, F.H.M., Nelen, A.J.M. and Geldof, G.D. (1992) Urban drainage, in *Drainage Design* (eds P. Smart and J.G. Herbertson), Blackie & Son.

WRc (2001) *Sewerage Rehabilitation Manual*, 4th edn, WRc.

23 Low-income communities

23.1 Introduction

Much has been written about the plight of the urban poor and the huge health burden that they bear. The World Health Organisation estimates that 12 000 000 men, women and, in particular, children die each year from water- and excreta-related diseases. A large proportion of those falling victim to these diseases live in urban areas (Mara, 1996b). The irony is that, although these diseases account for such a high proportion of illness, debility and death among the poor, and are a substantial cause of high infant mortality rates, they are preventable. As discussed in Chapter 1, urban health in *all* countries depends on the provision of basic water infrastructure.

Providing water and sanitation needed for increasing numbers of people with limited financial investment is an enormous challenge (a challenge taken up first in the International Drinking Water Supply and Sanitation Decade (1981–1990), then by Safe Water 2000 and Health for All 2000, and most recently by the WEHAB initiative as part of the World Summit on Sustainable Development held in Johannesburg in 2002). The greatest need can often be found in the peri-urban areas where slums and shanty-towns (so-called informal or transitional settlements) proliferate in many Third World cities. These are rarely given services prior to their establishment and subsequent provision places severe financial strain on already over-stretched government resources. Yet, if these basic services can be provided, the hope is that better public health will allow an upward spiral of social and economic development, leading to increased productivity, higher standards of living and improved quality of life (Okun and Ponghis, 1975).

An important part of the 'upward spiral' of development is provision of other services, such as storm drainage and solid waste collection, and organisation of community hygiene education. If these can be deployed, both water supply and sanitation services will, in turn, function more effectively and additional health benefits will accrue. In fact, an integrated whole of municipal services (see Fig. 23.1) will bring maximum benefits.

In addition to health benefits, sanitation is also valuable in giving dignity and privacy to people, and in providing a cleaner, more pleasant living environment. Similarly, good drainage is also valuable in reducing nuisance and economic loss due to flooding.

In this chapter, the emphasis will be placed on the two urban drainage services, sanitation and storm drainage, in the context of the financial constraints of low-income communities.

23.2 Health

First in public health importance are the many faeco-oral infections where the contaminated faeces of one person are transmitted via water, hands, insects, soil or plants to other individuals. These diseases include the well-known 'waterborne' diseases such as cholera and typhoid, but also the many common diarrhoeal diseases that particularly affect young children, contributing to malnutrition and death. In fact, these diarrhoeal diseases are often the single greatest cause of child mortality (Cairncross and Ouano, 1991).

The larvae of helminths like roundworms and hookworms may be transmitted from person to person when infected faeces are left on the

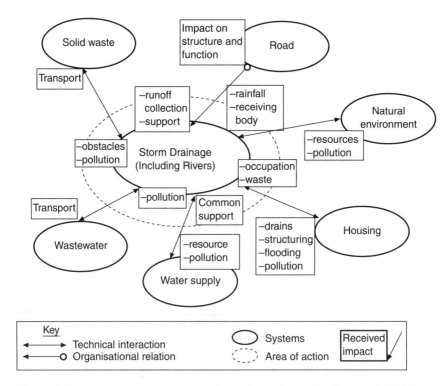

Fig. 23.1 Integrated municipal services (reproduced from Wondimu and Alfakih [1998] with permission of the authors)

ground. Children are particularly at-risk when playing or bathing in faecally-contaminated stormwater (see Fig. 23.2). Some roundworm eggs (e.g. *Ascaris*) remain viable for long periods.

A further group of diseases may be classified as 'water-related' rather than waterborne. For example, mosquitoes spread several different diseases. Their connection with water is that each species selects different types of water body in which to breed. Malaria is the best known of these diseases and is transmitted by the female *Anopheles* mosquito. These do not usually breed in heavily-polluted water, but can multiply in swamps, pools, puddles and other poorly-drained areas. *Culex* mosquitoes favour polluted water (e.g. pit latrines, blocked storm drains) and are a vector for filariasis which can lead to elephantiasis. Other mosquito species spread diseases such as dengue and yellow fever.

Another important disease is schistosomiasis (bilharzia), which can be transmitted in poorly-drained urban areas (Fig. 23.2). If standing stormwater becomes contaminated with egg-infected excreta, the microscopic larvae (miracidia) that hatch out can multiply in the bodies of small aquatic snails. From every infected snail, thousands of infective cercariae emerge and swim in the water. Local residents become infected through their skin when they stand or play in the water. Table 23.1 summarises the major diseases and their link to urban drainage.

Improved water supply, sanitation, drainage and hygiene education are important components in obstructing the transmission route of these diseases. Of course, in providing engineering works for this purpose, great care is needed to ensure that new breeding sites are not created inadvertently.

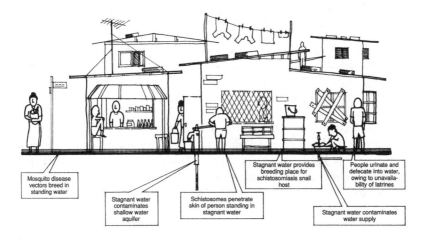

Fig. 23.2 Health implications of poor drainage (reproduced from Cairncross and Ouano [1991] with permission of WHO, Geneva)

Table 23.1 Classification of diseases linked to lack or precariousness of urban drainage (after Souza *et al.*, 2002)

Group		Disease
I	Diseases transmitted by flying vectors that can multiply in pools and wetlands	Urban yellow fever Dengue Filariasis Malaria
II	Diseases in which the etiological agent uses an intermediate aquatic host that can multiply in wetlands	Schistosomiasis
III	Diseases transmitted by direct contact of water or soil (with the presence of the hosts) – contamination is favoured by floods and wetlands	Leptospirosis
IV	Diseases transmitted by ingestion of water contaminated by etiological agents present in wetlands and floods that enter into water distribution systems	Typhoid fever (water) Cholera and other diarrhoea (water)
	Diseases transmitted by direct contact to contaminated soil – contamination is favoured by floods and wetlands	Hepatitis A (water) Ascaridiasis (water) Trichuriasis (water) Hookworm (water and soil)

23.3 Option selection

23.3.1 Sanitation

Alternative types of sanitation can be conveniently classified as 'on-site' or 'off-site'. In on-site methods, the excreta storage/treatment is in or near to the individual dwelling. Off-site systems remove the excreta from the dwelling for disposal. Additionally, systems can be designated as 'dry' or 'wet' with wet systems using water to transport the excreta (see Table 23.2).

The final choice between on- and off-site systems will normally be a financial one. In low-density, low-income settlements, on-site systems will almost certainly be the most cost-effective option. In higher density areas, the feasibility of using on-site methods becomes less, and some form of sewerage may be appropriate. In northeast Brazil, for example, Sinnatamby (1986) demonstrated that unconventional (shallow) sewerage is cost-effective for population densities exceeding about 160 hd/ha. Population densities in slums and shanty-towns can be much greater than this (2000 hd/ha is not uncommon).

Conventional sewerage, the most expensive option, will only be appropriate where property values are high and occupiers can pay for the full costs

Table 23.2 Classification of sanitation options

	Dry	Wet
On-site	Pit latrines	Pour-flush latrines
	VIP latrines	Septic tank systems
	Compost latrines	Aqua privies
Off-site	Bucket latrines	Conventional sewerage
	Vault latrines	Unconventional sewerage

involved. Suitable upgrade paths also need to be considered. However, each case is different and the merits of all the options need to be thoroughly evaluated in technical terms and by considering social, cultural, financial and institutional factors. This may involve considerably more consultation with community members than in conventional planning and design practice.

23.3.2 Storm drainage

Storm drainage options are more limited. The main classification is between 'closed' systems, relying on underground pipes, and 'open' systems requiring open channels (see Table 23.3).

In most situations, conventional piped drainage will not be an option, unless it is part of a simplified system. Open channels are widely used but need to be carefully designed, constructed and maintained.

23.4 On-site sanitation

On-site disposal is widely practised in low-income communities. This can be a perfectly satisfactory urban solution (even in crowded areas) provided the plot-size is large enough and subsoil conditions are suitable for disposal of the effluent to the ground without danger of contamination of nearby water sources (wells, for example).

The following sections contain an outline of the on-site options available. Design methods and typical details can be found in Pickford (1995) and Mara (1996b).

Table 23.3 Classification of drainage options

Open	Open channels
	Road-as-drain
Closed	Conventional piped drainage
	Dual drainage

23.4.1 Latrines

Pit latrine

The simplest form is the *pit latrine* (Fig. 23.3(a)). This consists of a squatting hole or plate directly above a pit in the ground into which the excreta falls. The conditions within the pit are anaerobic, promoting gaseous products (mainly CO_2 and CH_4 but also some malodorous gases) which escape into the atmosphere. The excreta gradually decompose and a solid residue accumulates in the pit bottom. Water, urine and other liquids infiltrate to the ground through the pit walls and base. When the residue reaches about 0.5 m from the top, the latrine either needs to be cleaned-out or abandoned. Pits are usually approximately 1 m in diameter (or 1 m square) and up to 3 m deep.

As the pit latrine works by allowing infiltration of liquids into the surrounding soil, it follows that there must be sufficient open space available on the individual plot, the actual size depending on the conductivity of the soil. However, pit latrines require only 1–2 m² of space, making them suitable sanitation options even in high-density areas (Reed, 1995).

VIP latrines

Ventilating pipes can be used to overcome the common complaints of bad smells and insect nuisance. The 'ventilated improved pit' (VIP) latrine

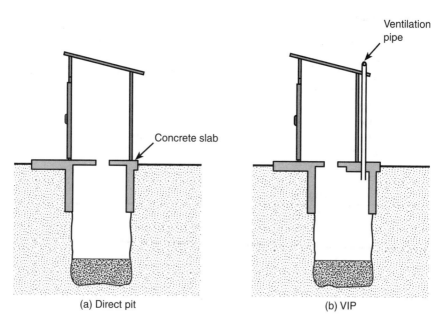

Fig. 23.3 Types of latrine

consists of a slightly offset pit with a vertical vent having flyproof netting at its exit (see Fig. 23.3(b)). The latrine is kept dark so any flies hatched in the pit are attracted to the light at the top of the vent and are trapped and die. The wind across the top of the vent causes low pressure and, therefore, an updraught extracting any foul odours. The pipe can also be painted black which helps heat up the air inside, causing it to rise and ventilate the pit. The shelter itself needs to be well-ventilated to allow a through-flow of air.

Permanent VIP latrines can also be built with two chambers (so called 'alternating twin pit' VIPs) where pits are filled and emptied alternatively. This allows safe manual emptying of 'old' sludge, but does, however, require more of the householder than other options (Tayler, 1996). Thus, VIP latrines constitute a very useful option, even in highly built-up, low-income areas.

Pour-flush latrines

Another solution to the odour and insect problem is to provide a pan and trap with a water seal above the pit. This has the additional benefit of removing the direct 'line-of-sight' between user and faeces below. Well-designed pans can be washed down with 1–3 l of water poured from a hand-held vessel.

Pour-flush latrines are widely and successfully used in low-income communities where there is a nearby water source, such as a well or stand-pipe.

Compost latrines

Composting faeces with vegetable wastes in an enlarged pit latrine or other chamber offers another method of on-site treatment. These require considerable care and continuous attention, and cannot currently be recommended. However, with further development, these should become a more realistic and sustainable option (see Chapter 23).

Communal latrines

Communal latrines are acceptable in some situations (e.g. city-centre areas), although this depends on the attitudes and habits of the local people. In some places, privacy is considered so important that local people refuse to use them. Where they are acceptable, and where arrangements can be made for frequent and regular cleaning, their cost is much lower than that of individual household latrines.

23.4.2 Septic tank systems

Septic tank systems consist of a tank and, ideally, a drainage field (see Fig. 23.4). The tank is a watertight, underground vessel that provides conditions suitable for the settlement, storage and (temperature-related) anaerobic decomposition of excreta. Sludge accumulates at the bottom of the tank and has to be emptied periodically. A hard crust of solidified grease and oil forms on the surface. Wastewater is fed directly to the tank, through which it flows, and then on to the drainage field. Direct discharge to a ditch, stream or open drain is not recommended, but is common practice in many Third World cities. The *drainage field* consists of a soakaway or sub-surface irrigation pipe system, which drains the effluent into the surrounding soil and provides additional treatment.

Tanks should have a minimum volume of 1 m³. The desludging interval should be short enough to ensure the tank does not become blocked but long enough to allow the benefits of anaerobic reduction in sludge volume. This will be typically 2–5 years.

Septic tanks require significantly more space than pit latrines. Depending on water use and soil conditions, this can range from 10–100 m² (Reed, 1995). They are also expensive to construct and operate, but have proved satisfactory in low- and high-income countries alike, especially in low-density housing areas (Butler and Payne, 1995).

23.4.3 Aqua privies

An *aqua privy* consists of a latrine set over a septic tank. The squatting plate is connected to a pipe that dips below the surface of the liquor in the chamber below. Overflowing liquor infiltrates into the soil through a drainage field and the water level in the tank is made up with small amounts of cleaning water. Regular desludging is required.

Aqua privies were popular but problems have always been encountered in maintaining the necessary water level in the tank and with faecal fouling of the drop-pipe. Both can cause odour nuisance and health hazard.

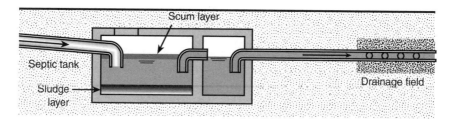

Fig. 23.4 Septic tank system

23.5 Off-site sanitation

23.5.1 Bucket latrines

Bucket systems are a traditional form of excreta removal. Waste is simply deposited into a container and the so-called 'nightsoil' collected on a daily basis. This method is still widely practised in urban areas, although it is objectionable from most points of view (including being hazardous to health), but at least it does remove excreta from the household.

23.5.2 Vault latrines

In principle, these are similar to bucket latrines except the squatting plate/seat is joined directly to a closed, watertight chamber where the excreta are deposited. Vaults need to be periodically emptied often by scoops or ladles. Objections similar to those for bucket latrines can be raised. The vault system is widely and successfully used in Japan where vacuum trucks empty the waste and transport it to be treated centrally. Similar systems connected to conventional WCs (*cesspools*) are also used in some remote properties in high-income countries. However, the need for very regular emptying makes this option expensive.

23.5.3 Conventional sewerage

As mentioned, the most expensive sanitation option is conventional sewerage. The advantages of using conventional sewerage (as described in the rest of this book) should by now be clear. However, the disadvantages for low-income communities are manifest: high cost, the need for an ample water supply, the difficulty of construction, operation and maintenance, and the potential for serious pollution at the outfall (unless expensive wastewater treatment is proposed). In addition, there are two other practical causes for concern.

Septicity

High temperatures accelerate decomposition and limit the amount of oxygen that can be dissolved in water, leading to the rapid development of anaerobic conditions. Such conditions can give rise to H_2S production, resulting in corrosion of cementitious materials (see Chapter 17).

Blockage

Blockage can be caused or exacerbated by:

- abuse of the system through ignorance
- use of traditional methods for anal cleansing e.g. leaves, rags, stones, newspaper
- use of traditional methods for pot cleansing e.g. sand, ash
- low water use
- too few sewer connections.

23.5.4 Unconventional sewerage

Simplified sewerage

Simplified sewerage (also known as shallow or condominial sewerage) is similar to conventional separate foul sewerage except it is reduced to the basics and less-conservative assumptions are used in its design. Thus, sewer diameters, depths and gradients are reduced compared with conventional systems, and locally-available materials are utilised. Hydraulic design is also similar to conventional foul sewers as described in Chapter 10 (see Table 23.4).

In addition to relaxation in the hydraulic design criteria, the following practical details are changed (Reed, 1995):

- conventional access points are replaced by ones of smaller diameter or rodding eyes
- access point spacing is increased
- more maintenance responsibility is taken on by residents
- layout is amended, in particular back-of-property collectors are used to minimise sewer length.

The latter point is characteristic of the variant known as condominial

Table 23.4 Typical unconventional sewerage design criteria

Criteria	Simplified	Settled
Per capita flow (l/d)	100	100
Peak factor (×DWF)	2	1.5
Minimum velocity (m/s)	0.5	0.3
Maximum prop. depth of flow	0.75	1.0
Minimum pipe size (mm)	100	75
Minimum slope	1/200	–
Minimum cover (mm)*	350–500	350–500

* Depending on location

sewerage. Stormwater drainage is needed to exclude runoff from such systems.

Simplified sewerage systems have most obvious application in high-density, low-income areas where space is at a premium and on-site solutions are inappropriate. However, their main advantage is their increased affordability. Some of the disadvantages of conventional systems also apply to simplified systems, although blockages are reported to be rare (Mara, 1996c).

Settled sewerage

Settled sewerage (also termed small-bore, solids-free or interceptor tank sewerage) consists of small diameter sewers connected to small tanks that collect individual household wastewater and capture much of the solid material. Solids accumulate in the tank and require periodic removal. The tank is provided to:

- settle-out heavier material which normally requires relatively high self-cleansing velocities
- settle out larger solids, grease and scum that might potentially block smaller pipes
- attenuate individual flow inputs to reduce the peak flow.

This allows small pipes to be provided at nominal fall. Indeed some designs (Otis and Mara, 1985) allow sections of adverse pipe gradient and pipes flowing surcharged, provided there is an overall positive hydraulic gradient.

As with simplified sewerage, hydraulic design is similar to that for conventional foul sewers but using the criteria summarised in Table 23.4. However, because surcharged pipes are allowed in some systems, extra care is needed in designing the system to ensure all tanks can empty under gravity. Mara (1996c) gives a worked example and further explanation. The tanks can be designed as single chamber septic tanks (see Section 23.4.2), although there are reports of much smaller tanks being used in certain areas (Tayler, 1996).

Settled sewerage may be the best option and the most economic choice where septic tanks are already in existence. Probably the main concern about these systems is their long-term performance. How well and for how long will they operate if tank desludging is neglected? What provision, if any, has been made for proper collection and safe disposal of the sludge?

23.6 Storm drainage

Provision of drainage for low-income communities often lags behind water supply and sanitation. This is a pity because, if flooding is frequent, water supply and sanitation will be difficult to install and ineffective in operation. As has already been argued, storm drainage also fulfils a disease control function.

23.6.1 Flooding

When using conventional storm drainage in developed countries, systems are designed to run full at relatively modest storm return periods (1 or 2 years) with the knowledge and expectation that flooding will occur even less frequently (e.g. once every 5–10 years). Little thought or engineering effort is, therefore, given to the case when flooding does occur. This is seen as system failure, over which the engineer no longer has control (see Section 11.2.2).

That situation can be contrasted with many low-income communities living with the monsoon. During the rainy season, high intensity storms take place frequently and any drainage system fails almost immediately from lack of capacity and blockage, and widespread flooding occurs. As storm drains are routinely contaminated with sullage and faeces, the flooding is with dilute wastewater. However, detailed study of some low-income householders' attitudes towards flooding is very revealing (see Box 23.1).

Box 23.1 Community attitudes in Indore, India

Flooding was ranked low in comparison to other risks and problems, such as improvements in job opportunities, provision of housing, mosquitoes and smelly back lanes.

A major concern mentioned by residents relates to the predictability of the flooding event. . . . In other words, even extensive inundation is bearable if expected. . . . Interventions aimed at ameliorating the effects of flooding should try to take account of these needs of the community to understand and adapt their coping strategies if necessary.

Residents in flat areas give equal, if not more, weight to the after effects of rains. . . . as compared to immediate effects. Water may stand for long periods in very flat areas. Respondents feel that this makes walking difficult and allows the breeding of mosquitoes. Faeces-contaminated mud caused by stormwater is seen as most problematic as it is also a perceived source of mosquitoes and noxious smells.

[In areas of the slum improvement programme . . .] residents had high expectations of drainage improvement when the projects were initiated and appear to have expected that flooding and inundations would cease or be reduced substantially. It is likely that the feeling that their expectations have not been met is in part an immediate sense of disappointment that flooding has not ceased altogether.

[Drainage] interventions would gain favour if residents understood and were clearly informed about the effects (good and bad) on the environmental risk which they perceive as inherently a natural event. This, of course, necessitates technical personnel being able to predict the consequences of technical interventions. Such a strategy might reduce the scale of expectation. (Stephens *et al.*, 1994)

So, in low-income communities subject to frequent flooding, even in improved areas, local people want to know most of the very things the drainage engineer cannot tell them! When will it flood and how long will it last, where will it flood and how high will the water rise (Kolsky *et al.*, 1996)?

23.6.2 Open drainage

Open channels

Open drains have a number of advantages when compared to closed pipes.

- They are cheaper to build as they are simpler and shallower than closed pipes.
- Blockage with refuse and washed-in sediment can be more easily monitored and cleaned out.
- They use available head more efficiently.
- Mosquito breeding is easier to control than in closed drains.

The simplest and cheapest drains are unlined channels along the roadside with water shedding from the road to the drain by positive fall. The sides of an unlined drain should not slope by more than 1:2 to ensure that they will be stable. If the gradient along the drain exceeds about 1:100, lining will usually be required to protect the channel from damage by scouring (Fig. 23.5(a)).

One of the keys to making open drains work is to maintain them properly. Drains will require cleaning at regular intervals (whatever the self-cleansing velocity specified) since material from the street inevitably gets washed or blown in. Kolsky *et al.* (1996) have demonstrated how the performance of the system is dramatically reduced by even small amounts of sediment in the system. Cleaning technique is also important. 'Sweepings' from drains are traditionally left to dry in roadside piles prior to collection. Unfortunately, this material can quickly find its way back to the drain from whence it came!

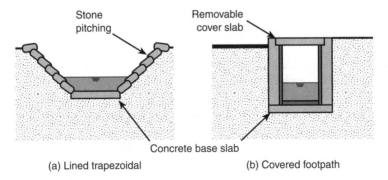

(a) Lined trapezoidal   (b) Covered footpath

Fig. 23.5 Typical open drainage channels

A compromise solution is to build channels (under the footpath, for example) with removable covers (see Fig. 23.5(b)) along their length. The covers discourage entry of rubbish and sediment, but can be removed for cleaning if necessary.

Design can be undertaken in a similar way to that of storm sewers described in Chapter 11. The major differences lie in the roughness of the channel being used and in ensuring velocities are not so high as to damage unlined channels (see Example 23.1).

Road-as-drain

Experience suggests that blockage of drains is likely to be a recurrent problem, and that frequent stormwater flow along the streets of low-income communities during the monsoon season is practically inevitable. Therefore, logic dictates that engineers should consider the deliberate and controlled routing of surface runoff over the streets of the slum (see Fig. 23.7). The street surface should therefore be depressed below the level of housing sites and not *vice versa*, as is usually the case. This may involve more extensive site grading, but will assure more reliable drainage than any system based on conduits (open or closed) that are subject to solids deposition. Roads as drains are also easier to maintain; street sweeping is easier than drain cleaning, and residents have an interest in keeping roads reasonably clear for access (Kolsky, 1998).

This is by no means a universal panacea, however! It is most appropriate in narrow streets where heavy vehicles do not pass at all and traffic is light. It will not be appropriate on steep streets (say >5%), and in flatter areas the micro-topography may well prevent surface routing of all flows. Road surfacing material needs to be carefully selected (e.g. concrete, compacted gravel or stone) to provide erosion protection.

23.6.3 Closed drainage

Conventional drainage

The main advantage of closed drains is that they do not take up surface space. They also reduce the risk of children playing in or falling into polluted water, and the possibility of vehicles damaging the drains or falling into them. Also, in principle, sediment entry to the system can be minimised with catchpits and gully pots. The main problem is blockage due to refuse or sediment that does enter, and then the difficulty of cleaning-out such material. Satisfactory simplified versions of storm drainage systems have not yet been developed.

As with conventional sewerage, conventional drainage using closed drains is the most expensive option available. Even though it is the most highly-engineered approach, uncritical transference of European or North

Example 23.1

An unlined (but vegetation free) trapezoidal open channel (geometry shown in Fig. 23.6) is to be built alongside a road of gradient of 1:100. If the design stormwater flow is estimated at 300 l/s, calculate the depth of flow in the channel and check its velocity. Use Manning's equation (8.23) and assume a roughness of $n = 0.025$.

Solution

A trapezoidal channel of bottom width $w = d/2$ and side slope 1:2 gives:
Cross-sectional area, A:

$$A = d(w + d \cot \alpha) = 2.5d^2$$

Hydraulic radius, R:

$$R = \frac{d(w + d \cot \alpha)}{w + \dfrac{2d}{\sin \alpha}} = 0.5d$$

Thus:

$$Q = \frac{1}{n} 2.5d^2 (0.5d)^{\frac{2}{3}} S_o^{\frac{1}{2}}$$

which for $Q = 0.3$ m³/s and $S_o = 0.01$ gives $d = 0.32$ m.
Velocity, $v = Q/A = 0.3 \times 2/(5 \times 0.32^2) = 1.2$ m/s
This velocity is unlikely to be high enough to self-cleanse, and may cause erosion of the unlined channel (Watkins and Fiddes, 1984).

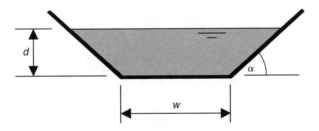

Fig. 23.6 Trapezoidal channel geometry (Example 23.1)

Fig. 23.7 Road-as-drain in Indore, India

American standards has littered developing countries with expensive infra-structure that does not work. Hence, piped drains should be built in low-income tropical areas only after very careful consideration of the other options.

Dual drainage

Dual major/minor drainage, as described in Chapter 11, also shows promise for low-income communities.

Problems

23.1 Explain the importance of taking an integrated approach to munici-pal service provision.
23.2 What are the main health-related issues concerned with poor sanita-tion and storm drainage?
23.3 Classify and compare the main sanitation options available to low-income communities.
23.4 Classify and compare the main storm drainage options available to low-income communities.
23.5 Compare and contrast the various types of pit latrine.
23.6 Discuss what is meant by 'simplified' and 'settled' sewerage and explain how they differ from conventional sewerage.
23.7 Assess the relative merits of open and closed storm drains.
23.8 For the channel cross-section described in Example 23.1, what gradi-ent is needed to reduce the flow velocity to 1.0 m/s to avoid erosion?
[1:140]

Key sources

Caincross, S. and Feachem R.G. (1993) *Environmental Health Engineering in the Tropics: An Introductory Text*, 2nd Edn, John Wiley & Sons.
Cairncross, S. and Ouano, E.A.R. (1991) *Surface Water Drainage for Low-income Communities*, World Health Organisation.
Mara, D.D. (ed.) (1996a) *Low-Cost Sewerage*, John Wiley & Sons.
Mara, D.D. (1996b) *Low-Cost Urban Sanitation*, John Wiley & Sons.
Mara, D.D., Sleigh, A. and Tayler, K. (2001) *PC-based simplified sewer design*, University of Leeds.

References

Butler, D. and Payne, J.A. (1995) Septic tanks: problems and practice. *Building and Environment*, 30(3), 419–425.
Kolsky, P. (1998) *Storm Drainage. An Engineering Guide to the Low-Cost Evalu-ation of System Performance*, Intermediate Technology Publications.
Kolsky, P.J., Parkinson, J. and Butler, D. (1996) Third world surface water

drainage: the effect of solids on performance. Chap 14 in *Low-cost Sewerage* (ed. D.D. Mara), John Wiley & Sons.

Mara, D.D. (1996c) Unconventional sewerage systems: their role in low-cost urban sanitation. Chap 2 in *Low-Cost Sewerage* (ed. D.D. Mara), John Wiley & Sons.

Okun, D.A. and Ponghis, G. (1975) *Community Wastewater Collection and Disposal*, World Health Organisation.

Otis, R.J. and Mara, D.D. (1985) *The Design of Small Bore Sewer Systems*, TAG Technical Note. 14, The World Bank.

Pickford, J. (1995) *Low-cost Sanitation. A Survey of Practical Experience*, Intermediate Technology Publications.

Reed, R.A. (1995) *Sustainable Sewerage. Guidelines for Community Schemes*, WEDC, Intermediate Technology Publications.

Sinnatamby, G.S. (1986) *The Design of Shallow Sewer Systems*, United Nations Centre for Human Settlements.

Souza, C., Bernardes, R. and Moraes, L. (2002) Brazil's modelling hopes. The public health perspective. *Water21*, October, 40–41.

Stephens, C., Pathnaik, R. and Lewin, S. (1994). *This is My Beautiful Home: Risk Perceptions Towards Flooding and Environment in Low Income Communities*, London School of Hygiene and Tropical Medicine.

Tayler, K. (1996) Low-cost sewerage systems in South Asia, Chap 4 in *Low-Cost Sewerage* (ed. D.D. Mara), John Wiley & Sons.

Watkins, L.H. and Fiddes, D. (1984) *Highway and Urban Hydrology in the Tropics*, Pentech Press.

Wondimu, A. and Alfakih, E. (1998) Urban drainage in Addis Ababa (Ethiopia): Existing situation and improvement ideas. *Fourth International Conference on Developments in Urban Drainage Modelling – UDM '98*, London, 823–830.

24 Towards sustainability

24.1 Introduction

24.1.1 Sustainable development

Sustainable development has been on the world agenda since the 1987
World Commission on Environment and Development. The outcome (the
Brundtland Report (World Commission on Environment and Development, 1987)) offered a viable alternative to the commonly-held view of
environmentalists in developed countries: that pursuit of economic growth
is incompatible with a responsible policy towards the environment. Such
an attitude was unacceptable in developing countries, where increasing
national wealth must be a primary aim, and where any global environmental policy that threatens growth would be seen as the rich countries
'pulling up the ladder after them'. In the context in which the phrase 'sustainable development' was coined, it was the inclusion of the word 'development' that was particularly significant: it affirmed the right of a country
to seek to develop – but only in a way that does not compromise
opportunities for the future.

Bruntland defined sustainable development as, 'that which meets the
needs and aspirations of the present generation without compromising the
ability of future generations to meet their own needs'.

Further reflection has shown that sustainable development must satisfy
environmental, economic and social criteria (Bruce, 1992).

- It must not damage or destroy the basic life support system of our
  planet: the air, water and soil, and the biological systems.
- It must be economically viable to provide a continuous flow of goods
  and services derived from the earth's natural resources.
- It requires developed social systems, at international, national, local
  and family levels, to ensure the equitable distribution of the benefits of
  the goods and services produced.

24.1.2 Sustainability

The emphasis on development is less necessary in developed countries, where the word 'sustainable' has become useful by itself in defining environmental policies. The main problem has been in defining 'sustainability'! However, use of the word has at least been effective in encouraging a broad view of environmental policy.

24.1.3 Implementation

Sustainable development became the central theme of the UN Earth Summit at Rio de Janeiro (United Nations, 1992) which called on governments to produce their own strategies for sustainable development, building on existing plans and policies.

The British Government published its national strategy, *Sustainable Development*, in 1995. In parallel to this, the Local Government Management Board published *Local Agenda 21. A Framework for Local Sustainability*, requiring local authorities to produce their own strategies. The 1995 Environment Act has confirmed the Government's commitment to sustainability. The Environment Agency for England and Wales and the Scottish Environmental Protection Agency are both legally required to promote sustainable development.

24.2 Sustainability in urban drainage

24.2.1 Objectives

The difficulty with the Brundtland definition of sustainable development is that, although its principles are widely accepted, it is not immediately clear how to put them into practice.

Butler and Parkinson (1997) have considered and argued what might be the technical objectives of 'sustainable urban drainage' and have proposed the following list in order of priority:

- maintenance of an effective public health barrier
- avoidance of local or distant flooding
- avoidance of local or distant degradation/pollution of the environment (water, soil, air)
- minimisation of the utilisation of natural resources (water, nutrients, energy, materials)
- reliability in the long term and adaptability to future (as yet unknown) requirements.

The list can be expanded to include the broader requirements of:

- community affordability
- social acceptability.

24.2.2 Strategies

The search for economically viable solutions with low water, energy and maintenance requirements which are both flexible to change and hygienically acceptable under changing conditions is a great challenge. Butler and Parkinson (1997) suggest three fundamental strategies that should be pursued:

- reduce the reliance on water as transport medium for waste
- avoid mixing industrial wastes with domestic wastewater
- avoid mixing storm runoff with wastewater.

They argue that, if priority were given to these strategies, many benefits would be immediately realised, even if they are only introduced singularly or incrementally. The strategies and their potential advantages and disadvantages are summarised in Table 24.1.

Water transport

Conventional urban drainage systems utilise water extremely inefficiently. Large quantities of water are abstracted and treated using expensive treatment technology to drinking standard, yet are subsequently used to flush faeces and urine in the WC (see Chapter 4). Not only is this wasteful of a precious resource, but it also promotes unnecessary mixing and dilution of wastes and contamination of previously unpolluted water. End-of-pipe treatment technology is then needed to extract the solid and dissolved pollutants from the liquid component of the waste flow to avoid receiving water pollution.

Household water consumption can be reduced substantially with minimal investment (National Rivers Authority, 1995). This may be achieved by installing low-flush toilets or possibly domestic water recycling systems (see below). The challenge is to ensure, for example, that existing systems will operate effectively with much lower dry weather flows (see Box 24.1). Alternative no-flow technologies for the conveyance of sanitary waste such as dry sanitation systems (Chapter 23) or some vacuum systems should be investigated as feasible technical options for the future.

Mixing of industrial and domestic wastes

The mixing of industrial effluents with domestic wastewaters can cause problems for conventional wastewater treatment resulting in the creation

Table 24.1 Strategies towards sustainable urban drainage

Component of wastewater	Problems	Proposed strategies	Potential advantages	Potential disadvantages
Carriage water	• unnecessary water consumption • dilution of wastes • requires expensive end-of-pipe treatment	• introduce water conservation techniques • re-use water • seek alternative means of waste conveyance	• conserves water resources • improves efficiency of treatment processes	• increases possibility of sedimentation in sewers • health hazards associated with water re-use
Industrial waste	• disrupts conventional biological treatment • increases cost of wastewater treatment • causes accumulation of toxic chemicals in the environment • renders organic wastes unsuitable for agricultural re-use	• remove from domestic waste streams • pre-treat, reduce concentration of problematic chemicals • promote alternative industrial processes using biodegradable substances	• improves treatability of wastewaters • improves quality of effluents and sludges • reduces environmental damage • saves costs associated with re-use of recovered chemicals	• costs associated with implementing new practice • lack of monitoring facilities • may promote illicit waste disposal
Stormwater	• requires large and expensive sewerage systems • transient flows disrupt treatment processes • discharge from overflows causes environmental damage • causes floods	• utilise overland drainage patterns • store and use stormwater as a water resource • provide infiltration ponds, percolation basins and permeable pavements • promote ecologically sensitive engineering, e.g. constructed wetlands	• reduces pollution from overflows • improves efficiency of treatment • recharges groundwater • reduces demand for potable water • reduces hydraulic capacity requirements of conduits	• decentralised facilities harder to monitor • increases space requirements • risk of groundwater contamination

Box 24.1 Droylsden, UK

Droylsden is a largely residential suburb of Manchester. It has a population of about 20 000 and an area of 420 ha. The majority of the catchment is urbanised, although there are substantial undeveloped areas adjacent to the River Medlock and its tributaries. The existing sewerage system is combined, with just a few newer developments having separate sewers. The system drains to a single outfall crossing the river before discharging into a large interceptor sewer.

A detailed theoretical exercise (Parkinson, 1999) was carried out to assess the implications on the existing sewers of introducing water-conserving devices throughout the catchment. To do this, a sewer simulation model was developed and verified using *in situ* measured data. Fig. 24.1 shows the relative frequency of velocities during dry weather in all the catchment sewers for the base-case of 9 l flush WCs. As illustrated, in this system, very few sewers reach velocities required for self-cleansing (a minimum of 0.6 m/s).

According to model predictions, application of lower flush WCs only has a small effect. However, the effect is interesting in that it increases the proportion of pipes that have lower velocities, and therefore decreases those with higher velocities. Thus, fewer sewers will reach the self-cleansing threshold.

More serious results may arise in smaller sewers, at the heads of systems, where it is expected that the transport of solids will be reduced with the potential for blockage formation.

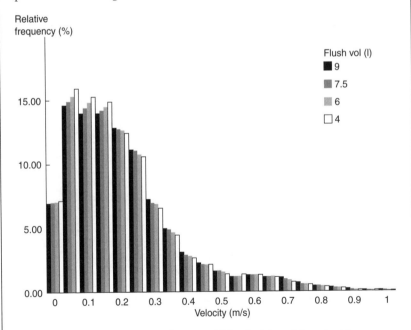

Fig. 24.1 Relative frequency of sewer DWF velocities with and without water-conserving WCs

of water pollution problems. The re-use potential of nutrients and minerals contained in sludge is diminished by small concentrations of industrial contaminants. In many cases, this results in the disposal of sludge to landfill sites or by incineration, and the waste of a valuable resource. Synthetic organic chemicals and heavy metals accumulate in the environment causing problems to ecosystems as they move through the food chain.

Removing the industrial waste component from municipal wastewater is, therefore, critically important to sustainable drainage strategies. Dealing with the waste as close to the location of production as possible reduces the problems associated with treating a highly complex mixture of substances, and increases the re-use potential of the more valuable components of waste (Hvitved-Jacobsen *et al.*, 1995). Where the complete isolation of an industrial waste stream is impractical, pre-treatment using best available technology to reduce concentrations of the undesirable wastes (be they refractory chemicals or excessively high concentrations of biodegradable waste) is required prior to discharge into the sewer.

Mixing of stormwater and wastewater

By disconnecting stormwater from overloaded combined systems, a number of benefits are achieved: CSO discharges are decreased, thus reducing the extent of environmental damage, and the problems caused at treatment works by peak, transient flows are reduced. The potential for use of rainwater for reuse or recreation is also increased (see Box 24.2). Generally, drainage systems do not exploit stormwater as a resource (for example by using it for toilet flushing and garden watering).

Isolating storm runoff from wastewater is not a new idea. As discussed in Chapter 1, separate systems were regarded in the 1950s as being the drainage systems of the future and, indeed, are still the current practice in most developed countries. Generally, these systems are more costly to construct, generate higher emissions during manufacture (Gigerl and Rosenwinkel, 1998), and yet have not delivered the expected environmental

---

Box 24.2 Edwinstowe, UK

Edwinstowe Youth Hostel consists of a 400 m² roof plan building with a 325 m² car park. The car park is constructed of permeable blocks allowing inflow of storm runoff to the sub-base, where it can be stored, with excess being overflowed to an infiltration trench. Roof runoff is also collected and stored.

The volume of storage available (34 m³) is approximately equal to the average monthly rainfall on the site and is adequate to supply the total WC flushing requirements of up to 33 people in the hostel (Pratt, 1999).

improvements. However, it seems that the overall strategy of separating the two components is not at fault, rather the method through which the strategy has been implemented.

More recent thinking advocates separation of storm runoff from other urban wastewaters (Andoh, 1994; Kollatsch, 1993; O'Loughlin, 1990) but, instead of routing stormwater directly into a piped system, the alternative approach recommends utilising the natural drainage patterns of the catchment, or using stormwater management techniques (SUDS) as described in Chapter 21. Many of these methods utilise on-site infiltration. Where the soil and the quality of the runoff permits, direct infiltration into the ground is preferable in order to recharge groundwater reserves. Where infiltration is not possible, the development of natural drainage patterns offers a range of opportunities for conservation, recreation and amenity, as well as providing basic flood and pollution control (Ellis, 1995).

24.3 Steps in the right direction

24.3.1 Domestic grey-water recycling

The reuse of grey-water (all the wastewater produced in a house except by the WC) potentially reduces the need to use potable water for non-potable applications, with the water effectively being used twice before discharge to the sewer. The major reuse potential is for WC flushing and garden watering, relieving demand on public water supplies and wastewater collection and treatment facilities.

The main elements of a grey-water reuse system are a collection and distribution pipe network, sufficient storage volume to balance inflows and outflows, and appropriate treatment to render the recycled water fit for the purpose.

If WC flushing and garden watering demand can be fully satisfied using recycled water, comparison with Table 4.1 indicates that a 36% reduction in demand will be realised. In theory, this demand could easily be met by reusing water first used for personal washing (26%) and clothes washing (12%). System modelling shows that 90% WC water saving efficiency can be achieved using a storage capacity of approximately 200 litres (Dixon et al., 1999).

Widespread adoption of such systems will inevitably depend on their cost. Retrofitting systems to individual houses is unlikely to be financially attractive, although systems built into new houses may well be. Larger scale application (e.g. hotels, office blocks) may prove to be more cost-effective.

24.3.2 Nutrient recycling

It is an important fact that, in principle, each person produces in their excreta enough nutrients to grow 250 kg of cereal per year, which is about enough for one person to live on (Drangert, 1998). Of course, some of these nutrients are currently used on agricultural land as sludge derived from WTPs (about 50% of sludge goes to land in the UK). However, significant quantities of nutrients are discharged through the WTP to the receiving water and so are effectively lost to agriculture. Also, industrial discharges add heavy metals to wastewater, rendering it less useable.

One way to make better (more sustainable) use of wastewater nutrients is to isolate and handle urine separately and use it as a fertiliser. One important reason for this strategy is the fact that urine is the source of around 30% of phosphorus and around 70% of nitrogen in wastewater (Fig. 24.2). Additional benefits include the fact that urine is relatively low in heavy metal concentration and also relatively easy to handle, and much less water is needed to flush it.

To isolate and reuse urine, special no-mix toilets are required, which can range from the simple to the sophisticated, with choice depending on the socio-economic culture. Box 24.3 describes a case-study application of urine-separation technology.

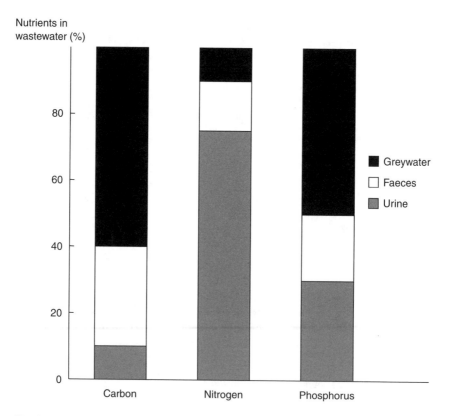

Fig. 24.2 Percentage source of nutrients in wastewater

Box 24.3 Björsbyn, Sweden

Björsbyn is an 'ecological' village built in 1994 in northern Sweden. It was designated as such to emphasise that the inhabitants' way of living is deliberately different from traditional urban dwellers, moving towards a more sustainable lifestyle. Interestingly, however, 'ecological concern' was a major reason for moving into the village for only two of the families (the main attraction was an appealing rural area within a reasonable distance from the city centre!). The village has urine-separating toilets, compost bins for biodegradable organic wastes at each house and a small outhouse for the collection of paper and glass separated from the solid wastes (Hanaeus *et al.*, 1997).

- Excreta are separated at source using no-mix toilets (Fig. 24.3). The toilets used were equipped with a small collection unit for urine with its own flush-water system (0.1 l per flush). The urine is collected separately and led through a sewer system to a collection tank with eight months' storage capacity. When the tanks are full, or urine is needed, they are emptied onto nearby farmland.
- The faeces and other grey water are collected and treated by septic tanks followed by infiltration beds. The sludge from the septic tanks is treated locally by a combined sludge drying, freezing and composting unit. After treatment, the sludge is used as a fertiliser and soil conditioner on agricultural land.
- Stormwater flows via ditches to the surrounding countryside where it is discharged and infiltrates to groundwater.

When studying this system, Hanaeus *et al.* (1997) concluded that the successful operation of a urine separation system is particularly dependent on complete separation (to avoid dilution or contamination of the urine). This can be promoted by well-designed toilets, careful design of the collection system and education of users.

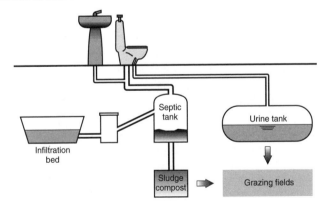

Fig. 24.3 The Björsbyn system (reproduced from Hanaeus *et al.* [1997] with permission of publishers Pergamon Press and copyright holders IAWQ)

24.3.3 Disposal of domestic sanitary waste

In some countries, and most noticeably the UK, it is common for the WC to be used as a disposal system for a variety of solid waste items (listed in Chapter 4). Each individual selects the method of disposal, in the privacy of the bathroom: the item is either dropped in the WC and flushed away, or placed in a bin for subsequent removal via the solid waste system.

Disposal of sanitary waste by the waterborne route is known to cause operational and environmental problems of small pipe blockage, screen ragging and blinding, and offence to the public if discharged to the environment. Common sense would suggest that this is an inefficient method of disposal – solid waste mixed and transported with wastewater requires an increase in effort and expense to re-separate.

A non-technical approach to reducing the problems caused by these wastes emphasises cultural rather than technical aspects. It is to create a change in public disposal practices: to reduce disposal by the waterborne route and increase disposal by the solid waste route. In the second half of the 1990s, there have been several public awareness campaigns drawing attention to the operational and environmental problems caused by waterborne disposal as described in Chapter 1.

Research by Ashley *et al.* (1999) investigated the advantages of encouraging a change in public disposal practice, and has shown that such a change improves the sustainability of the system as a whole. For public campaigns to be successful, there needs to be an emphasis on activity at local level.

24.4 Assessing sustainability

How can we assess whether one urban drainage option is more sustainable than another?

It is commonly stated that for a course of action to be sustainable it must simultaneously satisfy criteria in three areas: environmental, social and economic. The UK Government states (DETR, 1999a) that sustainable development 'means meeting four objectives at the same time, in the UK and the world as a whole:

- social progress which recognises the needs of everyone;
- effective protection of the environment;
- prudent use of natural resources; and
- maintenance of high and stable levels of economic growth and employment.'

To 'help people assess whether ... we are achieving the broader objectives of sustainable development' (DETR, 1999b), the UK Government uses a range of *indicators*. These include 15 headline indicators (for example, H9:

Emissions of greenhouse gases; H12: Rivers of good or fair quality), and a wide range of other core indicators.

At the level of assessing sustainability in relation to a choice between urban drainage options, a far more specifically focussed approach is clearly needed. One approach has been developed by a project named SWARD, Sustainable Water Industry Asset Resource Decisions (Ashley *et al.*, 2003). This project investigated the way asset investment decisions are made in the water industry, and how issues of sustainable development should be included. The decision support processes recommended are set out in seven stages:

1 define objectives
2 generate options
3 select criteria
4 collect data
5 analyse options
6 select and implement preferred option
7 monitor outcome.

The options, considered in this way, can be quite general (for example, alternative strategies for managing domestic sanitary waste) or quite specific (alternative locations for a wastewater treatment facility).

Candidate criteria are proposed under four headings: environmental, social, economic and technical. Recommended 'primary criteria' are listed in Table 24.2. Each primary criterion leads to a number of secondary

Table 24.2 SWARD primary criteria

Type	Primary criterion
Environmental	Resource utilisation Service provision Environmental impact
Social	Impact on risks to human health Acceptability to stakeholders Participation and responsibility Public awareness and understanding Social inclusion
Economic	Life-cycle costs Willingness to pay Affordability Financial risk exposure
Technical	Performance of the system Reliability Durability Flexibility and adaptability

Table 24.3 SWARD: secondary criteria and indicators under the primary criterion resource utilisation

Secondary criteria	Indicator
Water resource use	
– Withdrawal	Annual freshwater withdrawal ÷ annual available volume (%)
– River water quality	% of rivers of good or fair quality
– Nutrients in water	% of river length with greater than guideline nutrient concentrations
Land use	Land area used in km^2
Energy use	
– Energy for water supply	Energy use (kW.h/m^3)
– Energy for wastewater treatment	Energy use (kW.h/m^3)
Chemical use	
– WTP or on-site (herbicides)	Litres/year
Material use	
– aggregates, plastics, metals	Total material requirement (tonnes/year)

criteria, each with a recommended indicator to be used in the assessment. The proposed secondary criteria and indicators for one of the primary environmental criteria, *resource utilisation*, are given in Table 24.3.

Analysis and comparison of the options using the selected criteria and their indicators requires some form of multi-criterion analysis approach. Case studies using this general approach are described by Ashley *et al.* (2003).

Problems

24.1 Define 'sustainable development' and interpret its implications in terms of society, economics and the environment. Illustrate your answer with examples from urban drainage.
24.2 'A "sustainable city" is a contradiction in terms.' Discuss.
24.3 Explain what you understand by the term 'sustainable urban drainage'.
24.4 In what ways is conventional urban drainage unsustainable?
24.5 Describe three techniques or technologies that claim to be more sustainable, and cite evidence to support the claims.

Key sources

Aalderink, H., van Ierland, E., Klapwijk, B., Lettinga, G., Lexmond, M. and Terpstra, P. (eds) (1999) Options for closed water systems: sustainable water management. *Water Science and Technology*, **39**(5).
ASCE/UNESCO (1998) *Sustainability Criteria for Water Resource Systems*, American Society of Civil Engineers.

Beck, M.B., Chen, J., Saul, A.J. and Butler, D. (1994). Urban drainage in the 21st century: assessment of new technology on the basis of global material flows. *Water Science and Technology*, 30(2), 1–12.

Butler, D. and Parkinson, J. (1997) Towards sustainable urban drainage. *Water Science and Technology*, 35(9), 53–63.

Henze, M., Somlyody, L., Schilling, W. and Tyson, J. (eds) (1997) Sustainable sanitation. *Water Science and Technology*, 35(9).

Niemczynowicz, J. (1994) New aspects of urban drainage and pollution reduction towards sustainability. *Water Science and Technology*, 30(5), 269–277.

References

Andoh, R.Y.G. (1994) Urban drainage – the alternative approach, *Proceedings of 20th WEDC Conference*, Colombo, Sri Lanka.

Ashley, R.M., Souter, N., Butler, D., Davies, J., Dunkerley, J. and Hendry, S. (1999) Assessment of the sustainability of alternatives for the disposal of domestic sanitary waste. *Water Science and Technology*, 39(5), 251–258.

Ashley, R.M., Blackwood, D., Butler, D., Davies, J.W., Jowitt, P. and Smith, H. (2003) Sustainable decision making for the UK water industry. *Proceedings of the Institution of Civil Engineers, Engineering Sustainability*, 156 (ES1), 41–49.

Bruce, J.P. (1992) *Meteorology and Hydrology for Sustainable Development*, World Meterological Organization Report No 769.

Department of the Environment, Transport and the Regions (1999a) *A Better Quality of Life – a Strategy for Sustainable Development for the UK*. DETR, London.

Department of the Environment, Transport and the Regions (1999b) *Quality of Life Counts – Indicators for a Strategy for Sustainable Development for the United Kingdom: a Baseline Assessment*. DETR, London.

Dixon, A., Butler, D. and Fewkes, A.F. (1999) Water saving potential of domestic water reuse systems using greywater and rainwater in combination. *Water Science and Technology*, 39(5), 25–32.

Drangert, J.-O. (1998) Fighting the urine blindness to provide more sanitation options. *Water SA*, 24(2), 157–164.

Ellis, J.B. (1995) Integrated approaches for achieving sustainable development of urban storm drainage. *Water Science and Technology*, 32(1), 1–6.

Gigerl, T. and Rosenwinkel, K.-H. (1998) Life cycle assessment of sewer systems. Conceptual idea, *Fourth International Conference on Developments in Urban Drainage Modelling, UDM '98*, London, September, 673–680.

Gujer, W. (1996) Comparison of the performance of alternative urban sanitation concepts. *Environmental Research Forum*, 5–6, Transtec Publishers, 233–240.

Hanaeus, J., Hellstorm, D. and Johansson, E. (1997) A study of a urine separation system in an ecological village in northern Sweden. *Water Science and Technology*, 35(9), 153–160.

Kollatsch, D.-Th. (1993) Futuristic ideas to create a most efficient drainage system, *Proceedings of 6th International Conference on Urban Storm Drainage*, Niagara Falls, Canada, September, 1225–1230.

Hvitved-Jacobsen, T., Nielsen, P.H., Larsen, T. and Jensen, N. Aa. (eds) (1995) The sewer as a physical, chemical and biological reactor. *Water Science and Technology*, 31(7).

National Rivers Authority (1995) *Saving Water – Demand Management in the UK*, September.

O'Loughlin (1990) The issue of sewer separation, in *Proceedings of 5th International Conference on Urban Storm Drainage*, Osaka, Japan, 987–992.

Otterpohl, R., Albold, A. and Grottker, M. (1996) Integration of sanitation into natural cycles: a new concept for cities. *Environmental Research Forum*, **5–6**, Transtec Publishers, 227–232.

Otterpohl, R., Grottker, M. and Lange J. (1997) Sustainable water and waste management in urban areas. *Water Science and Technology*, **35**(9), 121–133.

Parkinson, J. (1999) *Modelling strategies for sustainable domestic wastewater management in a residential urban catchment*. Unpublished PhD thesis, Imperial College, University of London.

Pratt, C.J. (1999) Use of permeable, reservoir pavement constructions for stormwater treatment and storage re-use. *Water Science and Technology*, **39**(5), 145–151.

United Nations (1992) *Agenda 21: Program of Action for Sustainable Development*.

World Commission on Environment and Development (1987) *Our Common Future*, Oxford University Press.

Useful websites

Sewer system modelling
www.wapug.org.uk — Wastewater Planning User Group

Sewer system modelling software
www.epa.gov/ednnrmrl/swmm — SWMM development
www.wallingfordsoftware.com — Wallingford Software, including InfoWorks
www.dhisoftware.com — DHI software: MOUSE and MOUSE TRAP
www.sobek.nl — The SOBEK flow model
www.microdrainage.co.uk — Micro Drainage software: WinDes and WinDap

SUDS
www.ciria.org.uk/suds — CIRIA SUDS website
www.suds-sites.net — UK SUDS database

Organisations
www.met-office.gov.uk — Met. Office
www.environment-agency.gov.uk — Environment Agency
www.ukcip.org.uk — UK Climate Impacts Programme
www.ofwat.gov.uk — OFWAT (water industry regulator)

Research networks
www.sewnet.org — Sewer systems and processes network
www.watersave.uk.net — Water conservation and recycling
www.jiscmail.ac.uk/lists/urban-drainage.html — Urban drainage email discussion list

Index

eBooks – at www.eBookstore.tandf.co.uk

A library at your fingertips!

eBooks are electronic versions of printed books. You can store them on your PC/laptop or browse them online.

They have advantages for anyone needing rapid access to a wide variety of published, copyright information.

eBooks can help your research by enabling you to bookmark chapters, annotate text and use instant searches to find specific words or phrases. Several eBook files would fit on even a small laptop or PDA.

NEW: Save money by eSubscribing: cheap, online access to any eBook for as long as you need it.

Annual subscription packages

We now offer special low-cost bulk subscriptions to packages of eBooks in certain subject areas. These are available to libraries or to individuals.

For more information please contact webmaster.ebooks@tandf.co.uk

We're continually developing the eBook concept, so keep up to date by visiting the website.

www.eBookstore.tandf.co.uk